FAY RUMSEY

School of Environmental and Resource Management
State University of New York/Syracuse

WILLIAM A. DUERR

Division of Forestry and Wildlife Resources
Virginia Polytechnic Institute and State University

SOCIAL SCIENCES IN FORESTRY:

A Book of Readings

1975

W. B. SAUNDERS COMPANY

Philadelphia, London, Toronto

W. B. Saunders Company: West Washington Square
Philadelphia, PA 19105

12 Dyott Street
London, WC1A 1DB

833 Oxford Street
Toronto, Ontario M8Z 5T9, Canada

Library of Congress Cataloging in Publication Data

Rumsey, Fay, comp.

Social sciences in forestry.

1. Forests and forestry—Economic aspects—United States—Addresses, essays, lectures. 2. Lumber trade—United States—Addresses, essays, lectures. 3. Wood-using industries—United States—Addresses, essays, lectures. 4. Forest management—Addresses, essays, lectures. I. Duerr, William Allen, 1911- joint comp. II. Title.

HD9755.R85 338.1′7′490973 74–24518

ISBN 0–7216–7839–4

Social Sciences in Forestry: A Book of Readings ISBN 0-7216-7839-4

Last digit is the print number: 9 8 7 6 5 4 3 2 1

PREFACE

This book is the outcome of a project conceived a decade ago. Users of Mrs. Rumsey's bibliographies in forestry economics had inquired whether *readings* might not appropriately flow as a by-product of the work. The inquirers were teachers of forestry economics and management in a scattering of United States and Canadian forestry schools. Their vision was that readings could serve them in two ways: (1) as the set of topics for an informal graduate or senior discussion course in economic thought and principles related to forestry, and (2) as a supplementary literature source for a formal textbook course in forestry economics.

We circularized the schools for opinions of a readings project and received both encouragement and helpful suggestions. A tentative set of articles and excerpts was used by the junior editor with his class of graduate students in the fall term of 1970. Subsequent classes, from spring 1971 through fall 1972, reviewed and discussed the selections and proposed alternatives. Meanwhile, colleagues at the schools were given review drafts of material and responded constructively. The present draft results from all this consultation and criticism.

The editors are most grateful to the students and colleagues for their help. We also thank the authors and publishers for their permission to use the works here included. And of those involved in the production itself, we especially appreciate the typing assistance given us by Martha Melfi and Doris Wrightsman and services in the print shop performed by Ellen Gural—all of the College of Environmental Science and Forestry at the State University of New York—and secretarial help from Amy McKee, of Virginia Polytechnic Institute and State University.

What is the philosophy, the set of criteria for including or excluding items for this book? Does it consider relevance to a pre-set area of interest? orthodoxy of viewpoint? conformance to rules of logic? coverage of a body of principles? prestige of authors? understandability of their writing? Clearly, a mixed criterion governs: a subjective one, not easy to explain or defend. Perhaps the following is what it turned out to be: Include that group of readings which will give mature students the most educational experience in the economics and other social sciences of forestry—provoke their thinking, inspire their critical faculties, help them see the sides of issues, give them some excellence to admire, some faults to discover, some uncertainties to ponder.

The book is organized into five parts, consisting of social-science applications to successive aspects or subdivisions of forestry. As between the parts, so within each of them, the story proceeds from the more general subjects to the more specific. Most of the social sciences are represented among the readings. First emphasis is given to economics, partly because it is so given in the literature as a whole and partly because this is the field with which both the editors and their reviewers are most familiar.

Part One, on the economics and other social sciences of forestry at large, is organized so as to bring out both the subject's historical development and its concern for a range of interdisciplinary combinations. The opening item is a historical treatment by Edward I. Kotok, a national forestry leader during the formative years of social-science applications. It is followed by a review of economic theory which serves to establish some of forestry's social characteristics. The story then moves successively into areas bordering upon cultural anthropology (Items 3 to 5), social ecology (5 and 6), psychology (7 and 8), education (8 to 10), sociology (9 and 10), political science (11 to 13), and management science (14 to 16). As the first item concerns the history of forestry, so the last concerns its growing tip.

Part Two comprises articles on the main classes of inputs into forestry. (a) The forest owner is the subject of the scholarly opening item, by the late Professor Keniston, who reviews the literature and advocates more application of sociology and psychology in ownership studies. (b) Next, forest land and its use are dealt with in an imaginative theoretical article. (c) Capital and its investment are discussed in a series of four items from the viewpoint both of society and of the individual firm. (d) Finally, labor, a topic much neglected in the American forestry literature, is represented by a general article on problems and a specific analysis of productivity.

Part Three concerns forest resource management. It opens with three items on multiple use and general land-management considerations. From there it moves to specific commodities. There is one all-too-brief article on forest recreation and one on upstream watershed management. The remaining items deal mostly with timber: the applications of economic and related thought to decision making, goal setting, forest regulation, and the weighing of alternative silvicultural practices.

Part Four touches upon the subject of wood-using industry. It commences with an article on industrial interpretation of forest inventories, which thus forms a bridge of sorts with the previous part of the book. Next come three specialized analyses: of the lumber industry, of plywood, and of pulp and paper. Part Four closes with an item on an analytical tool for evaluating industrial raw material: conversion surplus. Stressing, as it does, the end product and the market, this item leads readily into the final part of the book.

Part Five concerns, in a sense, the market: quantities, prices, and trade of forest resources, in this case primarily timber. The series commences with a classic item—perhaps the best-known article in forestry economics—written by Vaux and Zivnuska on the subject of production goals. The following two items also treat production or consumption. Next come two articles on price, the first of which, also a classic, is one of many attempts—and apparently the first largely successful one—to explain the long upward trend in the constant-dollar price of lumber. The work is that of Zaremba. Part Five ends with two articles on trade: in logs and in lumber futures.

CONTENTS

Part One. FORESTRY AT LARGE

Part One

FORESTRY AT LARGE

1

FORESTERS AND ECONOMICS

Edward I. Kotok

First published in *Society of American Foresters, Division of Forest Economics and Policy, Proceedings;* pages 88–94. 1958.

BIOGRAPHY

Mr. Kotok (1888–1966) was Deputy Chief in Charge of Research when he retired from the Forest Service in 1951 to become the first leader of the FAO program in Santiago, Chile. During his 4 years there, he was instrumental in helping that country to draft new forestry legislation, to establish a new forestry school, and to put the FAO program on a solid footing. His long career with the Forest Service commenced after he was graduated from the forestry program at the University of Michigan in 1911. As a research worker, he made many significant professional contributions, beginning with his 1924 publications on forest fire. As Director of the California Forest Experiment Station, he organized and led the early research in watershed management, forestry economics, and fire control. His next-to-last position in the Service was that of Deputy Chief for State and Private Forestry. He remained an active leader and officer in national and international forestry organizations until his death.

EDITORS' SUMMARY

The early development of the forestry economics field was strongly related to current problems, such as those of water conservation and flood control. It is interesting to compare professional foresters and forestry economists with other professionals such as military and agricultural engineers, agricultural economists, and soil conservationists.

QUESTIONS TO CONSIDER

(1) Mr. Kotok characterizes forestry economics as having arisen in response to urgent public questions of the day. Is this kind of origin characteristic of other applied social sciences? of "general" social sciences? of all sciences? of all areas of human interest?

(2) Mr. Kotok explains that the earliest workers in forestry economics were foresters, and

the later workers, foresters who studied economics. Why have there been so few who were economists, or economists who studied forestry? Which of the two alternatives would have been preferable?

(3) How do you explain the emphasis placed upon flood control by Mr. Kotok in his history of the forestry economics field?

(4) Mr. Kotok refers to the problem of choosing the *sound* alternative for reaching a given goal. How do you interpret *sound*? How does one decide what is sound?

(5) How does the forestry economist contend with the problem that he is concerned with human behavior but that economic theory provides only fragmentary and imperfect insight into human behavior?

* * *

American foresters entered the field of forest economics with the very beginning of the Bureau of Forestry in the U.S. Department of Agriculture. Studies were conducted on specific areas to determine the economic justification for specific management plans. Probable costs and returns for given systems of silviculture and management were evaluated. Many of us will recall Graves', Frothingham's and Zon's bulletins that dealt with the possibilities for forestry on private properties. As a matter of fact, these bulletins were used as texts in the forestry schools of that day. The economic evaluations used by these early authors were drawn chiefly from formulations included in German forestry texts. These early economic studies were written by foresters and not economists.

EARLY WORK IN ECONOMICS

With the establishment of the U.S. Forest Service and the national forest system, a division was set up in Washington charged with gathering statistical data dealing with forestry and forest products. Kellogg and Hall, the earliest chiefs of this division, used these data to venture into the field of economic forecasts.

Whatever courses in economics were given in the early forest schools, at Cornell, Yale, Michigan, Nebraska, and Harvard, were in conjunction with management studies. Whatever economic methodology was introduced followed closely Germanic forestry texts. Here again, it should be noted that forest economics as such in the early forestry schools was taught by men trained in general forestry.

The first forest economics textbook in the United States was Fernow's *Economics of Forestry*. This text could hardly be considered adequate today either in content or treatment. Nevertheless, whatever forest economics was taught was based on this text.

Foresters also entered the field of forest economics in inquiries dealing with general forest conditions. Beginning with Gifford Pinchot's term as chief of the Forest Service, the Congress, through resolutions, requested the Forest Service to survey and report on specific forest problems. The reports prepared in conformity with these resolutions generally included economic criteria and economic speculation. The state of forest conditions, the treatment of forests, national requirements for forest products, and inferences as to the impact of forestry on the national economic welfare were appraised. It is to be noted that these reports were prepared by foresters and not by trained economists. Whatever shortcomings these reports may have had in their economic methodology, they had an effective influence in actions by the Congress.

The Forest Service undertook economic inquiries as part of other programs. For example, the Forest Taxation Inquiry involved many economic aspects. Thus the Forest Service, for the first time, in 1925 employed a distinguished economist to head the inquiry, Professor Fred R. Fairchild of Yale.

During World War I, Forest Service personnel developed information on supplies and requirements of forest products needed in the war effort. These short studies had many economic implications.

In 1926 the Division of Research created a unit of Economic Investigation with R. E. Marsh in charge. Here again, the personnel of this group were not economists but foresters.

The Forest Service enlarged its activities in economic research with the passage of the McSweeney-McNary Act in 1928. For the first time provision was made for a comprehensive program of forest research including economics as one of its key divisions in its organizational setup. Work in forest economics was provided at the Washington level and at each forest and range experiment station. It was recognized that the staffing of these new economic research divisions would require trained forest economists. Because of this, a number of Forest Service men were encouraged to take advanced degrees in economics. Forestry schools, on their part, added chairs in economics and offered advanced degrees in that subject. By 1935, as a result of this effort, a few forest economists became available both for service in the Forest Service and professorships in forestry schools. The present leaders in the economics field, both in the government and in the schools, have been largely recruited from this early crop of doctorates in economics.

At this point it might be well to review the results of some of these early efforts by the Forest Service in the economics field. The earliest studies in management of specific tracts attempted to show through economic criteria that private forestry in the long pull would pay. R. S. Kellogg's review of the timber supply of the United States and subsequent reports including the United States Congress Joint Congressional Report of 1938 were largely overall economic appraisals of our forest problems. The tone of the reports reflected the Forest Service concern as to the weak position of our forest situation and the need for strong national forest policy. These reports pled for greater congressional and public support along the whole forestry front. They preached the need for a bountiful supply of timber resources, the maintenance of our forests in fullest productivity, predicating these suggestions on the belief in an expanding economy.

These findings and recommendations of the Forest Service did not go unchallenged. Organized industry, through its associations, registered objections to the inference that forest lands were being destructively exploited or that the industry was not equal to the task of putting its own house in order. The critics seriously questioned the predictions of a future imbalance of supply and demand in timber or in forest areas.

The overall effect of these early Forest Service economic studies might, in the light of historic perspective, be stated as follows:

The Forest Service concern as to the situation had a definite impact on the executive and legislative branches of our government, which was translated into positive and substantial support in affirmative legislation and increased federal appropriations. As a corollary it might be stated that these inquiries also stimulated individual state action on forestry programs. There were some industrial leaders who gave more considered attention to the forestry aspects of their enterprises, even though they believed the Forest Service reports to be slanted and biased.

Up to 1938, then, we have an illustration where inquiries with strong economic implications were largely prepared by foresters untrained in economics. Perhaps the economic criteria and methodology employed in these inquiries would not meet the exacting methodology of a formal, orthodox discipline. Nevertheless, these economic prognostications, with all their limitations, served an important part in the formulation of national forest policy.

If we examine some of the criticisms of that period, we find these charges against the inquiries: The studies were made not to formulate policy but to prove that established or proposed policies were warranted by imposed economic assertions. The veracity of the data and the assumptions of the premises were questioned. However, one cannot find any criticism as to the economic methodology used except objections to the long time factor used in projecting forecasts.

Beginning with the Franklin D. Roosevelt Administration we find forestry economics entering into many fields of government action. I shall cover a few to illustrate this trend.

Among the more significant programs was the Flood Control Act of 1936. This important act formally and specifically placed part of the responsibility for flood control reports and action in the Department of Agriculture. By law, the department was given the responsibility of employing economic justification for its recommendations. In the department there was a division of responsibility. The land-action agencies, namely, the Soil Conservation Service and the Forest Service, were assigned the job of making surveys and carrying out the action program. The responsibility for the economic evaluation was assigned to the Bureau of Agricultural Economics. This division of responsibility for surveys and economic evaluation led to confusions and conflicts. It has an important bearing on the part forestry should play in flood control. The economic criteria of the forestry items in the survey were subjected to the viewpoints of agricultural economists, not forest economists. The inherent philosophy of the agricultural economists who participated in these programs was unusually conservative as compared to the philosophy of forest economists. Forest economists were challenged as being propagandists and unrealistic. The forest economists employed by the Forest Service in its part of the flood control program formulation were largely thwarted in impressing their point of view.

This conflict between foresters and agricultural economists was never satisfactorily resolved. The question of whether or not there are any economic scientific criteria sufficiently measurable to be useful in estimating the public interest was never answered. Thus, this was not a clash between economic methodologies but a clash between economic philosophies.

In previous studies by foresters in the governmental services, economic theory and philosophy revolved around the concept of inherent public interest working in a free enterprise system. It predicated that the overall public good transcended all other values in the long-term point of view. Now the foresters found themselves in a conflict where economic yardsticks were to be applied to a minutia of choices and where a discount factor in evaluation practically de-emphasized the long-term public interest.

THE WATER PROBLEM

In a world of conflict, beset with innumerable problems, it would be dangerous to single out any one field of activity as a focal point of our national concern. However one can safely venture the guess that the problems revolving around water, namely, the control of floods, the harnessing of water for useful purposes, and the prevention of the wasting of agricultural soils, loom high among the vexing problems confronting our nation.

The cost for carrying out long-term programs of water conservation will reach astronomical figures. Water conservation is needed in every region of our country.

Correlation of Water Uses

The control of floods cannot be treated in a vacuum. The activity is generally associated not only with controlling the destructiveness of moving water but also in harnessing and using it for agricultural and industrial needs. This multiple role of water presents complexities, conflicts of interests and program direction. It would be a rather simple matter to develop a program solely in flood control to protect a given area. The economic criteria for justifying such a program could be easily established. But when a flood control program must be considered with other objectives, such as irrigation and navigation, power development and distribution, economic criteria become confused. What further complicates the problem because of this multiple use of water is that a simple program for flood control must frequently be radically changed to fit into the requirements of a complex multiple use program.

Alternative Solutions

Flood control problems are confronted with other inherent difficulties. We have been three-quarters of a century in the business of controlling floods. We have built up a large body of information both about floods and their control. Yet, at the present time, the technicians in this field seem miles apart as to methodology in treatment of the physical factors that can be accurately measured. We still argue the questions of big dams versus small dams, dikes and revetment of streams, and overflow areas. The place that vegetative cover plays in flood control is not universally accepted, particularly by the engineering groups charged with the responsibility of harnessing water and controlling floods. Thus, even under the new federal Flood Control Act, land treatment as a measure designed for flood control is reluctantly included in current programs. If there is lack of agreement within the technical aspects of the program, the economic criteria for their evaluation, particularly in the land-treatment phases, have even more divergent points of view. The premises of foresters and agriculturists are constantly challenged as to the extent to which specific land treatment effectively modifies the characteristics of water-flow. Unfortunately, the major data dealing with this premise are less than 25 years old, and are largely drawn from isolated experimental plots. Foresters, then, have the job not only of converting the challenges to their premises, but must also establish the economic evaluations with which to equate these premises.

UNDERLYING POLITICAL AND SOCIAL PHILOSOPHY

In the general field of economic theory much has been written as to the extent that political and social philosophies must be applied to economic evaluations. It may not be necessary at this point to fully review the conflicts that have arisen from the widely differing viewpoints of the protagonists of these theories. Nor is it necessary to raise the moot question as to the extent to which politics and philosophy are involved in the economic inquiries. I do want, however, to indicate that the political and social philosophy of those who engage in flood control surveys and correlated activities must inevitably have an important bearing, even if the orthodox methodology of economics is used in arriving at solutions. It might be well, then, to consider with complete candor the general political and social philosophy of the professional groups who participate in the activities of flood control.

Foresters

Foresters are generally optimists, even though they may approach a problem conservatively. They believe in an expanding national economy. They accept surpluses of raw materials as a benign index of a nation's wealth. They believe in an expanding economy because the population will increase. They believe in an increasing standard of living. Foresters, from the beginning of their training, even if they later become forest economists, are taught methods of dealing with long-term intervals. The very nature of the crop raised compels them to accept the long-term approach. Foresters, therefore, use a time factor both consciously and unconsciously that involves an impact on the future and future generations.

Foresters believe that in handling forest properties, whether publicly or privately held, the general common good inescapably must be considered. Foresters deal with forests as an ecological society, although sometimes they do get lost in the woods because of the trees. They unconsciously recognize interrelationships and symbiotic dependencies of a collective group in nature. This underlying philosophy in dealing with trees has undoubtedly influenced their approach to society. It is not difficult, therefore, for foresters to accept Secretary Wilson's aphorism "The greatest good to the greatest number in the long run."

Foresters are taught that, regardless of the silvicultural or management plan devised for a forest property, the inherent capital of the property must not be diminished. In fact, it is stressed that good forest practice predicates that the inherent capital value must be augmented through time. This means that the basic capital, the soil and its fertility, cannot be sacrificed in any method of exploitation without some compensating work to restore it to its initial value. These professional tenets are a potent guide to a forester's thinking. When translated into a philosophic concept it predicates that forest properties are held in trust for future generations. This applies regardless of the character of the ownership of the land. This is not blind worship of the forest per se. It is a recognition that a forest, with its ever-changing ecological sequences, is a heritage of Man that can be used without being despoiled. In a sense, while the forest may be temporarily in the possession of an individual or in the ownership of the body politic, it must be considered a part of the total economic capital of the nation.

The American forester, in contrast to his European colleague, covers a variety of subject matter in his training. While the Old World forester's training emphasizes crop production, we, here, include knowledge of harvesting, manufacturing, and marketing. Thus, the American forester must anticipate the market requirements in the markets of his region and of the nation. He therefore thinks of the forest property under his charge as part and parcel of a vast and broad economic area.

The American forester must consider the ancillary values such as watershed conditions, recreational opportunities, and fish and wildlife. He thus is trained to give attention to on-site and off-site benefits, and the values these may have for the locality and the region, irrespective of whether the owner of the forest negotiates these values or not. When calculating costs and returns for a given management plan for a forest property, the ancillary values are rarely included as an asset for profit return.

These are the professional tenets with which an American forester is indoctrinated in his schooling. Those who enter into public service receive additional dosages of this indoctrination. Now this same forester, even though he receives later specialized and advanced training in economics warranting the title of forest economist, cannot wholly dissociate himself from the political and social philosophy which is part and parcel of the forestry profession itself.

I have outlined the professional training of a forester and the philosophic content of the tenets of his profession. We might now consider the other professional groups that participate in flood control and conservation. It would be presumptuous to try to psychoanalyze the motives underlying the philosophy of any professional group other than one's own. Nevertheless, through the policies that they support and through their publications, one can draw some general inferences as to their underlying philosophies.

U. S. Army Corps of Engineers

This important professional group has had a major part to play in flood control for over half a century. The army engineer was primarily trained as a soldier at West Point. Engineering formed but a small part of his total curriculum. A good many men, however, who were assigned to the Engineering Corps received specialized training in first-class engineering schools. These men assigned to the civilian activities of the corps had the benefit of on-the-job training. But as all good soldiers, they were imbued with certain basic philosophic concepts. Discipline, chain of command, high integrity, and consideration of overall public interest, permeated their approach to the job. Above all, a given assignment had to meet the specifications set up and the time schedule for accomplishment. In their schooling they received very little instruction in economics. Normally, their choice of action was determined by other than economic criteria.

The Army Engineers entered the water field by way of responsibilities assigned them by the Congress for maintaining navigability of streams and harbors. In conjunction with this work their activities were later extended to include flood control of our major river systems. It is only in very recent years that they have entered the full field of multiple use of water development, including even power and irrigation. In the field of flood control the economic evaluations used in their flood surveys have been set by legislative acts.

The charge is made that the Army Engineers first plan and set the specifications for a flood control project and then adjust the economic evaluations required by law in order to justify their engineering recommendations. This may be an unfair charge. But it is undoubtedly true that the engineering features are given first consideration. Undoubtedly this group has employed economic criteria in determining choices between alternative engineering plans. As the Army Engineers have entered into the multiple use of water, the importance of economic evaluation has been specifically recognized. To meet this need they have followed the general economic models used by the U.S. Reclamation Service.

At this point it is well to note that there was a long period during which the army engineer gave little credence to any value in what we now call upstream engineering and land treatment in ameliorating flood conditions. Since the passage of the Flood Control Act of 1936 making the Department of Agriculture a partner in flood control, the Army Engineers have looked more sympathetically upon the part that upstream measures must play in flood control.

Without being too critical, I venture to state that in the formulation of Army Engineer flood control plans there is little concern as to the need for maintaining the inherent values of the soil, the base for our agriculture and forestry. In their flood control program their emphasis has been mainly the protection of the large urban centers.

Agricultural Engineers

This professional group has been in the field of water conservation and use for a long time. It has had the responsibility for leading and guiding the work of the U.S. Reclamation Service. Their first and most important task was to bring irrigation to the arid West. In progressive steps it has entered the whole field of multiple use of water, and has carried forward important federal projects.

The agricultural engineer has had basic training in engineering as his major discipline, and with it, often, wide training in agricultural subject matter. In the beginning, when he was concerned with irrigation only, it was natural that his philosophy should be influenced by the effect of his work on a farmer. The questions he had to answer were: Will the farmer be able to produce a salable crop and will he be able to make payment for the land and water charges? In the U.S. Reclamation Service these agricultural engineers were assisted by agricultural economists in making economic evaluations for a given project in order to determine its feasibility. Frequently the agricultural engineer himself undertook the economic evaluation.

The charge has been made that reclamation projects were undertaken without giving consideration to the impact additional crops would have on the general national market situation. It is further charged that in the anxiety to create a new project, economic devices were used for justification by introducing subventions in abnormally low interest rates and long periods for repayment. Even with these subventions many of the projects had to be given further federal financial support. Be that as it may, the basic philosophy of agricultural engineers was to help settle the arid West. That was their mission, economics or no economics. And it was accomplished.

Agricultural Economists

My purpose in discussing this group is that they had a part to play in formulating public policy in the water conservation field, in the federal departments, and in many states that have their own water development agencies.

I have not been able to trace just how agricultural economics branches off from the older classical economic discipline. I assume that the underlying reason for such a separation was the belief that an economist should have knowledge of the activities and material on which his economic theorizing was based. Normally, then, an agricultural economist has been trained in the agricultural sciences and in economics as a major. There are, of course, many exceptions to this generalization.

The general tendency, as this new discipline developed, was to draw away from studies of purely philosophic and social content to concrete and specific problems of the farmer and agricultural groups. It placed an enormous emphasis on farming as a business. Farm management studies were designed which would help a given farmer determine economic choices in handling his business. It is true that their activities progressively expanded into wider areas of economic inquiry. This general trend is still going on and its impact is felt in the consideration of overall complexes of our general economy. This discipline has contributed much in developing techniques and methodology for economic evaluations in agricultural economics. It is from this group that the federal and state agencies have drawn agricultural economists in the field of water conservation and flood control.

In recent years much has been published by agricultural economists covering the subject matter of water conservation and flood control. In perusing these writings one can draw some idea of the social and philosophic criteria which permeate their thinking. As in all generalizations, one must be careful not to assume that all within the profession are in accord. It is quite otherwise.

Certain attitudes, however, are discernible. For example, when agricultural economists consider flood control they are philosophically concerned in how a given program will affect a farmer and rural life. The old farm management theory of costs and returns intrudes itself when they consider a flood control program. Once they fall into this line of thinking, economic evaluations are constricted by certain limitations. The span of the time element falls within the narrow limits of a farmer's working lifetime. The interest rates used reflect more nearly what a farmer would have to pay on a relatively short time interval. The choice in the use of money approaches what a farmer would do if he had to make the choice, rather than what the body politic must do in a long-term point of view. This group shies away from meeting the whole question of social and political implications in a given program. They do not want to commit the present generations to any stringent obligations for the requirements and needs of generations yet unborn. They are not too clear as to the extent to which federal and state obligations must of necessity be applied in protecting the general welfare if it extends beyond a very limited time span.

Soil Conservationists

Within the last 20 years a new professional group has been emerging—the soil conservationists. They are a phenomenon of our great depression of the thirties.

The origin of this profession starts with the founding of the Soil Conservation Service. The profession includes more than one discipline. Engineers, a gamut of agricultural scientists and practitioners, foresters and economists, all work together for a single purpose—conserving our soil wealth.

The Soil Conservation Service was born in an economic depression, in strife and conflict with established agencies in the field of agriculture. Fortunately, its leadership was in the hands of protagonists who could not be diverted from one primary purpose—the necessity for stopping the wasteful erosion of our precious soil by using land practices which ultimately would build up soil values. Theirs was a long-term philosophy. Theirs was the belief that we hold our agricultural lands in trust for future generations. They were not overly concerned with precise economic evaluations, nor with the fact that the body politic would have to underwrite a substantial part of the costs for their programs. In a sense it was a daring and startling new philosophy of conservation, but in a general way it followed the pattern established by Gifford Pinchot and Theodore Roosevelt when the conservation movement in forestry was initiated in 1905. Conflicts and contests have a way of molding and shaping philosophies and of bringing the professional groups that support such movements together with a high sense of public responsibility.

Under the new Flood Control Act the Soil Conservation Service was given the leading responsibility for the Department of Agriculture's part in flood control programs. In carrying out its assignment for the land-treatment phases of the upstream programs it, too, has been confronted with the restricted limitations that the agricultural economists of the department set up for evaluation of the programs. Frequently the Forest Service and the Soil Conservation Service have jointly been faced with the economic restrictions imposed by the agricultural economists and have jointly sought a wider and more favorable interpretation in economic evaluation of the overall public interest.

Policy Makers

I have listed the professional groups that have the official responsibility for recommending flood control and water conservation programs and their philosophic attitudes that influence their approach to such programs. These professional groups, too, through educational facilities, publicity and propaganda, have the responsibility of molding public opinion. In this educational process these groups are often charged with being biased bureaucrats, feathering their own nests, dreamy-eyed idealists, and, in some bitter conflicts, sharper and unkinder names.

In the last analysis, however, our water conservation programs rest squarely on the responsibility vested in the executive and legislative branches of our government. The philosophy of these political groups is predicated on the viewpoints of public opinion and the powerful pressure of special interest groups. Since public opinion fluctuates and the conflicts within special interest groups give varying political power to one or another, the ultimate philosophy that guides the political branches of government reflects these variations.

So, our water conservation programs depend upon the mood and attitudes of the body politic to the extent to which it makes itself felt through the executive and legislative branches of our government. In this general process of building up a national program for water conservation there are overall trends that resist the vagaries of temporary political fluctuations. For shorter periods of time there may be and often are radical changes in policy approach. The question can be raised whether the general voting public is well enough informed in the complicated field of water conservation to arrive at sound judgments as to the position they should support. There is merit to this question. However, we are fortunate that our democracy often produces leaders who prophetically espouse a cause in the public interest and who stimulate response from the public itself. Within the legislative branch of government we frequently find men of strong belief and influence who are willing to support a cause against great odds. One can name such outstanding men as the late Senators Norris of Nebraska, Hiram Johnson of California, and McNary of Oregon. I am hesitant in naming some equally strong present day protagonists of water conservation now in the present Congress.

It is not surprising to find in the conservation field an outstanding individual who epitomizes and personifies the essence of this movement. Thus, we have had a Gifford Pinchot and a Hugh Bennett, in times past when the whole conservation idea suddenly crystallized. There is room for such leadership.

UPSTREAM PROGRAM DETERMINATION

In discussing water conservation it is convenient to artificially segregate the upstream aspect from the downstream. The upstream generally covers the land treatment phases while the downstream carries the multiple-use aspects of a program.

Normally, the upstream features are subject to little controversy excepting for the economic evaluation aspects. The agencies who carry forward upstream measures adjudicate their differences and resolve the problem of division of work with reasonable harmony. The difficulties in the upstream program proposals arise from the conflict between the philosophic viewpoints of the conservationists and the agricultural economists. These issues often confront them: First, agreement as to the premises upon which evaluation should be made; second, the extent to which the classical and orthodox use of the time factor as it affects discounts, interest rates, and other minutiae of an economist's procedure, is applicable.

No one will quarrel that good economic methodology for evaluation is a requisite for good work. The underlying difficulties between the conservationists and the economists, however, go deeper, determined largely as to recognition of the public interest in the long run. Such sharp differences cannot be easily compromised or resolved.

DOWNSTREAM PROGRAM DETERMINATION

Downstream programs are complex and full of conflicts. They include power and irrigation, sources of revenue, and profit. Along with these revenue-producing features we have flood control, navigation, and public recreation. In the revenue-producing areas we have conflicts between federal and state governments as well as aggressive private corporations. There is less pressure for the flood control and navigation phases which are non–revenue producers. Political stakes are high, frequently determining the selection and election of members of the Congress. Since downstream programs are usually under the control of federal agencies, we find that within the government there are also further conflicts between agencies as to jurisdiction in programming and implementation. Our concern here is to analyze the extent to which the economist entered into this area of conflict.

In the last decade an exhaustive literature has been produced which covers the economic aspects and implications of downstream programs. While much has been written as to the economic methodology used, the major concern has been the problem of formulating criteria for the allocation of costs between the revenue producing features and those that are non–revenue producing. Throughout this literature one finds a tendency for special pleading. The power groups want to allocate to irrigation a fair share of their true costs; the irrigationists, in contrast, want just the reverse. But both of these groups agree to maximize allocations of costs to the non–revenue producing features in terms of the public interest at stake. This in effect places the flood control and navigation parts of the program in a favorable position because both the revenue-producing groups support these regardless of any economic justification for them. The charge has been made that the proponents for downstream programs manipulate their economics to suit their needs. There is, however, a general trend to work out a uniform formula for estimation, evaluation, and allocation of cost by agreed upon criteria and methodology.

THE TASK AHEAD

The forest economist and the economist who deals generally in the field of conservation can play important roles in the formulation of our water conservation program. Their major contributions will lie largely in the area of upstream flood control. The downstream programs will for a long time to come be considered by other economists who are primarily concerned with direct revenue-producing developments. The conservationists will face the task of seeing that in the downstream programs the flood control features as they apply to non-urban areas are fully considered.

In approaching the economic formulation of upstream measures one should be more concerned with the philosophic and social premises that the economist employs rather than with economic methodology as such. I fear, speaking of the forest economist, that his desire to conform with economic orthodoxy will divert him from his major task.

Since many of the forest economists have been trained under agricultural economists the forest economist must reorient himself as to the inherent differences between the agricultural short-term viewpoint and the conservationist's long-term point of view. In my reading I have found very little philosophic content in the literature of the agricultural economist. They have for the most part been largely concerned with methodology. In contrast, I have found far richer material dealing with the social and philosophic yardsticks in the contributions of classical economists. We should not forget that if short-term economic philosophy had prevailed there would have been no Louisiana Purchase or Alaska Purchase, and a great share of the development of our West. We find ourselves in the same dilemma—shall we approach the conservation program in upstream engineering by strict calculated short-term economics or shall we commit another Seward's Folly?

THE PRESENT SITUATION

We might pause to examine the general contributions of forest economists since they were added to our professional group.

Duerr, in 1948, summarized the economic articles included in the *Journal of Forestry* for a ten-year period. He noted that the contributions covered the fields of production, supply, demand, tenure of lands, and management of forest products. There was a complete lack of consideration of the economic implications of the public interest. The field of flood control, from the economic angle, was not treated. I have reviewed the *Journal* for the 1948–1958 period and find that, except for one issue in 1951, the contributions in economics followed the same pattern noted by Duerr for the earlier period. For the decade I examined, covering 43 articles, only three dealt with watershed management as a factor in flood control. One issue for 1951 did consider some aspects of the public interest in forestry.

THE FOREST SURVEY

The inauguration of the national *Forest Survey,* because of the nature of the task, stimulated a series of economic studies. Substantial data dealing with forest land areas, stands and growth, by regions and subregions, furnished unusually valuable material which could be subjected to economic appraisals. The last publication in this field, *Timber Resources for America's Future,* is an excellent illustration of the kinds of studies to which I refer. It might be well to note here that the major leaders in this last study came from our first crop of trained forest economists, although many other forestry specialists participated.

SUMMARY

May I be permitted as one of the passing generation to make this observation: Techniques, methodology, skills in the use of material, all are important in scientific inquiry. But when we touch an area in which human behavior and human response are involved we must depend upon some underlying assumptions which reflect with reasonable accuracy the hopes and aspirations of a people. We are, therefore, challenged in trying to develop a national forest policy, to adduce a sound economic philosophy and a sound social philosophy, which will reflect the true worth of the public interest.

Let us assume we can have agreement as to the premises in a given economic philosophy. Let us further assume that there is agreement as to the public interest at stake in a given proposition. Within these limits we still may have choices for alternatives in reaching the goal. Here the methodology of economic theory must be provided to arrive at a sound equation which will indicate the most favorable choice for action.

The question is whether the methodology now largely employed by agricultural economists is applicable to conservation economics, or whether new ground must be plowed which will furnish a more accurate and useful tool to serve our needs.

This generation of forest economists must boldly and daringly venture into a new approach towards the whole problem of conservation economics. I wish them success.

2

SOME ASPECTS OF THE ECONOMIC THEORY OF FORESTRY

John A. Zivnuska

First published in *Land Economics* 25(2):165–171. May 1949.

BIOGRAPHY

Mr. Zivnuska, a native Californian with academic degrees from his state's university and from the University of Minnesota, is Professor of Forestry in the College of Natural Resources at Berkeley. He is known for his perceptive and constructive critical commentary upon forestry and conservation problems and his profession's approach to them. He is known, too, as a research worker (long-term demand, economics of the wood-using industry, forest policy, taxation, international forestry), a widely traveled visiting lecturer, and a consultant to industry and government. He is a fellow of the Society of American Foresters and a member of such broadly diverse organizations as the Society of Range Management and the Sierra Club.

EDITORS' SUMMARY

There is need for more theoretical studies of forestry economics in light of the special characteristics of forestry (timber) technology, notably (1) the great length of the production period and (2) the fact that trees are both wood factory and wood product.

QUESTIONS TO CONSIDER

(1) What is meant by the *period of production?* Do you agree with Mr. Zivnuska in his emphasis upon the long period of production as a major economic characteristic of forestry? Are there industries with even longer production periods? industries with notably short periods?

(2) Do you think that forestry production periods in the United States have been growing shorter or longer in recent decades? What will happen in the future? How do you explain the change?

(3) What are the actual consequences of the fact that trees are both wood factory and wood product?

(4) Like Mr. Kotok, Mr. Zivnuska uses the term *sound*—in the expression *sound forest economy* (page 17). What is a sound forest economy? By what means might such an economy be achieved or approached? Is *laissez-faire* policy the answer? Why or why not?

(5) If Mr. Zivnuska were writing about the economic theory of forestry today, what topics would you expect him to emphasize that he did not emphasize in 1949?

* * *

Forested areas have for some years been one of the primary zones for work in land economics and land use planning. Such studies as well as most studies in the economics of forestry have been centered almost exclusively on the institutional aspects of the forest problem. Studies of tenure, ownership pattern, taxation, marketing, governmental costs, and so on, have done much to explain the nature of the forest problem and to indicate means of encouraging the development of sound forest practice. However, theoretical studies of the economic bases of forestry have been largely overlooked. It appears probable that such studies could be of substantial value in clarifying the problem with which we are attempting to deal and in developing a model within which to evaluate various plans and procedures. This paper is intended as a preliminary exploration of this field and as an indication of the types of problems which are involved.

* * *

I

Under existing institutions, primary reliance for the development of a sound forest economy on [the] . . . vast privately-owned area [in the United States] is placed on forces of an essentially *laissez-faire* nature. While the theories of monopoly, of bilateral monopoly, and of monopolistic and imperfect competition all are of value in understanding particular phases of the forest economy, the general economic theory of forestry can well be developed within the context of the theory of competition.

One of the basic assumptions of *laissez-faire* economics is that the individual or group will normally attempt to select the course of action that will maximize present net worth in terms of money. This implies that, under existing institutions and with certain well-known minor exceptions, commercial forestry will be practiced to the extent that entrepreneurs anticipate a return to resources and efforts devoted to forestry which will have a present net worth at least equal to that for equivalent expenditures in alternative fields of economic activity. The key to the forestry entrepreneur's action, then, is the same as that for other entrepreneurs: a comparison of anticipated production cost with anticipated sale price. Thus in overall conception the economic theory of forestry is in accord with orthodox theory. However, in its detailed development certain distinctive characteristics of forestry enterprises become of great significance in modifying the conclusions which may be drawn.

One of the most significant and distinctive characteristics of forestry is the extreme length of the production process. The production of forest trees is essentially a biological process the length of which is determined by a combination of the natural characteristics of the several species, the site on which the forest crop is grown, the requirements of the industrial uses to which the trees are to be put, and the intensity of the silvicultural treatment. In the majority of cases the production period of a forest tree falls within the range of 50 to 150 years, although in exceptional cases both longer and shorter production periods are found. This length of the production period, which is not approached by any other industrial or agricultural production process, is the basis of many of the characteristic problems of forestry enterprises.

A second significant characteristic of forestry lies in the fact that a productive forest is simultaneously the "factory" used in producing wood and a storehouse of marketable raw material. Thus a wood inventory many times larger than annual sales must be carried if production is to be maintained. According to studies of the U.S. Forest Service[1] the board-foot volume of the present inventory of timber on commercial forest lands is over thirty times larger than the annual cut. Assuming an active management and a proper distribution of age classes it is estimated that, to achieve sustained yield of forest products, a ratio of board-foot volume of growing stock (inventory) to yield (annual production) approximating 23 to 1 on a national basis would be required. Regionally, this ratio would vary from about 17 to 1 in the South to 45 to 1 in the Rocky Mountain region.

Due to the length of the production period, additional resources and labor applied to forest production at the present time will not achieve their full expression in terms of the quantity of standing timber until after many decades. Conversely, if productive efforts applied to forestry are reduced the resulting effect on the quantity of standing timber will again require many years to develop fully. At the same time it is possible to achieve rapid adjustment in the quantity of timber actually offered in the market. Reservation price plays an important role in timber marketing. Although occasionally pathological or insect conditions may necessitate prompt cutting, it is ordinarily possible to hold old-growth timber on the stump over a fairly long period of years without substantial change in the stand, while with young-growth timber such as would be produced by forestry enterprises, withholding the timber from the market will usually result in both an increase in volume and an improvement in quality. Conversely, the characteristic high ratio of inventory to production makes it possible to offer on the market during a short period a volume very much larger than the production during that period.

Thus, these two characteristics of forestry combine to create an anomalous condition for a produced good in which the physical quantity of standing timber is subject to only slow change over a long period of time while the effective supply offered on the market is subject to pronounced short-period variation. The significance of these characteristics can be clearly illustrated by a consideration of the application of conventional competitive price theory in explaining stumpage value.

II

Stumpage value is the value of the timber as it stands uncut in the forest. The accepted methods of stumpage appraisal are forms of residual determination.[2] The usual conception of stumpage values has been concisely summarized by Steer,[3] who states:

> The theoretical market value of standing timber is the residue left when the total cost of manufacturing and marketing the forest products fabricated therefrom plus a reasonable margin of profit is subtracted from the total price received from the sale of the manufactured forest products. . . . Thus stumpage realization values, theoretically at least, fluctuate up or down directly as does the selling price of the product and inversely as the costs of logging, manufacturing, and merchandising.

While the use of a residual technique in appraisal does not necessarily involve such a theoretical justification, it is apparent in discussions of stumpage values that such values are considered to be economic residuals in nature. This is particularly apparent in the conception of differential stumpage values. For example, Chapman and Meyer[4] state:

> As in the case of land rent, this stumpage value depends on the margin between the prices of the various qualities of products obtainable from the timber and total operating costs and profits incidental to its removal and manufacture.

Since timber, unlike land, is not indestructible, the conception is essentially that of a royalty rather than a rent.

During a period in which timber is treated as a mine, such a theoretical conception of stumpage appears wholly adequate. However, once it is recognized that timber can be a crop—that is, a material which can be produced through the application of economic effort—it becomes necessary to modify this conception to allow for the influence on value of production costs.

This problem has been recognized by Marquis,[5] who distinguishes between two types of timber, "virgin timber" which is "costless" and "produced timber" which is "the result of measurable cost." He then holds that these two types occupy different niches in the theory of value.

> The value of virgin timber, like the value of agricultural land, is derived from the value of its products. . . . A different method of determining value is involved when stumpage is produced as a conscious effort. . . . The value of the lumber produced is still the most important determinant of the demand for stumpage, but costs affect the supply of timber on the market, and through their influence on supply, the value of timber.

Accordingly, Marquis holds that the stumpage value of "virgin timber" is entirely product-price determined whereas the stumpage value of "produced timber" is product-price determining:

> If the logs for manufacture come from a cultivated forest, the cost of producing lumber or other products properly includes the cost of the logs. The cost of such stumpage is one of the determinants of the value of stumpage, which in turn is the cost to the lumberman and one of the determinants of the price of lumber. On the other hand, the cost of logs from a virgin forest is not properly included in the costs of manufacturing lumber when such manufacturing cost is considered as a determinant of supply and an influence on lumber prices. . . . Stumpage that derives its value from the value of the product cannot be reckoned as a cost in determining this same price of the product.

Such a distinction between "virgin timber" and "produced timber" is highly artificial and unsound. The remaining areas of virgin timber have been in fee ownership for many years. Substantial investments in costs of protection, administration, and carrying charges are involved. There has been produced within them through conscious effort what might be termed a utility of time which is as real as a utility of form or of place. On the other hand, many of the second-growth forests which now exist have developed with no conscious effort other than that involved in the claim to ownership. Thus it becomes impossible to distinguish between the two in terms of conscious effort.

Furthermore, the fact that timber is a crop means that the existing timber, both "virgin" and "produced," can be replaced by the expenditure of productive effort. In the determination of market price, historical costs are not relevant. The only alternative to selling an existing stock of goods in a given market at whatever price is offered for it is to sell it in a different market—different either geographically or in time. Thus the cost element in the determination of market price is essentially the seller's anticipations of future prices. The relevant costs of production are future costs, for it is these which bear on the anticipations of sellers and buyers. Once the possibility of producing timber is granted, such future costs exert their influence equally whether the extant crop is virgin or produced.

Thus in the theoretical determination of the market price of stumpage the influence of costs of production on the position and shape of the supply curve is the same for timber of a particular quality, whether it is virgin or second-growth. Further, the usual statement of competitive price theory that the market price *tends* to equal the normal price applies to all types of timber. However, in evaluating the significance of this tendency it is essential to consider both the relative importance of virgin timber in the economy's timber supply and the distinctive economic characteristics of forestry enterprises.

In a forest economy dominated by virgin timber it is obvious that economic forces have had but limited opportunity to adjust the quantity of timber so as to bring market price into equality with normal price and so with the cost of production. Such a situation is exemplified by past experience in the United States during which market price appears to have been substantially below normal price and entrepreneurs have not been attracted to forestry enterprises since stumpage could be bought more cheaply than it could be produced.

As the virgin timber is liquidated and the forest economy shifts increasing to second- and third-growth timber, the opportunity for economic forces to affect the quantity of timber increases and it becomes theoretically possible that market price will come into equality with normal price and so with the value of the economic effort required to produce the timber. However, the significant question is the extent to which this will actually occur. In the absence of institutional obstacles a substantial degree of equality between market price and normal price can be expected under conditions of a relatively stationary economy, for goods which are subject to rapid changes in production, or for some combination of these characteristics. Against these requirements the actual situation for forestry is seen to be one of a highly dynamic economy with rapidly changing markets and a production process which is sluggish in response to changes in productive effort to a degree unknown in other economic enterprises. As a result, and aside from all questions of institutional influences, any equality between market price and normal price for stumpage can be expected to be both fortuitous and transitory.

As Alfred Marshall[6] has clearly pointed out:

As a general rule, the shorter the period which we are considering, the greater must be

> the share of our attention which is given to the influence of demand on value; and the longer the period, the more important will be the influence of cost of production on value.

In interpreting this statement, "shorter" and "longer" are best thought of as being relative to the length of the production cycle. As has been emphasized above, the usual period of economic decision and economic activity is extremely short as contrasted with the length of the production period of forest trees. For example, assume that, as is the usual practice, a year is considered an appropriate unit of time for the analysis of market price phenomena of timber. For most manufactured products such a time unit would be a multiple of the production period and for most agricultural products, with the exception of orchard-type crops, it would be at least equal to the production period. For timber, however, the usual production period is anywhere from fifty to one hundred fifty times this long. Thus, for any period of time ordinarily used in price analysis the pricing of stumpage is, in the Marshallian sense, an extremely short-period phenomenon.

As a result, demand is the primary determinant of stumpage price. It appears, therefore, that even after the liquidation of the virgin forests, for any usual period of economic analysis, such as a year, stumpage prices will be largely determined by the price of forest products while the costs of production of stumpage will be of limited significance in determining the price of forest products. While there will be some indeterminate element of product-price determination in stumpage prices, it can be expected to be wholly overshadowed by the forces of residual determination. This follows from the disequilibrium of market price and normal price which can be expected in an economy characterized by shifting demand during a period of time too short to permit any substantial change in production. Any equality of the two which might actually occur would be essentially coincidental. Theoretically, at least, the effect of this residual determination may be either negative or positive. In other words, market price may be either forced below or held above normal price. It appears possible that this latter situation now prevails in some segments of the forest economy and may be more common in the future.

It might also be noted here that this same type of situation results in a much less satisfactory basis for entrepreneurial expectations in forestry than in other enterprises. In most enterprises the experience of one or two decades covers several to many production cycles and provides a fairly adequate basis for expectations as to production changes needed and to be expected in response to changes in demand. In forestry, however, a corresponding experience base covers only a fraction of a production cycle and is a proportionately less satisfactory basis for expectations.

Thus the simple fact of the length of the production period of forest trees makes the two-way mobility assumption of the competitive adjustment invalid except in terms of very long periods of years. The resulting difficulty is greatly magnified by the extremely high ratio of inventory to annual sales resulting from the fact that trees are the "factory" as well as the product. For example, if it be assumed that an equilibrium price situation had been achieved and was then followed by a decline in demand, the resulting excess of production pressing on the market would be multiplied many times by the necessity of reducing growing stock to reestablish the equilibrium situation. This in turn would lead to a pronounced depression of stumpage prices below costs of production which would continue over a prolonged period while the sluggish production adjustment took place. On the other hand, if demand rose following the establishment of an equilibrium, the need for increased production would be multiplied many times by the necessity of increasing growing stock to increase production. Prices would rise above production costs and a prolonged period of deficient timber supplies would ensue.

It might be held that this type of problem is not unique to the field of forestry but occurs to some degree in almost all forms of economic activity. This is, of course, true. For any form of production there is a lag in the reestablishment of equilibrium following a shift in demand, the length of which depends on both institutional factors and the type of production process. The difference is one of degree. However, the magnitude of the degree is such as to place forestry in a distinctive economic context. Aside from all institutional problems, as long as demand is unstable pronounced conditions of disequilibrium between market and normal price prolonged far beyond the usual period of economic action and planning can be expected to develop simply as a result of the inherent nature of forest production.

This basic problem is both partially hidden and made the more complex by the possibility of extreme flexibility in volume of cut which is combined with the extreme sluggishness in response of the quantity of timber extant to changes in productive effort. When shifts in demand occur, the exigencies of the moment can be met readily by either holding timber on the stump or overcutting growth. If the shift in demand is temporary, no serious problem arises. Indeed, it is only through some combination of these procedures that the problem of cyclical shifts in demand can be met by forestry enterprises. However, if the shift in demand represents a permanent change the extent of the maladjustment is compounded by this short-period adjustment. If demand has declined, the short-period adjustment increases growing stock while the long period need is for reduction of growing stock. On the other hand, and probably the more serious of the two, if demand has increased, the short-period adjustment reduces growing stock while the long period need is for increase in growing stock.

III

The implications of these economic consequences of the simple biological facts of tree growth extend well beyond the analysis of the theoretical aspects of stumpage value. Under a *laissez-faire* philosophy the criterion of proper allocation of productive effort is that of equal returns to equal productive effort in all lines, since under such conditions there would be no possibility of increasing total returns through shifting a unit of effort from one type of activity to another. Reliance is then placed on the action of competitive forces to bring about such an allocation. When, for any particular good, market price is either higher or lower than normal price, it is evident that this desired allocation of resources and effort has not been achieved.

Thus the same type of analysis that indicates the probability of a substantial discrepancy between market price and normal price of stumpage also indicates the probability of a substantial failure to allocate the optimum effort to forestry enterprises. Aside from all institutional problems, competitive economic forces cannot be expected to maintain a proper allocation of resources to forestry simply due to the problems arising from the inherent nature of the forestry production process in a dynamic economy. The actual allocation may be either greater or less than the optimum.

In a modified *laissez-faire* economy social action in the realm of production is usually undertaken either (1) to modify or control existing institutions so as either to enable economic forces to lead to the competitive adjustment or to simulate the results of the competitive adjustment, or (2) to modify the results of the competitive adjustment in those instances in which the economic consequences conflict with particular social conceptions. In the development of social action in the field of forestry through land use planning programs and other devices it is important to realize that the inherent economic characteristics of forestry production place a definite limit on the results which can be achieved through manipulation of the institutional environment. At the same time these problems emphasize the importance of developing a favorable institutional environment, since even at best a full adjustment of forest production to needs cannot be expected.

The significance of this condition to forest policy is obvious. It is not the purpose of this paper to develop these implications. However, it should be made clear that this analysis does not necessarily lead to the conclusion that more centralized planning is needed since it is frequently held that it is under conditions leading to prolonged and difficult adjustments to unforeseen changes in circumstances that a dispersion of decision is most desirable.

In the development of land use planning for forested areas it is essential to study both the institutional problems and the economics of the production processes being considered. A clear conception of the problem is the necessary base for adequate analysis and action. Further development of the economic theory of forestry holds promise of contributing much to such a clarification in the conception of the forest problem.

* * *

REFERENCES

1. U.S. Forest Service. Gaging the Timber Resource of the United States, Rep. 1 from A Reappraisal of the Forest Situation, 1946.
2. For details of these procedures, see H. H. Chapman and Walter H. Meyer, Forest Valuation. New York: McGraw-Hill Book Co., 1947. Pages 361–471; and D. M. Matthews, Management of American Forests. New York: McGraw-Hill Book Co., 1935. Pages 275–336.
3. Steer, Henry B. 1938. Stumpage prices of privately owned timber in the United States, U.S. Dept. Agriculture Tech. Bul. 626. Page 8. It should be noted, however, that Steer places major emphasis on institutional factors in the actual determination of sale price, stating that "the relationship between the fair market stumpage value of many tracts of timber and the price for which they have been sold has, during the period covered by this investigation, been so slight as to be practically negligible."
4. Chapman and Meyer. Work cited. Pages 53–54.
5. Marquis, Ralph W. 1939. Economics of Private Forestry. New York: McGraw-Hill Book Co. Pages 27–37.
6. Marshall, Alfred. 1920. Principles of Economics, 8th ed. London: Macmillan and Co., Ltd. Page 349.

3

BROMIDES AND FOLKLORE IN FOREST ECONOMICS

Ralph W. Marquis

First published in *Society of American Foresters, Division of Forest Economics, Proceedings, 1947;* pages 76–81. 1948.

BIOGRAPHY

Mr. Marquis (1903–1966), an economist by education, was adopted, early in his career, into the forestry profession and became one of the American pioneers in forestry economics. While a Ph.D. candidate at the University of Wisconsin, he developed an absorbing interest in the economics of private forestry. His book on this subject, published in 1939, was instrumental in his recruitment into the Branch of Research of the Forest Service, where his first assignment was to study the national demands for wood products and his last was to direct the Northeastern Forest Experiment Station near Philadelphia. Among his honors was an invitation, shortly before his death, to review German forestry activities as a guest of the German government. As an adoptive forester, he retained a degree of detached concern for the profession, typified in the paragraphs here reproduced.

EDITORS' SUMMARY

The traditional forestry doctrines, which show the heavy influence of silvicultural thinking upon the profession, are unfounded and frequently mistaken. Examples are an abhorrence of "idle" land and a belief in unlimited demand for wood. To make forestry economics a social science worthy of the name requires more research.

QUESTIONS TO CONSIDER

(1) Mr. Marquis, like many of the early forestry economists, approached the subject of traditional, or *folklore,* forestry by poking fun at it. Weigh the value of this approach to teaching against that of alternative approaches.

(2) Are "folklore and bromides" to be found in economics? in other social sciences? in other sciences? What are some examples? What is the distinction between "superstition and mythology" on the one hand and "truth" on the other (page 29)?

(3) How may the field of silviculture be conceived and structured so as to escape the faults of anachronism and heterotopy attributed to it by Mr. Marquis?

(4) Like both of the preceding authors, Mr. Marquis uses the term *sound:* He espouses the development of a *sound body of economic doctrine* (page 26). How do you interpret Mr. Marquis's point? Do you think that there is something about the field of social sciences in forestry that produces preoccupation with soundness, or is the use of the concept by our first three writers a mere coincidence?

* * *

One of the most distinctive of all human characteristics is the constant search for knowledge, the quest for answers to all the questions that arise from natural phenomena and from human behavior. One of the less commendable attributes of too many members of the human race is the ease with which their curiosity can be satisfied.

The history of science, philosophy, and primitive religion abounds with examples of the fact that whenever accurate knowledge about any subject continues to be lacking for an extended period of time, something is created to take its place. Many rules of thumb devised to establish operating procedures were adopted not because they worked satisfactorily but because they filled an intellectual void. Now, after long usage, they are accepted as authoritative.

Today, with a vast store of scientific knowledge and what we prefer to regard as a high level of intelligence, we look with tolerance on the superstitions and folklore of our ancestors. We appreciate the ingenuity often displayed in arriving at a wrong answer. We try to understand the mental inertia and the fear of the unfamiliar that caused backward peoples to resist the encroachments of scientific thinking.

But are we equally appreciative of some of the ready-made answers that today take the place of rational thinking and scientific investigation? Do we realize that our actions have been guided for many years by pronouncements made by the medicine men of forestry? Are we fully conscious of the extent to which we accept without question the traditional and sometimes fallacious beliefs that have established and continue to establish our attitudes toward forest economics? Are not many of us still planting our seeds of thought in the dark of the moon, and grinding up unicorns' horns to ward off the things we fear in the future?

This is not to imply that there is no constructive thinking being done in the field of forest economics today. Both in public service and in private industry a healthy scientific attitude can be found, and many sincere attempts are being made to understand and straighten out tangled relations. But because studies in forest economics have been slow to develop, we have accepted doctrines and beliefs that have never been tested adequately. So firmly established is this body of bromide and folklore that it is almost heresy to question it, just as it was once heresy to question the movement of the sun around the earth or the displeasure of the gods that caused crops to fail and animals to die.

Until our ready-made answers have been tested, analytically or in practice, no one can be sure whether they are true or false. But there is an open field for questioning them. The purpose of this paper is to consider a few of the beliefs that have been repeated often enough to gain a certain degree of authority and to question the basis on which they have been accepted.

The dominance of the silviculturists in forestry has been a major factor retarding the development of a sound body of economic doctrine. The belief that silvicultural perfection is the ultimate goal of forestry, and the preemptive assumption that what is silviculturally good is economically sound, often threaten to relegate forest economics to a minor role or to complete extinction. This belief has been expressed in a variety of ways. One of the earliest was the advocacy of European methods of management for the forests of the United States. Though a few die-hards still favor this, most of us have come to realize that a system of management designed to fit conditions of wood scarcity, relatively high land values, high density of population, and cheap labor could not be economically successful when none of these conditions exist.

But our actions are still influenced by the traditional emphasis on silviculture. From habit rather than as the result of rational processes we are inclined to conceive the most desirable forest to be one with fully stocked stands of a preferred species, rather than one that will provide the maximum returns in the form of watershed protection, game refuges, community support, useful products, and direct financial return. We fight what often seems to be a losing battle to protect a favored but disappearing species against insects, disease, fire, and cutting. We fail to consider the fact that protection from fire often encourages the less desirable species, or that the forest of useful species is often actually the product of long periods of uncontrolled natural fire. Only after years of fighting blister rust and fire, with expenditures running into millions of dollars, do we finally begin to question the economic justification of these efforts. Perhaps we should be devoting more effort to making the most of the so-called inferior species that seem destined to replace some of our old favorites.

The silvicultural bias leads us not to an advocacy of cultural practices that may be uneconomic, but to a belief that it is wasteful to leave any land idle if forests would grow on it. Nature's opinion of a vacuum is nothing more than a mild dislike compared to an ardent forester's abhorrence of an idle acre. We are frequently told that a certain number of acres are in need of planting. Why? To provide cover, to assure future supplies of timber, or because this land has no other use? If planting is advocated simply to prevent the waste of land, we need a definition of waste. It cannot be held wasteful in an economic sense to leave land idle if the costs of cultivation could not be equaled by the value of the tangible and intangible products resulting from cultivation. It is economically wasteful to devote capital, labor and land to a submarginal enterprise. The economic goal of more production with less effort requires that our resources be so allocated as to achieve maximum productivity. A submarginal enterprise by definition does not accomplish this.

Part of the doctrine that has guided the thinking of many forestry protagonists is the belief that there will be profitable markets for all the timber that can be produced. If we err in our attempt to produce the correct quantity of timber for the future an excess would probably be more desirable than a deficit as a matter of public policy. But public policy should be justified on the basis of possible emergency or public security rather than by unsupported assumptions of future demand. The statement that future markets will take all the timber that can be produced may be true. Or it may only be wishful thinking. It is not true simply because it has been stated by some of our high priests and repeated over and over again. Nor can all the reasons advanced in support of this doctrine be accepted without question.

The favorite argument used to support this belief is that we are destined to enjoy a rising standard of living, and that a rising standard means a higher per-capita consumption of forest products. It is granted that if consumption does increase the economy will benefit from a greater use of replaceable resources and a lessened depletion of nonrenewable resources. But it does not necessarily follow that individual preferences and market demands will conform to the national interest. The coincidence of high living standards and large per-capita wood consumption in several European countries is not conclusive evidence of any causal relationship. The supply of wood in relation to other materials may have more effect than standards of living on per-capita consumption. Certainly there is enough evidence of a preference for nonwood products in areas of high living standards within our own country to require further exploration of this doctrine before it can be fully accepted.

The doctrine of unlimited demand is also supported by the belief in the indispensability of wood. It is true that wood is now necessary for railway ties, for newsprint, for bobbins and shuttles, and for crates. There is a preference for wood over other materials for many other uses. But the indispensability of wood has been challenged many times. New materials and new methods are constantly threatening markets for forest products. The idea of indispensability, much as we may like it, is another that is subject to question.

This idea of indispensability is weakened by the collateral argument that there would be an almost unlimited demand for forest products if prices were low. The implication is that price may be the principal basis for the preference of wood. The argument is not realistic unless we assume that production costs in forestry and forest products industries will be reduced considerably, and more so than costs for competing materials. The argument is not sound unless we assume a very elastic demand for forest products. These assumptions are blithely made, but are they valid? Has anyone demonstrated how stumpage, logging, manufacturing, selling, and transportation costs in the forests industries are to be cut so that the products of wood can undersell substitute materials? Has anyone proved conclusively that lower prices of lumber, for example, would greatly stimulate its use in building, in crating, in furniture, in railway cars, in motor vehicles, or in any of its other uses? To what extent does the cost of newsprint affect its uses? These assumptions, part of our folklore, need to be examined carefully.

The use of dogma, doctrine, bromides, and half truths reaches its peak in the condemnation of past logging practices. The "ruthless destruction," the "greed" of the operators, the creation of "ghost towns," are all familiar to the reader of forestry literature. Today we know that the economy would be in much better shape if past practices in logging had been different. But do we take into account the social pressures that induced the type of logging we now condemn—the legend of inexhaustibility, the demand for lands cleared for farming, the need of forest products to build an empire, and the ever-present need in a pioneer community to convert plentiful and cheap raw material into capital, the scarce factor? In attempting to improve present day logging practices is our approach the most intelligent possible? Vituperative condemnation may not be the best way to win the logger over to the cause of forestry. Do we understand fully the social and economic factors now at work, and do we do everything we can to modify them in the interest of our forestry goals?

The idea of sustained yield has been glorified by those who condemn forest destruction. But failure to define sustained yield precisely has encouraged its loose usage by forest owners and operators. Without complete and clear definition, sustained yield is almost meaningless. The forester may mean a high level of intensive cultural management, but the owner can use the term to describe a liquidation operation if he is holding and protecting his cutover lands. The public is taught that sustained yield is necessary to preserve our forest economy, and is lulled into a sense of security when it is told that millions of acres in private ownership are on a sustained-yield basis.

There are many cases of words or phrases used in forestry acquiring good or bad connotations. Through continued loose usage an expression may come to convey meanings that were not originally intended. Just as "sustained yield" is a good term, "clearcutting" is a bad one, even though clearcutting is an accepted silvicultural practice under certain conditions. "Light cutting" and "selective cutting" have a good connotation that is entirely unjustified until they are defined. "High-grading" is a bad term, but high-grading and light cutting, because of the loose way in which these terms are used, may be employed to describe the same method of cutting. One thing forest economics and forestry in general need to take them beyond the stage of folklore is a set of definitions for these generally used terms and rigid adherence to their proper usage. Only by this means can our discussions convey definite meaning instead of spreading confusion.

The cause of forestry has not been furthered by the bit of folklore that private forestry will not pay. It has been emphasized that forestry represents a long-term investment, possibly attractive to corporations, governments, or other long-lived institutions, but not to the individual. The less attractive methods of forest culture, notably bare land planting with compound interest a large element of cost, have been used to support the argument. The best means of refuting it will be a few demonstrations of financial success.

The complementary doctrine is that public forestry will pay. But many of the advocates of public forestry have relied on broad generalizations about social benefits to support their claims, and have abandoned economic analysis. It is granted that many of the indirect and intangible social returns from public forests cannot be measured accurately in monetary terms. But would not our thinking be advanced if those returns that can be measured were expressed in dollars and weighed against the costs, so that we could compare a smaller unrecovered cost with the unmeasurable returns?

In discussions of forest economics the statement is frequently made that because of the long time required to grow timber of merchantable size, supply cannot readily be adjusted to demand. The timber owner is contrasted with the farmer, who within a year can adjust his acreage of corn or potatoes to the market. The apparent disadvantage in growing timber results from a misconception of the economic meaning of supply, and a confusion of supply with gross stocks or volume in existence. Supply in the market sense means the quantities offered at a range of prices at any time. Actually the market supply of timber may be extremely flexible, and readily adjusted to changes in demand. Growing trees can be harvested and supplied on the market before their normal financial maturity if demand increases and log prices rise. Conversely timber can be withheld from a depressed market without the danger of loss that affects many perishables, and with an increment in volume and quality that often will equal the cost of storage on the stump. The sale of timber at distress prices and the withholding of timber on a rising market are not evidences of an inflexible supply. The same conditions characterize the markets for annual crops and manufactured products.

It is true, of course, that the normal or long-run supply of timber cannot be adjusted as readily as market supply. This fact has been recognized in our suspicion of the doctrine that future demand cannot exceed future supply. But here, as I have pointed out, we need the most careful analysis of trends in the consumption of wood and the best estimates of probable future consumption, rather than an unsupported assumption that there can never be a surplus. At this point industry and public agencies are in greatest disagreement.

Perhaps these illustrations are sufficient to support the thesis that slogans and doctrines, often untested, play an important part in our thinking and actions in forestry. There are a few more that may be mentioned briefly.

There is a widely held belief that all the timber cut in any area should be manufactured into finished products in that area, to support local employment and community life. This doctrine fails to consider the advantages of exchange, or the law of comparative costs which operates as effectively among communities and regions of a nation as among nations. It is reasonably certain that the national economy would suffer from a full application of the doctrine of manufacture-at-home. It is doubtful if all individual timber-producing communities would benefit from it. To advocate this doctrine as a step toward social and economic progress is unwise until the full effects of its application on other communities and regions have been considered.

Then there is confusion in our thinking about big and small business in forestry. Large ownerships are said to be absorbing all the forest land, particularly in the West. They are tending toward monopoly, and driving out the small owners and operators who represent the American Way of Life. But we are also told repeatedly that small ownerships are unstable, and that continuity of ownership is necessary for sustained-yield operation. Thus bromide is pitted against bromide! What is the optimum size of a forest operating unit? Slogans will not answer that question.

The doctrines I have chosen to question are not universally accepted. Sometimes we suspect that they are not always believed by the persons who give voice to them. But they, and many others like them, are often used to support a policy or justify a position because full knowledge is lacking or because it is easier to parrot the words of someone else than it is to think. Perhaps many of these beliefs are sound. They cannot all be, for many are contradictory. All of them are open to question until study or experience has proved them true or untrue.

It is one thing to indicate that there has been much fallacious thinking in forest economics and an avidity to accept untested doctrines. It is another thing to remedy the situation. We need both a new approach to the economic problems of forestry and more factual material to analyze. Perhaps the most logical place to develop a scientific approach is in our forestry schools. Periodically the curricula of these schools are subjected to review, and courses of study are recommended. But how much attention is paid to the subject matter in these courses? It is in the classroom that our future foresters receive their indoctrination in folklore. It is in the same classroom that they should be encouraged and trained to question this folklore and to search for the truth.

The teachers in our forestry schools need not, however, shoulder the entire responsibility for the perpetuation of beliefs that may be false. They have so little to work with! The obvious need, if forest economics is to become a real social science, is more research. This research must be more than just a compilation of superficial facts. Every bit of folklore that remains in our thinking must be torn apart and analyzed thoroughly. Only by doing this can we progress from superstition and mythology to truth.

4

THE ROLE OF FAITH IN FOREST RESOURCE MANAGEMENT

William A. Duerr
Jean B. Duerr

First published in *Man and the Ecosystem;* pages 45–55. City Printers, Inc., Burlington, Vermont. 95 pages. 1971. Presented as a Zeltzerman Lecture, University of Vermont, 1968.

BIOGRAPHY

Mr. Duerr is Thomas M. Brooks Professor of Forestry at Virginia Polytechnic Institute and State University, Blacksburg. Before his appointment to this position in 1972, he was for 20 years at Syracuse, New York, where he served as Chairman of the Forestry Economics Department in the Forestry College and toward the end of the period was appointed Distinguished Professor in the State University of New York. His earlier career was with the Forest Service's Branch of Research, at a number of the regional experiment stations and in Washington, D.C. One of the founders and once Chairman of the Division of Forestry Economics and Policy in the Society of American Foresters and author or editor of textbooks, a laboratory manual, and a research guide, Mr. Duerr is a member of the first generation of American foresters who sought, through postgraduate study, to become qualified also in the field of management and economics.

Mrs. Duerr's interest in, and deep appreciation for, the social-science aspects of forestry stems from 30 years of marriage to a forestry economist. Her research for the report reprinted here included 3 months of study and observation in Germany and extensive reading in the American professional literature: all issues of *Forestry Quarterly* and *Journal of Forestry*, 1902 through 1967. She serves as collaborator with the research and editorial team of the Society of American Foresters engaged in developing an approach to forest-management teaching and a supportive textbook, *Forest Resource Management, Decision-Making Principles and Cases.*

EDITORS' SUMMARY

The uncertainties of resource management increase as knowledge grows. Meeting uncertainty successfully requires not only science but also faith—i.e., a body of doctrine, a set of tenets, constituting the life philosophy of the professional subculture. The challenge is to keep this subculture changing in step with the world's needs.

QUESTIONS TO CONSIDER

(1) It can be argued that cultural tenets of faith not only supplement science, as the Duerrs contend, but underlie science, so that everything in human experience (except instinct?) rests on faith. That is, culture not merely comprises learned behavior, but also determines learned behavior: It determines our social science and even our mathematics. What do you think of this argument?

(2) If forestry is a culture (subculture), is each other profession also a culture? Are the different professions the same culture or different ones? Cite examples.

(3) What, precisely, is meant by *uncertainty?*

(4) How might you explain the origin of the doctrine of sustained yield?

(5) Why are foresters taught to speak of the "volume" of wood in a tree but not ordinarily the "quantity" of wood?

(6) Take, as an example, a forest tract with which you are familiar. List the tenets of faith that are being followed in the management of the tract.

* * *

Forest resource management is all those decisions, or plans, that are made and carried out for forest land. The decisions concern how to use the land and its many potentials. Decisions include who shall own the land, what purposes it shall be put to, and what revenues and other benefits it shall be asked to produce for its owners and for the public. There are decisions about the goals of management and about the mix of products to be cultivated in pursuit of goals: water, timber, recreational services, derivatives such as the fruit or sap of trees, livestock forage, minerals, wildlife, materials taken from lesser vegetation, or other products. There are decisions about how to treat the ground: whether, for example, to mine it and whether to fertilize it and what improvements to establish on it, such as roads, trails, and recreational developments. There are decisions about the living things: when and how to crop and harvest them, to reproduce them, to cultivate and protect them. And there are decisions about surrounding arrangements: personnel and their organization, marketing the products, financing, public relations, and so on.

RESOURCE MANAGEMENT FACES MANY UNCERTAINTIES

Resource management decisions all relate to the future, and for forestry, many relate to the long future. Think what far-off horizons may be in view when one decides to buy a forest or to plant a certain kind of tree or to engage in flood-control measures or to set the annual harvest at a certain level. Such horizons are at best dimly seen, veiled in the manager's uncertainty. His uncertainty stems from the rapidity and unpredictability of the change he foresees, compounded over the length of time in view. Even for relatively short-term decisions such as those concerning the wood-using plant or the recreational improvements, the decision maker faces much uncertainty amidst today's swift and swerving change: By the time his investments are ready to pay out, what sort of world will this be? What inventions and revolutions, what new fashions and consumers' ways, will have given his judgments the lie?

It may seem as though the rapid growth of science should offset the upsurge of uncertainty. Science enables us not only to predict more accurately, but also to devise more ingenious schemes for hedging and even for controlling the future. But in fact, the expansion of science appears to heighten our uncertainty. Our store of knowledge is a little enclave in the universe of the knowable. The borders of this enclave are manned by defenders who are pushing their lines forward and outward. The greater their success, the more extended their lines of defense and the more gigantic the opposing arrays of the unknown. And so, assuming that the knowable is infinite, science alone cannot subdue uncertainty: Science alone—knowledge, fact, reason alone—cannot suffice in decision making for resource management.

Put the point in still another way. The process of reasoning-out a decision can be idealized in some such terms as these: The decision maker identifies his alternative courses of action and places on each a net value appropriate to his goals. He then chooses the highest-valued course, reviewing his method from time to time in light of its results. Now, in the face of change and uncertainty, the manager cannot know all about his alternatives and their probable consequences and the values of these and so is not able fully to deliberate his decision. *For we know in part and we prophesy in part*. Reasoning alone does not suffice in resource management.

Consider, then, this thesis: that to face uncertainty successfully, to manage resources successfully, requires two types of armament: science on the one hand and faith on the other. Every decision is achieved with some combination of the two. The manager carries out such factual evaluation of alternatives as his knowledge permits, within the constraints set by his faith. Let us dwell, now, on the latter: What is faith? What is its role in management?

CULTURE IS A RESOURCE-MANAGEMENT SYSTEM

From the beginning of our lives, each of us learns ways of behaving and thinking. At home, we are taught how to sit and to walk, to use eating utensils and otherwise perform at table, to think that certain foods are appropriate and others not. We start to learn a language. We acquire ideas about clothing, play, work, and the rest; about religion, morals, aesthetics, life, and property; about behavior toward others within the family and toward neighbors. Above all, we form an attitude toward those normal persons who share with us the same cultural patterns and those queer ones who do not. We continue the learning process in school. As adults, we are still subjected to enculturation. We are always being exposed to new things and new ideas and attitudes. Now, however, we have left the shelter of our parents' house: It is no longer possible or suitable to accept all the innovations that come our way. Often we must form judgments and make choices and thus take an active part in shaping the culture by which we ourselves, until now, have been shaped.

Our culture, whether it is that of New England or of some other part of the Western world or that of southern China or of a remote pocket from the Stone Age, will invariably exhibit certain institutions. Language, for example, will be one. A second will be social structure; a third, provisions for making a living, including a technology and economy; a fourth, aesthetic ideas joined with artistic activities. A fifth cultural institution—and this is the one to be stressed—is a philosophy of life: goals, religion, tenets of faith regarding some of life's great issues and grave uncertainties. These five will serve as illustrative. If a people has cultural institutions different from those of other peoples, here is evidence that *a* culture exists.

What functions does a culture perform? What survival values does it have?

One function of a culture: It lets the individual enjoy a basic human want: a sense of his identity. He is enabled to say proudly, "This is my language and yours; through our language we know who we are, and we are kindred." Out of his identity arise a person's goals and values, and these, in turn, are his inspiration.

Another function of a culture: It promotes the survival of the group. Through family, community, and government, a degree of harmony is assured. Individual behavior is bent to the group standard; one can predict what the other fellow is going to do (for instance, drive always, or almost always, on a particular side of the street); and thus social life is made possible. Culture incorporates the code of behavior that has proved successful: ways of obtaining such basic essentials as food, clothing, and television sets; the banning of theft and of in-group murder; the taboo against walking beneath a ladder.

Still another function: In fostering the survival of the group, culture favors the individual's survival as well. It promotes his physical health by prescribing the right food and personal hygiene. It promotes his mental health by making others' behavior more understandable and predictable and thus lessening the little maladjustments of daily life. One can enter a revolving door without wondering if it is a trap; he can take the other man's extended hand in the fairly secure knowledge that it is a greeting and not a judo trick.

A CULTURAL SYSTEM HELPS COPE WITH UNCERTAINTY

Finally—and this is the principal summary point about the functions of a culture: It provides its members with a store of ready-made goals, values, and predictions, such as those mentioned, all of which can be used for deliberating decisions in short order. Beyond this, it provides a host of decisions themselves, which are taken as a matter of custom. In brief, by supporting deliberated decisions and furnishing customary decisions, culture plays a large role in the management of every sort of resource.

A culture's ready-made choices allow the decision maker to cope, almost without realizing it, with many uncertainties. Indeed, each person's early enculturation results in putting most of his behavior below the level of conscious thought. Decisions regarding the occasion and manner for washing his hands, holding his spoon, wearing his hat, speaking his mind are taken in large measure automatically. What a blessing! He is thus enabled to keep his sanity and to find time for other than routines.

Each of the cultural institutions serves an area of our decision making and lessens our uncertainty in this area. The great bulk of our questions about language are ready-answered for us, and about authority and about making a living and about what is beautiful and what ugly. Our philosophy of life—our goals, religion, our tenets of faith—offer us guidance through uncertainty in many of our largest personal decisions. Faith instructs us about our obligations to our families, giving us immediate answers to many an otherwise puzzling question. It guides us in our relations with our employers and fellow workers. It takes us firmly by the hand as we contemplate the mystery of death.

Now to get back to forestry—to the role of faith in forest resource management. Consider the following argument: Forest managers have a subculture within our general culture; the forestry culture exhibits all the institutions that one looks for in any culture. The forestry culture contains the same kinds of ready-made predictions, goals, values, and decisions for forest resource management as does our general culture for our thinking and our behavior at large. The life's philosophy of the forestry culture—the goals, the religion—generates the culture's tenets of faith, and faith in this sense plays a major role in resource management amid change and uncertainty. Let us examine this premise. It applies to forests and foresters, and it applies also, in some degree, to every resource and every profession.

FORESTRY CONSTITUTES A CULTURAL SYSTEM

Each occupational group is more or less clearly distinguished from others. Its distinctiveness arises in the life and work which its members share with one another but not with outsiders. Its distinctiveness arises, too, in the technology which its members employ but others do not, and thus in its common and peculiar problems, words, and manner of thinking. Again, distinctiveness is a function of the group's devices for assuring continuity and the storage of ideas and their transmission from generation to generation of members. We all recognize that farmers, with their relative isolation and personal freedom, present their own aspect. Miners who live in the shadow of catastrophe develop distinctive ways. So do members of the regular army, with their comparative release from decision making. The old-time chimney sweep wore a top hat; the old-time lumberjack wore a soft felt with the crown unindented; and both were marked in many regards more fundamental than these.

So it is with the occupational group charged with forest resource management—that is, the profession of forestry. Check off the cultural institutions and see to what extent forestry may qualify as a culture.

Language? This is perhaps the foremost criterion. Yes, there is a distinctive tongue—learned, not at mother's knee, but at professor's. Foresters enjoy conversing in their own technical language, much to the boredom of laymen who overhear. The Society of American Foresters publishes a *Forestry Terminology,* and an international group is now at work on an analogous compilation. These are dictionaries of Forestrese.

Social structure? Is there among foresters a counterpart of the family?—of government? Undoubtedly. The family is the class: of forestry students, of new enrollees in the Forest Service, of industrial foresters. As for government, there is the professional society. Its executive secretary is its benevolent dictator.

A way to get a living: a technology, an economy? No need to elaborate here.

Aesthetic activities: forms of art and a concept of what is beautiful and pleasing? There are clear examples of this cultural institution, notably in the idea of the aesthetic

Figure 4–1. The aesthetic forest.

forest, which is something to be striven for and maintained. The aesthetic forest is a congested aggregate of aesthetic trees, which are tall woody plants with long, nearly cylindrical, limb-free stems and just enough crown of foliage to keep them growing satisfactorily (Figure 4–1). That is, they are an efficient timber-producing machine.

FORESTRY CULTURE INCORPORATES A BODY OF FAITH

A philosophy of life: goals in life, religious beliefs? In this last of the five cultural institutions it may seem that we have finally run out of forestry parallels. On the contrary, it is here that the greatest abundance of analogies can be drawn. It is here that the professional culture reaches its fullest development. The forestry philosophy of life—that is, the professional philosophy, or religion—is contained in the profession's cultural tenets of faith. These tenets are the profession's response to the immense uncertainties of forest management. They acquired their identity as received doctrine during the formation of classical forestry in Germany in the sixteenth, seventeenth, and eighteenth centuries.

A word about classical forestry. It represented a combining of pragmatic and idealistic forces, of authoritarianism and deep personal humility: ingredients which have endured in the professional heritage to the present. The German states were a landlocked, war-possessed, timber-dependent region dedicated to self-sufficiency. Resource depletion had often threatened community well-being. Many laws had been ordained, expressing the social imperative of conservation and designating the forester as its official, quasi-military instrument. The Founding Fathers, with Teutonic and professorial attention to detail, propounded rules and formulas for every step of the timber-growing process. To achieve a regular, heavy, and enduring flow of wood from the forest was their prime preoccupation, and bitter were the arguments over alternative schemes for reaching this goal of "sustained yield." But amidst these academic wars, gentler voices were often to be heard. Listen to the words of Heinrich von Cotta, born in Thuringia in 1763. It was Bernhard Fernow, the German father of American forestry, who translated the words into English in 1902:

> I am a child of the forest; no roof covers the spot where I was born. Old oaks and beeches shade its solitude, and grass grows upon it. The first song I heard was of the birds of the forest; my first surroundings were trees. Thus my birth determined my calling.[1]

To return, now, to the cultural tenets of forestry, let us identify the *classical* tenets: those brought to the United States from Europe at the beginning of this century. They have been handed down—some unchanged, and most, essentially so—from professional generation to generation in the universities and government bureaus, over hundreds of years. The forestry tenets serve all those functions which are served by the tenets of any culture. They serve conspicuously to mold the professional person so that he may become an appreciative spokesman and instrument for the profession's goals. Thus one identifies the tenets by studying the professional journals and observing the enculturation of foresters. The classical forestry tenets may be grouped into four major doctrines.

FOUR MAJOR DOCTRINES EXPRESS CLASSICAL FAITH

Doctrine of Timber Primacy: Timber is the chief product of the forest; all else that comes from the forest is by-product, of secondary interest: water, forage, wildlife, and the rest, including recreation. Indeed, people are a nuisance in the forest. Wood is, and will always be, a necessity, for it has no true substitutes. Its consumption is assured; its consumers may be taken for granted. In fact, there is going to be a shortage of timber, and the central problem in forest management is the biological and engineering problem of growing more timber.

Doctrine of Sustained Yield: To fulfill our obligation to our descendants and to stabilize our communities, each generation should sustain its resources at a high level and hand them along undiminished. The sustained yield of timber is an aspect of man's most fundamental need: to sustain life itself.

Doctrine of the Long Run: Nature moves and changes slowly; she takes a long time to accomplish such purposes as the growing of timber. Society must adapt itself to this fact. Be patient. Curb the selfish, short-sighted interests such as those of private enterprise and notably small enterprise. Look to the past. The future will be like the past, and, indeed, should be like the past.

Doctrine of Absolute Standards: The forest is a living thing with its own ends and its means for attaining them under natural law. The successful manager, regardless of his forest's location or ownership, finds his goals and guides in the forest itself, by looking there, and by listening to what the forest tells him. People are not to be trusted in such matters. Thus the manager plants those kinds of trees which are best suited to the site. He carries the stock of growing timber which nature has shown she is capable of carrying. He aims to produce wood of high quality in maximum quantity, and this again is the aim of the forest. Needless to say, idle land is a cost to society.

Every one of the classical tenets of faith contains ready-made decisions for the forest resource manager. In every one, the manager hears the voice of the Founding Fathers admonishing him out of the profession's long experience. True, he is sometimes inclined to talk back, and this is a point worth pursuing. But for the moment think only of those tenets which retain acceptance. Take an illustration of faith's role in forest resource management.

THE FOREST MANAGER MAKES GOOD USE OF HIS FAITH

Suppose that you are at the head of a government bureau which manages public forest land. Consider your obligations respecting the timber resource. You are charged with the awesome responsibility for husbanding and servicing and bringing into use a vast stock of growing timber. You have the even more frightening task of living in some degree of harmony with legislators; public administrators; the wood-using industry; recreationist, sportsman, and conservation groups; and many others, including the members of your own organization. You are faced with the obligation to decide how much of your timber you will make available for harvest, decade by decade, into the future. How will you go about taking such a decision?

Your reason insists that the only sensible approach is to regulate the outflow of wood in such a way that over time the net benefits, in some sense, will be as large as possible. But here are staggering difficulties. Immense quantities of timber may be harvestable today, and their harvest may be essential, let us say, to rapid regional development. But then what of tomorrow, when the outflow would have to be severely curtailed because stocks were running low? Tomorrow's losses might greatly exceed today's gains, and tomorrow's citizens would call down imprecations on your memory for your bad judgment. On the other hand, suppose you were to hold back the timber stock at present, counting on a heightened use of it when timber should become scarcer and more valuable to society. But meantime, let us suppose, the world changes; the wood industry dwindles away; the values of your carefuly hoarded timber shrink and fall; and society condemns you for a fool. What can you do? To what security can you cling in the face of the void?

You can cling to the security of the faith you learned at your mother's knee—that is, at the knee of the Founders. You can do what virtually every public forest administrator in the Western world has done for the past few centuries: take shelter in a policy of sustained yield. Under this policy, you will manage the forest to the goal of producing the same, hopefully large, physical quantity of wood period by period into the indefinite future. Under this policy, you will find peace with yourself.

SUSTAINED YIELD IS THE FOCAL POINT OF FAITH

Just after the turn of the present century, while Bernhard Fernow was interpreting German forestry to Americans, the British cultural anthropologist, W. H. R. Rivers, was completing studies of the Toda people, whose villages occupied the hills east of Calcutta.[2] His findings furnish us with an illuminating digression.

The Todas, reported Rivers, took their orientation and their life's meaning from the focal element in their culture: the buffalo dairy. This pastoral people used and depended upon buffaloes at every turn: for wealth, for food, for religious gratification. The men devoted most of their labor to their buffaloes and dairies. The animals or their milk figured in the rituals of birth, childhood, marriage, death, and burial. Buffaloes and buffalo dairies were prominent in the Toda mythology and in the Toda concept of the ways of their gods.

How the buffalo dairy provided focus for the Toda culture is illustrated in the religion of this people, in which the buffalo was a central figure. The Todas performed religious ceremonies at nearly every important point in the buffaloes' lives. They based their religious rituals upon the milking and churning and other ordinary operations of the dairy, which was, then, their temple.

How the buffalo dairy provided cultural focus is illustrated, too, by elaboration and variation in the dairy institution. Buffaloes, milk, and dairies were all classified according to degree of sanctity. The more sacred the dairy-temple, the more fixed and complex the ritual, the more prestigious the dairyman-priest, the more lavish his induction ceremonies, and the more stringent the rules governing his daily life. Each of the two divisions of the Toda people developed distinctive dairy and herd characteristics, and within each division were dairies in which the ritual developed in special directions.

As with the buffalo dairy among Todas, so with sustained yield among forest land managers: It provides the point of cultural focus. Not only is it the cultural element most stressed, but also it is the element upon which the greatest effort has been lavished toward variation and elaboration. The Founding Fathers concentrated upon systems for harvesting, regenerating, and regulating the forest; early, they developed a great diversity of systems. Their descendants have written variations without end upon the theme of timber regulation: methods, problems, guiding rules—all directed to the goal of sustaining yield.

Of course, sustained-yield doctrine, viewed boldly, is ridiculous. Perhaps in early Germany, communities wished to be sustained; but in the modern world, they want growth and accept change as part of it. And even if they did desire stability, surely the processing of the same number of tons of wood each year would offer poor promise of it: Technology changes, and with it, the economic significance of a ton of wood. And still again, looking at the doctrine of sustained yield, one asks oneself why he should be so concerned about his descendants, who show every promise of being better-heeled than he.

The answer to such pangs of reason is this: We can approach many a resource-management question mainly on the basis of fact, by the scientific method, and the number of such questions is growing fast. The questions are typically those of intermediate scope, short-run questions at the operational level, and they can well be deliberated by the systems analysts, operations researchers, and other applied mathematicians, backed up by the computers. But the littlest questions we may decide mostly on the basis of habit or custom. And for the large and the long-run questions—the numbers of which, too, are growing fast—we may well rely primarily on faith because we have few facts: The variables are numerous and the predictions uncertain.

Yes, sustained yield and the other doctrines are absurd. Yet we want guides; and where fact is too dear, we take our guides on faith: *Credimus quia absurdum.*

HOW CAN WE BRING FAITH ABREAST OF CHANGE?

But here arises a dilemma: How can we enjoy the *security* of faith, which is a function of its strong rooting and its constancy, and at the same time escape the attendant *penalty:* that the rooting of faith is in the past, that it looks backward and not at tomorrow's world? This is a question of powerful import for any profession which gains strength, but loses adaptability, in its reliance on faith. The issue, stated constructively, is this: How can the tenets of faith be made responsive to change and still retain the strength that derives from their position as a part of culture? The issue is cultural evolution, a major issue for any people or any profession. Take forest resource management.

There is much evidence of drift from the classical forestry culture, and some even of revolution. Opinion polls[3] of land-management students and of practitioners invariably show some mixture of agreement and disagreement regarding every one of the classical tenets, a good indication that change is in progress. Indeed, the problem often is not to achieve change, but to have new tenets at hand as the old die off. A profession which replaces its tenets with vacuums runs the same risk as any other people in these circumstances: over-rejection of heritage and loss of identity. "Who am I?" is a question which a few forest land managers have always asked, and which today many are asking.

Those classical tenets which our *experience* has proved wrong are being cast off. (The emphasis here is upon experience, as a reminder that we are speaking of the past and only conjecturing the future.) For example, with clear public insistence upon the use of forest lands for recreation, water production, and other non-timber purposes, the Doctrine of Timber Primacy has become heavily undermined. Sixty years ago, in the professional forestry journals of the United States, as many as a fifth of the major articles were explicitly supporting timber primacy, while none was speaking out against it. Today, after an era of remarkably consistent change, one finds the proportions approximately reversed (Figure 4–2).

Where experience has shown the need for new areas of faith, these are being built. An instance is found in private industrial forestry, a field of management which

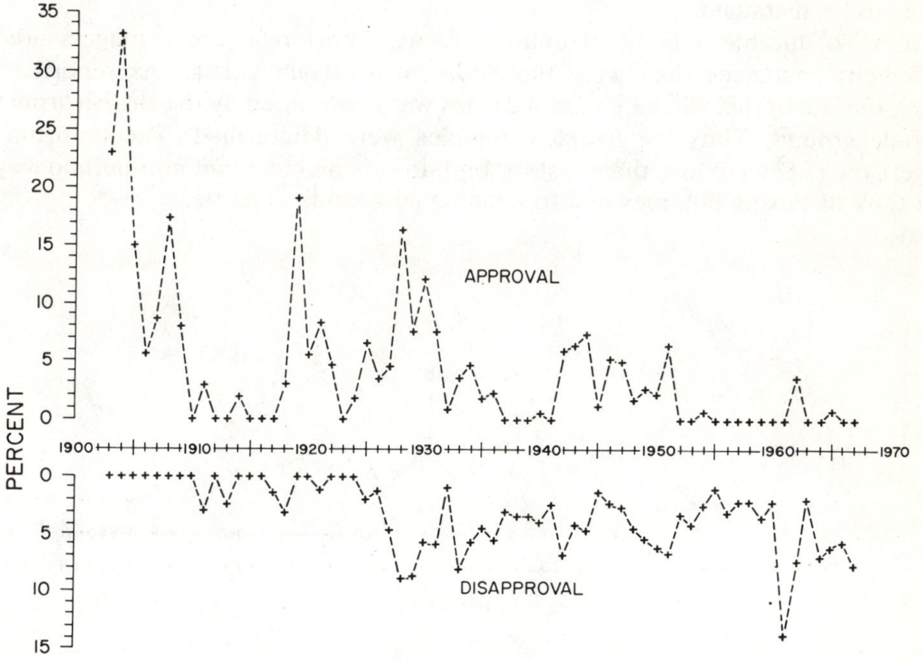

Figure 4–2. How writers in the *Forestry Quarterly* (1903–1916) and *Journal of Forestry* (1917–1967) have judged the Doctrine of Timber Primacy: Percentage of all major articles published each year, which expressed approval or disapproval of the Doctrine.

did not exist in the sphere and time of the Founders. Among industrial forestry's many special articles of faith are, for example—

> Forestry pays.
>
> Young-growth timber will be more valuable than old growth ever was.
>
> A pulp company should control enough forest land to supply at least half of its wood requirements.
>
> For guidance in forest management, look to the consumers of your products.

But where *experience* is unavailable, as it inevitably is for the longer-run questions about forest resource management, the classical tenets endure. Here the Doctrine of Sustained Yield is a good illustration. Through all the years of American forestry, there has been little or no dissent from this Doctrine. At the same time, there has never been impassioned defense: There was no need for it (Figure 4–3). In our day, God has been challenged and termed dead; not so sustained yield.

The viability of the sustained-yield idea—and these remarks hold also, to a degree, for the other enduring doctrines—rests in its powerful mixture of simplicity and realism, of definitiveness and adaptability. In order to live, any guide to management must call, in understandable terms, for data that the manager will be able to find. The sustained-yield principle, under this rule, is faultless. In concerns timber, the most readily measurable of the forest products. It states with clarity unmistakable instructions and supplies an inspiring argument for following them. It asks for information—on timber inventory, growth, and removal—which is the forester's stock-in-trade. And with all its freedom from equivocation, it still allows the manager latitude for judging and for changing (notably, for increasing) the level at which the yield is to be sustained.

With so durable a point of cultural focus, forest resource managers are in happier circumstances than were the Todas in a certain village. As reported by Rivers, the site of this village's sacred dairies was confiscated by the British army for a parade ground. Thus the people's temples were demolished. The unfortunate villagers then began to lose their zest for buffalo care and to adopt non-buffalo ways. They took to raising potatoes and to keeping new kinds of cattle.

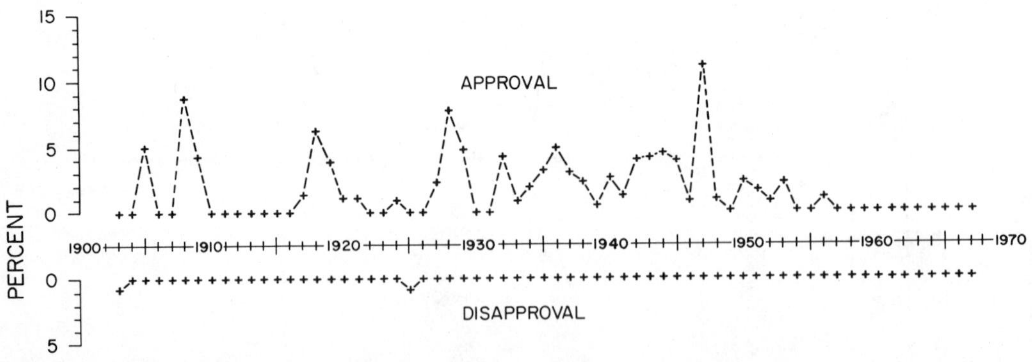

Figure 4–3. The Doctrine of Sustained Yield as viewed by writers in the *Forestry Quarterly* (1903–1916) and *Journal of Forestry* (1917–1967): Percentage of all major articles published each year, which expressed approval or disapproval of the Doctrine.

FAITH CAN BE REPLACED, REVISED, NEWLY GENERATED

Sustained yield, unlike the buffalo dairy, has never suffered demolition. What, indeed, could be put in its stead? In the absence of guiding doctrine, there is only the void: spiritual anarchy: the eating of potatoes. One wants something new ready to replace the old.

Considering that faith will always play a large part in forest resource management, what can we do to improve its performance?

Recognize the existence of faith. Understand its role in management.

Identify and analyze the tenets of faith. Codify them. Make them explicit, so as to expose them to the light of inquiry. Encourage debate over them.

Through research, supplement faith with fact, and where possible, even replace faith with fact. At the same time, discover the emerging questions in respect to which new faith will be wanted.

Experiment with the use of alternative articles of faith in resource management. Bring to mind the adult, who no longer shelters in his childhood home. Now he cannot merely accept enculturation: He must also help shape it. In the experimentation, the public administrators, being ultrasensitive to insecurity, are necessarily handicapped. Surely private industry can take leadership. And the universities can contribute fresh ideas.

As the old buildings are razed in the process of cultural renewal, start the new construction close behind. Idle land may not be a cost to society, but the bare ground can be a disaster to a profession.

* * *

REFERENCES

1. Forestry Quarterly. October 1902. Page 5.
2. Rivers, W. H. R. 1906. The Todas. London and New York: Macmillan and Co., Ltd.
3. See, for example, Bond, Robert S., and Joseph C. Mawson: Some attitudes of student and professional foresters about forestry. Jour. of Forestry, March 1968. Pages 181–186.

5

THE CONSERVATION ETHIC

Aldo Leopold

First published in *Journal of Forestry* 31(6):635–643. October 1933.

BIOGRAPHY

Mr. Leopold (1887–1948) was an Iowan of wide-ranging interests and talents. Educated at Yale University, with degrees both from the Sheffield Scientific School and from the School of Forestry, he followed a two-phase professional career. In the first phase, he worked with the federal Forest Service for 20 years, mainly on resource management in the Southwest, though his last position was as Associate Director of the Forest Products Laboratory, Madison, Wisconsin. In the second phase, he joined the University of Wisconsin to become the first professor of game management in the United States. He is remembered as a teacher and, more widely, as a writer who could express the penetrating insights of an ecologist in the graceful style of a poet. He authored a game-management textbook and pioneering works on game policy, soil erosion, and the value of wilderness. His essays, which voiced his concern for nature and for man's relation to land, were collected into two books: *A Sand County Almanac* and *Round River.*

EDITORS' SUMMARY

Pressures, imbalances, and changes in this machine age require that man recognize, study, and act appropriately to his interdependence with his environment: other animals, plants, and land. Examples abound of past despoliation of resources. Past abuses inspire us to create a more stable environment in the future.

QUESTIONS TO CONSIDER

(1) List the cultural tenets of faith that are held by Mr. Leopold. Then envisage a world in which these tenets are obeyed. Describe some of the principal aspects of forestry and society in that world and some of the principal features of social science and the study of forestry. What would be the role of the forestry economist?

(2) The anti-Malthusian argument is that while larger populations mean more mouths to feed, they also mean more heads and hands for inventing, working, and producing an abundant livelihood. How do you view this argument?

(3) Have efforts been made to improve the depleted resource conditions in the Southwest described by Mr. Leopold? What form have they taken? How successful have they been?

(4) Analyze Mr. Leopold's indictment of "economics." What does he mean by the term? To what degree can forestry-economic thought be considered as compatible with conservation thought as enunciated by Mr. Leopold?

(5) Show that Mr. Leopold's approach to resource problems is a *systems* approach.

* * *

A harmonious relation to land is more intricate, and of more consequence to civilization, than the historians of its progress seem to realize. Civilization is not, as they often assume, the enslavement of a stable and constant earth. It is a state of mutual and interdependent cooperation between human animals, other animals, plants, and soils, which may be disrupted at any moment by failure of any of them. Land despoliation has evicted nations, and can on occasion do it again. As long as six virgin continents awaited the plow, this was perhaps no tragic matter—eviction from one piece of soil could be recouped by despoiling another. But there are now wars and rumors of wars which foretell the impending saturation of the earth's best soils and climates. It thus becomes a matter of some importance at least to ourselves, that our dominion, once gained, be self-perpetuating rather than self-destructive.

This instability of our land relation calls for example. I will sketch a single aspect of it: the plant succession as a factor in history.

In the years following the Revolution, three groups were contending for control of the Mississippi Valley: the native Indians, the French and English traders, and American settlers. Historians wonder what would have happened if the English at Detroit had thrown a little more weight into the Indian side of those tipsy scales which decided the outcome of the Colonial migration into the cane lands of Kentucky. Yet who ever wondered why the cane lands, when subjected to the particular mixture of forces represented by the cow, plow, fire, and axe of the pioneer, became bluegrass? What if the plant succession inherent in this "dark and bloody ground" had under the impact of these forces, given us some worthless sedge, shrub, or weed? Would Boone and Kenton have held out? Would there have been any overflow into Ohio? Any Louisiana Purchase? Any transcontinental union of new states? Any Civil War? Any machine age? Any depression? The subsequent drama of American history, here and elsewhere, hung in large degree on the reaction of particular soils to the impact of particular forces exerted by a particular kind and degree of human occupation. No statesman-biologist selected those forces, nor foresaw their effects. That chain of events which on the Fourth of July we call our National Destiny hung on a "fortuitous concourse of elements," the interplay of which we now dimly decipher *by hindsight only*.

Contrast Kentucky with what hindsight tells us about the Southwest. The impact of occupancy here brought no bluegrass, nor other plant fitted to withstand the bumps and buffetings of misuse. Most of these soils, when grazed, reverted through a successive series of more and more worthless grasses, shrubs, and weeds to a condition of unstable equilibrium. Each recession of plant type bred erosion; each increment to erosion bred a further recession of plants. The result today is a progressive and mutual deterioration, not only of plants and soils, but of the animal community subsisting thereon. The early settlers did not expect this; on the cienegas of central New Mexico some even cut artificial gullies to hasten it. So subtle has been its progress that few people know anything about it. It is not discussed at polite tea tables or go-getting luncheon clubs, but only in the arid halls of science.

All civilizations seem to have been conditioned upon whether the plant succession, under the impact of occupancy, gave a stable and habitable assortment of vegetative types, or an unstable and uninhabitable assortment. The swampy forests of Caesar's Gaul were utterly changed by human use—for the better. Moses' land of milk and honey were utterly changed—for the worse. Both changes are the unpremeditated resultant of the impact between ecological and economic forces. We now decipher these reactions retrospectively. What could possibly be more important than to foresee and control them?

We of the machine age admire ourselves for our mechanical ingenuity; we harness cars to the solar energy impounded in carboniferous forests; we fly in mechanical birds; we make the ether carry our words or even our pictures. But are these not in one sense mere parlor tricks compared with our utter ineptitude in keeping land fit to live upon? Our engineering has attained the pearly gates of a near millennium, but our applied biology still lives in nomads' tents of the stone age. If our system of land use happens to be self-perpetuating, we stay. If it happens to be self-destructive we move, like Abraham, to pastures new.

Do I overdraw this paradox? I think not. Consider the transcontinental airplane which plies the skyways of the Southwest—a symbol of its final conquest. What does it see? A score of mountain valleys which were green gems of fertility when first described by Coronado, Espejo, Pattie, Abert, Sitgreaves, and Couzens. What are they now? Sandbars, wastes of cobbles and burroweed, a path for torrents. Rivers which Pattie says were clear, now muddy sewers for the wasting fertility of an empire. A "Public Domain," once a velvet carpet of rich buffalo-grass and grama, now an illimitable waste of rattlesnake-bush and tumbleweed, too impoverished to be accepted as a gift by the states within which it lies. Why? Because cows eat brush when the grass is gone, and thus postpone the penalties of overutilization. Because certain grasses, when grazed too closely to bear seed stalks, are weakened and give way to inferior grasses, and these to inferior shrubs, and these to weeds, and these to naked earth. Because rain which spatters upon vegetated soil stays clear and sinks, while rain which spatters upon devegetated soil seals its interstices with colloidal mud and hence must run away as floods, cutting the heart out of the country as it goes. Are these phenomena any more difficult to foresee than the paths of stars which science deciphers without the error of a single second? Which is the more important to the permanence and welfare of civilization?

I do not here berate the astronomer for his precocity, but rather the ecologist for his lack of it. The days of his cloistered sequestration are over;

> "Whether you will or not,
> You are a king, Tristram, for you are one
> Of the time-tested few that leave the world,
> When they are gone, not the same place it was.
> Mark what you leave."

Unforeseen ecological reactions not only make or break history in a few exceptional enterprises—they condition, circumscribe, delimit, and warp all enterprises, both economic and cultural, that pertain to land. In the cornbelt, after grazing and plowing out all the cover in the interests of "clean farming," we grew tearful about wildlife, and spent several decades passing laws for its restoration. We were like Canute commanding the tide. Only recently has research made it clear that the implements for restoration lie not in the legislature, but in the farmer's toolshed. Barbed wire and brains are doing what law alone failed to do.

In other instances we take credit for shaking down apples which were, in all probability, ecological windfalls. In the Lake States and the Northeast lumbering, pulping, and fire accidentally created some scores of millions of acres of new second growth. At the proper stage we find these thickets full of deer. For this we naively thank the wisdom of our game laws.

In short, the reaction of land to occupancy determines the nature and duration of civilization. In arid climates the land may be destroyed. In all climates the plant succession determines what economic activities can be supported. Their nature and intensity in turn determine not only the domestic but also the wild plant and animal life, the scenery, and the whole face of nature. We inherit the earth, but within the limits of the soil and the plant succession we also *rebuild* the earth—without plan, without knowledge of its properties, and without understanding of the increasingly coarse and powerful tools which science has placed at our disposal. We are remodeling the Alhambra with a steam-shovel.

ECOLOGY AND ECONOMICS

The conservation movement is, at the very least, an assertion that these interactions between man and land are too important to be left to chance, even that sacred variety of chance known as economic law.

We have three possible controls: legislation, self interest, and ethics. Before we can know where and how they will work, we must first understand the reactions. Such understanding arises only from research. At the present moment research, inadequate as it is, has nevertheless piled up a large store of facts which our land using industries are unwilling, or (they claim) unable, to apply. Why? A review of three sample fields will be attempted.

Soil science has so far relied on self interest as the motive for conservation. The landholder is told that it pays to conserve his soil and its fertility. On good farms this economic formula has improved land practice, but on poorer soils vast abuses still proceed unchecked. Public acquisition of submarginal soils is being urged as a remedy for their misuse. It has been applied to some extent, but it often comes too late to check erosion, and can hardly hope more than to ameliorate a phenomenon involving in some degree *every square foot* on the continent. Legislative compulsion might work on the best soils where it is least needed, but it seems hopeless on poor soils where the existing economic setup hardly permits even uncontrolled private enterprise to make a profit. We must face the fact that, by and large, no defensible relationship between man and the soil of his nativity is as yet in sight.

Forestry exhibits another tragedy—or comedy—of *Homo sapiens,* astride the runaway Juggernaut of his own building, trying to be decent to his environment. A new profession was trained in the confident expectation that the shrinkage in virgin timber would, as a matter of self interest, bring an expansion of timber cropping. Foresters are cropping timber on certain parcels of poor land which happen to be public, but on the great bulk of private holdings they have accomplished little. Economics won't let them. Why? He would be bold indeed who claimed to know the whole answer, but these parts of it seem agreed upon; modern transport prevents profitable tree cropping in cut-out regions until virgin stands in all others are first exhausted; substitutes for lumber have undermined confidence in the future need for it; carrying charges on stumpage reserves are so high as to force perennial liquidation, overproduction, depressed prices, and an appalling wastage of unmarketable grades which must be cut to get the higher grades; the mind of the forest owner lacks the point of view underlying sustained yield; the low wage standards on which European forestry rests do not obtain in America.

A few tentative gropings toward industrial forestry were visible before 1929, but these have been mostly swept away by the depression, with the net result that forty years of "campaigning" have left us only such actual tree cropping as is underwritten by public treasuries. Only a blind man could see in this the beginnings of an orderly and harmonious use of the forest resource.

There are those who would remedy this failure by legislative compulsion of private owners. Can a landholder be successfully compelled to raise any crop, let alone a complex long-time crop like a forest, on land the private possession of which is, for the moment at least, a liability? Compulsion would merely hasten the avalanche of tax-delinquent land titles now being dumped into the public lap.

Another and larger group seeks a remedy in more public ownership. Doubtless we need it—we are getting it whether we need it or not—but how far can it go? We cannot dodge the fact that the forest problem, like the soil problem, *is coextensive with the map of the United States*. How far can we tax other lands and industries to maintain forest lands and industries artificially? How confidently can we set out to run a hundred-yard dash with a twenty foot rope tying our ankle to the starting point? Well, we are bravely "getting set," anyhow.

The trend in wildlife conservation is possibly more encouraging than in either soils or forests. It has suddenly become apparent that farmers, out of self interest can be induced to crop game. Game crops are in demand, staple crops are not. For farm species, therefore, the immediate future is relatively bright. Forest game has profited to some extent by the accidental establishment of new habitat following the decline of forest industries. Migratory game, on the other hand, has lost heavily through drainage and overshooting; its future is black because motives of self interest do not apply to the private cropping of birds so mobile that they "belong" to everybody, and hence to nobody. Only governments have interests coextensive with their annual movements, and the divided counsels of conservationists give governments ample alibi for doing little. Governments could crop migratory birds because their marshy habitat is cheap and concentrated, but we get only an annual crop of new hearings on how to divide the fast-dwindling remnant.

These three fields of conservation, while but fractions of the whole, suffice to illustrate the welter of conflicting forces, facts, and opinions which so far comprise the result of the effort to harmonize our machine civilization with the land whence comes its sustenance. We have accomplished little, but we should have learned much. What?

I can see clearly only two things:

First, that the economic cards are stacked against some of the most important reforms in land use.

Second, that the scheme to circumvent this obstacle by public ownership, while highly desirable and good as far as it goes, can never go far enough. Many will take issue on this, but the issue is between two conflicting conceptions of the end towards which we are working.

One regards conservation as a kind of sacrificial offering, made for us vicariously by bureaus, on lands nobody wants for other purposes, in propitiation for the atrocities which still prevail everywhere else. We have made a real start on this kind of conservation and we can carry it as far as the tax string on our leg will reach. Obviously, though, it conserves our self respect better than our land. Many excellent people accept it, either because they despair of anything better, or because they fail to see the *universality of the reactions needing control.* That is to say their ecological education is not yet sufficient.

The other concept supports the public program, but regards it as merely extension, teaching, demonstration, an initial nucleus, a means to an end, but not the end itself. The real end is a *universal symbiosis with land,* economic and esthetic, public and private. To this school of thought public ownership is a patch but not a program.

Are we, then, limited to patchwork until such time as Mr. Babbitt has taken his Ph.D. in ecology and esthetics? Or do the new economic formulae offer a shortcut to harmony with our environment?

THE ECONOMIC ISMS

As nearly as I can see, all the new isms—Socialism, Communism, Fascism, and especially the late but not lamented Technocracy—outdo even Capitalism itself in their preoccupation with one thing: The distribution of more machine-made commodities to more people. They all proceed on the theory that if we can all keep warm and full, and all own a Ford and a radio, the good life will follow. Their programs differ only in ways to mobilize machines to this end. Though they despise each other, they are all, in respect of this objective, as identically alike as peas in a pod. They are competitive apostles of a single creed: *salvation by machinery.*

We are here concerned, not with their proposals for adjusting men and machinery to goods, but rather with their lack of any vital proposal for adjusting men and machines to land. To conservationists they offer only the old familiar palliatives: Public ownership and private compulsion. If these are insufficient now, by what magic are they to become sufficient after we change our collective label?

Let us apply economic reasoning to a sample problem and see where it takes us. As already pointed out, there is a huge area which the economist calls submarginal, because it has a minus value for exploitation. In its once-virgin condition, however, it could be "skinned" at a profit. It has been, and as a result erosion is washing it away. What shall we do about it?

By all the accepted tenets of current economics and science we ought to say "let her wash." Why? Because staple land crops are overproduced, our population curve is flattening out, science is still raising the yields from better lands, we are spending millions from the public treasury to retire unneeded acreage, and here is nature offering to do the same things free of charge; why not let her do it? This, I say, is

economic reasonong. *Yet no man has so spoken.* I cannot help reading a meaning into this fact. To me it means that the average citizen shares in some degree the intuitive and instantaneous contempt with which the conservationist would regard such an attitude. We can, it seems, stomach the burning or plowing under of overproduced cotton, coffee, or corn, but the destruction of mother earth, however "submarginal," touches something deeper, some subeconomic stratum of the human intelligence wherein lies that something—perhaps the essence of civilization which Wilson called "the decent opinion of mankind."

THE CONSERVATION MOVEMENT

We are confronted, then, by a contradiction. To build a better motor we tap the uttermost powers of the human brain; to build a better countryside we throw dice. Political systems take no cognizance of this disparity, offer no sufficient remedy. There is, however, a dormant but widespread consciousness that the destruction of land, and of the living things upon it, is wrong. A new minority has espoused an idea called conservation which tends to assert this as a positive principle. Does it contain seeds which are likely to grow?

Its own devotees, I confess, often give apparent grounds for skepticism. We have, as an extreme example, the cult of the barbless hook, which acquired self-esteem by a self-imposed limitation of armaments in catching fish. The limitation is commendable, but the illusion that it has something to do with salvation is as naive as some of the primitive taboos and mortifications which still adhere to religious sects. Such excrescences seem to indicate the whereabouts of a moral problem, however irrelevant they be in either defining or solving it. Then there is the conservation booster, who of late has been rewriting the conservation ticket in terms of "tourist bait." He exhorts us to "conserve outdoor Wisconsin" because if we don't the motorist on vacation will streak through Michigan, leaving us only a cloud of dust. Is Mr. Babbitt trumping up hard-boiled reasons to serve as a screen for doing what he thinks is right? His tenacity suggests that he is after something more than tourists. Have he and other thousands of "conservation workers" labored through all these barren decades fired by a dream of augmenting the sales of sandwiches and gasoline? I think not. Some of the people have hitched their wagon to a star—and that is something.

Any wagon so hitched offers the discerning politician a quick ride to glory. His agility in hopping up and seizing the reins adds little dignity to the cause, but it does add the testimony of his political nose to an important question: is this conservation something people really want? The political objective, to be sure, is often some trivial tinkering with the laws, some useless appropriation, or some pasting of pretty labels on ugly realities. How often, though, does any political action portray the real depth of the idea behind it? For political consumption a new thought must always be reduced to a posture or a phrase. It has happened before that great ideas were heralded by growing pains in the body politic, semi-comic to those onlookers not yet infected by them. The insignificance of what we conservationists, in our political capacity, say and do, does not detract from the significance of our persistent desire to do something. To turn this desire into productive channels is the task of time, and ecology.

The recent trend in wildlife conservation shows the direction in which ideas are evolving. At the inception of the movement fifty years ago, its underlying thesis was to save species from extermination. The means to this end were a series of restrictive enactments. The duty of the individual was to cherish and extend these enactments, and to see that his neighbor obeyed them. The whole structure was negative and prohibitory. It assumed land to be a constant in the ecological equation. Gunpowder and blood lust were the variables needing control.

There is now being superimposed on this a positive and affirmatory ideology, the thesis of which is to prevent the deterioration of environment. The means to this end is research. The duty of the individual is to apply its findings to land, and to encourage his neighbor to do likewise. The soils and the plant succession are recognized as the basic variables which determine plant and animal life, both wild and domesticated, and likewise the quality and quantity of human satisfactions to be derived. Gunpowder is relegated to the status of a tool for harvesting one of these satisfactions. Blood lust is a source of motive power, like sex in social organization. Only one constant is assumed, and that is common to both equations: the love of nature.

This new idea is so far regarded as merely a new and promising means to better hunting and fishing, but its potential uses are much larger. To explain this, let us go back to the basic thesis—the preservation of fauna and flora.

Why do species become extinct? Because they first become rare. Why do they become rare? Because of shrinkage in the particular environments which their particular adaptations enable them to inhabit. Can such shrinkage be controlled? Yes, once the specifications are known. How known? Through ecological research. How controlled? By modifying the environment with those same tools and skills already used in agriculture and forestry.

Given, then, the knowledge and the desire, this idea of controlled wild culture or "management" can be applied not only to quail and trout, but to *any living thing* from bloodroots to Bell's vireos. Within the limits imposed by the plant succession, the soil, the size of the property, and the gamut of the seasons, the landholder can "raise" any wild plant, fish, bird, or mammal he wants to. A rare bird or flower need remain no rarer than the people willing to venture their skill in *building it a habitat*. Nor need we visualize this as a new diversion for the idle rich. The average dolled-up estate merely proves what we will some day learn to acknowledge: that bread and beauty grow best together. Their harmonious integration can make farming not only a business but an art; the land not only a food factory but an instrument for self expression, on which each can play music of his own choosing.

It is well to ponder the sweep of this thing. It offers us nothing less than a renaissance—a new creative stage—in the oldest, and potentially the most universal, of all the fine arts. "Landscaping," for ages dissociated from economic land use, has suffered that dwarfing and distortion which always attends the relegation of esthetic or spiritual functions to parks and parlors. Hence it is hard for us to visualize a creative art of land beauty which is the prerogative, not of esthetic priests but of dirt farmers, which deals not with plants but with biota, and which wields not only spade and pruning shears, but also draws rein on those invisible forces which determine the presence or absence of plants and animals. Yet such is this thing which lies to hand, if we want it. In it are the seeds of change, including, perhaps, a rebirth of that social dignity which ought to inhere in land ownership, but which, for the moment, has passed to inferior professions, and which the current processes of land skinning hardly deserve. In it, too, are perhaps the seeds of a new fellowship in land, a new solidarity in all men privileged to plow, a realization of Whitman's dream to *"plant companionship as thick as trees along all the rivers of America."* What bitter parody of such companionship, and trees, and rivers is offered to this our generation!

I will not belabor the pipe dream. It is no prediction, but merely an assertion that the idea of controlled environment contains colors and brushes wherewith society may some day paint a new and possibly a better picture of itself. Granted a community in which the combined beauty and utility of land determines the social status of its owner, and we will see a speedy dissolution of the economic obstacles which now beset conservation. Economic laws may be permanent, but their impact reflects what people want, which in turn reflects what they know and what they are. The economic setup at any one moment is in some measure the result, as well as the cause, of the then prevailing standard of living. Such standards change. For example: some people discriminate against manufactured goods produced by child labor or other antisocial processes. They have learned some of the abuses of machinery, and are willing to use their custom as a leverage for betterment. Social pressures have also been exerted to modify ecological processes which happened to be simple enough for people to understand—witness the very effective boycott of bird skins for millinery ornament. We need postulate only a little further advance in ecological education to visualize the application of like pressures to other conservation problems.

For example: the lumberman who is now unable to practice forestry because the public is turning to synthetic boards may then be able to sell man-grown lumber "to keep the mountains green." Again: certain wools are produced by gutting the public domain; couldn't their competitors, who lead their sheep in greener pastures, so label their product? Must we view forever the irony of educating our sons with paper, the offal of which pollutes the rivers which they need quite as badly as books? Would not many people pay an extra penny for a "clean" newspaper? Government may some day busy itself with the legitimacy of labels used by land industries to distinguish conservation products, rather than with the attempt to operate their lands for them.

I neither predict nor advocate these particular pressures—their wisdom or unwisdom is beyond my knowledge. I do assert that these abuses are just as real, and their correction every whit as urgent, as was the killing of egrets for hats. *They differ only in the number of links composing the ecological chain of cause and effect.* In egrets there were one or two links, which the mass mind saw, believed, and acted upon. In these others there are many links; people do not see them, nor believe us who do. The ultimate issue, in conservation as in other social problems, is whether the mass mind *wants to* extend its powers of comprehending the world in which it lives, or granted the desire, *has the capacity to do so*. Ortega, in his *Revolt of the Masses,* has pointed the first question with devastating lucidity. The geneticists are gradually, with trepidations, coming to grips with the second. I do not know the answer to either. I simply affirm that a sufficiently enlightened society, by changing its wants and tolerances, can change the economic factors bearing on land. It can be said of nations, as of individuals: "as a man thinketh, so is he."

It may seem idle to project such imaginary elaborations of culture at a time when millions lack even the means of physical existence. Some may feel for it the same honest horror as the Senator from Michigan who lately arraigned Congress for protecting migratory birds at a time when fellow humans lacked bread. The trouble with such deadly parallels is we can never be sure which is cause and which is effect. It is not inconceivable that the wave phenomena which have lately upset everything from banks to crime rates might be less troublesome if the human medium in which they run *readjusted its tensions*. The stampede is an attribute of animals interested solely in grass.

6

ENVIRONMENT: NEW IMPERATIVES FOR FOREST POLICY

Carl H. Reidel

First published in *Journal of Forestry* 69(5):266–270. May 1971.

BIOGRAPHY

Mr. Reidel is Professor of Forestry and Director of the Environmental Program in the University of Vermont, a position which he has held since 1972. He had been Bullard Fellow at Harvard, and, before that, Assistant Professor of Political Science and Assistant Director of the Center for Environmental Studies at Williams College, information officer and ranger in the intermountain West—this varied career having been contained within 15 years. Educated at the University of Minnesota and at Harvard—a person whose wide and active interests are implied in such memberships as of the Board of Directors of the American Forestry Association and of the Trustees of the National Parks and Conservation Association—he has served forestry, in recent years, as an interpreter of the profession's context: its social ecology.

EDITORS' SUMMARY

To be effective, the forestry profession must understand the meaning and implications of the environmental movement. It is not simply an extension of the conservation movement, but a set of related aspirations that focus on worldwide interdependencies. Forestry must be viewed as part of a system requiring broad interprofessional and interagency cooperation.

QUESTIONS TO CONSIDER

(1) Offer your own definition of the conservation idea and of the environmental idea. What differences do you find between the two ideas? What differences can you discover in the way the two movements originated and developed?

(2) Read pages 322–323 in Gifford Pinchot's *Breaking New Ground,* in which the author

describes what he views as the origin of the conservation idea. Does this affect your answer to the first part of Question 1?

(3) Were there figures in the environmental movement analogous to Roosevelt and Pinchot in conservation? How do you explain any differences that you observe?

(4) What does Mr. Reidel mean when he says (page 56) that environmentalists view their values as infinitely great? What does a value of infinity imply for resource allocation?

(5) Select a wood-using industry in which you are interested, and sketch a study (a statement of problem, purpose, and procedure) designed to produce a social plan for the industry. The study should aim to describe production goals and measures consistent with the philosophy expressed in the latter part of Mr. Reidel's article.

* * *

American forestry is in trouble. . . . What has gone wrong? How is it that policies lauded as progressive less than a decade ago are suddenly almost universally maligned outside our immediate profession—by the industries we serve and by the conservationists with whom we share a common heritage? Part of the answer lies in our interpretation of, and response to, the social phenomenon that has been labeled the "environmental movement."

ENVIRONMENT: THE MEANING AND THE MOVEMENT

"Environment" is a word that has gained more obtuse and diverse meanings in less time than almost any word in our language. "Religion" and "politics" pale before "environment" as loaded words. Only a fool would attempt to define it, but perhaps such foolishness will give us some perspective on its meaning and significance for forestry.

In the classical sense environment is simply "the aggregate of surrounding things, conditions, or influences." In that sense we know full well what people mean when they talk about the environmental crises. Our "surrounding things" are deteriorating at an alarming rate. We face a very real crisis in terms of air, water, world resources, food, and the quality of life on this planet. Some would suggest that the question is not even "quality," but survival itself. To recite the dimensions of this crisis in order to prove that a crisis does, in fact, exist would be to insult your intelligence. The reality and gravity of the environmental crisis have been too well documented by eminent scientists around the world to debate the question here. Furthermore, I don't believe it is necessary to specify all the dimensions of the crisis itself. The most vital question is the nature of public perception of the crisis and, most importantly, people's response to that perception. In other words, it is the environmental *movement* with which we must deal in formulating policy.

Like the word "environment," the movement bearing its name is varied and complex. It ranges from traditional garden club conservation to extreme radical emotionalism. It includes concern with atmospheric physics, organic gardening, traffic on Manhattan, wilderness preservation, and nonreturnable bottles. It reaches across disciplinary, professional, and political boundaries. It is popular and yet esoteric. But there is a basic theme; a core philosophy. It is this theme that we must seek to understand and interpret in order to shape responsive policy. While I neither claim to fully understand the theme, nor to be able to adequately interpret its meaning for forest policy, I shall try to highlight what I consider some key points. In so doing I will only attempt to characterize the best in the movement—that to which we should respond.

A Holistic Approach

As a point of departure, we must shed the notion that the environmental "thing" is merely an extension and outgrowth of what we reverently call the "conservation movement." It is not! Perhaps conservation and environment share some common heritage, but environment is more than another wave of conservation.

Perhaps the clearest theme is an overriding holistic approach or "world view," as its younger adherents would say. This movement is integrated in an intuitive way. The concepts of ecology are applied to the entire social, biological, and physical environment in a metaphorical sense. Intrinsic values are given relative consideration as they are integrated into the total man/nature system. This, I believe, is a major point of departure from traditional conservation. Where the old conservation mourns the loss of an eagle for its intrinsic value alone, or perhaps as a vague symbol of paradise lost, the environmentalist mourns the eagle not only for these reasons but also in recognition of the fact that its death is a clear warning of systemic disruption. The eagle's death, like the death of the 19th century coal miner's canary, is a signal that something is wrong in the biological system *and* quite likely in the political, economic, and social realms as well. This holistic approach was well stated recently by Barry Commoner. He stated the laws of ecology as three proverbs, which you will recognize as three statements of the same principle: (1) all things are interconnected; (2) everything goes somewhere; and (3) there is no such thing as a free lunch.

This holistic approach reaches back into the life style of the advocate as well. Where traditional conservation was often launched from the position of wealth or bureaucracy, the environmental movement is clearly a cultural revolution seeking changes in basic attitudes and social behavior. This is clearest in the younger members of the movement for whom the environmental crisis is inseparable from every other aspect of life. Warren Bennis, author of *Planning Change* and of *Changing Organization,* characterizes the younger generation this way:

> Their morality is person centered. It values technology, only if technology serves personal growth and social goals. The new culture is willing to tolerate inefficiency for the sake of [human] development . . . [it] treasures smallness, human scale, and cultural pluralism.[1]

Renaissance or Revolution

It is perhaps the younger generation's involvement in the environmental movement that is most significant—not only because people under 25 constitute half of the population and will *be* the future, but because they are willing to make a more total commitment.

William B. Boyd of the University of California believes that:

> A generation of youth that rates involvement higher than status will not choose either the same job or the same clothes as its father. Neither the carrot nor the whip will drive this new generation. When we reach the moment when the quality of life, rather than its preservation becomes our true concern, then the attitude of modern youth will have come into its own.[2]

In the words of a member of that generation, a participant at the November UNESCO meeting in San Francisco:

> We will stop the destruction of this planet even at the cost of our own futures, careers, and blood. The situation is simply like that. If you are not going to live for the earth, what are you going to live for?[3]

These words cannot be taken lightly. Youth is determined to seek change and they will take us either to renaissance or to revolution. The choice, I would argue, is ours, depending upon our response. What they are asking for, however, is much more than a shift in marginal policies. They are challenging fundamental tenets of our society—the basic assumptions of our economic and political systems. That is a major departure from traditional conservation!

These challenges do not, however, come only from youth, although I believe they hold these ideas more generally than does the rest of society. In recent months Paul Ehrlich, Barry Commoner, Rene Dubos, and a host of eminent social and natural scientists have also advocated the need for fundamental change and personal commitment. Nor is the movement limited to youth and scientists. It is a broad cultural movement, despite the fact that it has originated in great part from an understanding of the world gained from rigorous scientific analysis. In its quest for the wholeness of life, it accommodates paradox and pluralism because its perspective encompasses the total system and its approach is integrated. Although lacking in the precision and clarity of objectives we would like, these themes are not unlike the concepts of systems theory, or what we recognize as rigorous ecosystems analysis. They are also what Aldo Leopold meant when he asked for a new land ethic.

In summary, the environmental movement is *not* just another conservation movement—it is a deep cultural change that challenges the strongly entrenched assumption that the greatest good is served by endlessly increasing production and consumption. As a cultural movement it asks not for better regulation of the system, but a transformation of it—a reordering of priorities and policies. Unlike some conservationists who would retreat from the complexity of modern technological society, the environmentalist seeks a transformation of that society based on a sensitivity to ecological reality and humane values. And finally, this movement is more than a mere critique of the present technoeconomic system; it offers an integrative conceptual alternative.

The themes we find in the environmental movement are not clear and explicit, but I would contend that they are clear enough to provide justification for a serious examination of forest policy and even to suggest certain imperatives for the future.

FOREST POLICY

Economic Analysis and Policy Formation

The tensions between economics and ecology are clearly intensifying with the increasing tempo of the environmental movement. The clash is sharpest between the applied, or popular, expressions of these disciplines. Both camps claim to be the true holders of the ecos—to be the best keepers of the earth household. The message of the environmentalists was well stated by *Saturday Review* editor Norman Cousins in an editorial entitled "Needed: A New Dream":

> The dream is dated. For nothing could be more dangerous today to the human future than if the American Standard were to be achieved in country after country. . . . A tidal wave of prosperity would sweep across the globe, but within a few years the Earth would become uninhabitable.[4]

The American Dream has been just that—a dream. It has been sustained by a technological system that has borrowed heavily from the world ecosystem—a debt we are beginning to pay.

John Fischer, in the April 1970 issue of *Harper's,* suggested the "profoundly subversive idea" that "our prime national goal should be to reach zero growth rate as soon as possible—in people, in GNP, and in our consumption of everything."[5] In the July 1970 issue of *American Forests,* Van Trumbell suggested in similar fashion that: "Endless growth is impossible. The population must stop somewhere and the Gross National Product also."[6]

These cries for a zero growth rate are clear indications of a growing public suspicion that the economic housekeepers are not doing the full job. There is ample evidence they have been sweeping a lot under the rug. Our faulty economic bookkeeping has encouraged and sustained spiraling patterns of production and consumption with severe environmental effects. Synthetics replace natural fibers; chemicals replace organic fertilizers and biological pest controls; disposable products replace those easily recycled—not because they are environmentally *or economically* sound, but because they are profitable in the short run due to our faulty cost/benefit accounting. About such misleading economic strategies Stephen Spurr states:

> Cost-benefits ratios have fallen into disrepute because they have too often been manipulated to justify a project which is wanted for political reasons . . . the principle is sound—provided, first, that the intangible values to mankind are properly introduced into the formula. . . .[7]

Quantifying Intangibles

There are two reasons why crucial environmental externalities have not been introduced into the formula. First, economists have simply been unable to get adequate data from ecologists and other natural or social scientists to plug into the formula. This is a two-edged sword, however; economists demand quantification beyond the limits of reason while environmentalists view all their values as inalienable rights with infinite value. Both sides are unrealistic.

"Quality environment" is no less concrete an idea than "freedom" or "greatest good," and we have successfully built an economic system within the constraints of the Bill of Rights. On the other hand, scientists can give us increasingly better estimates of the value of specialized soil organisms or of the impact of DDT on the productive capabilities of the natural system, as well as estimates of the social costs of pollution.

There are real problems in attempting to value inputs to an economic analysis which includes apparently intangible values, but using those problems as excuses for reinforcing traditional stands—by either side—is a "copout" of the first order. However, the problem of quantifying intangibles is not the primary reason for the failure of our economic analyses to account for environmental externalities. There is, I believe, a more basic problem.

A Total Perspective

An environmentally sensitive economic analysis must begin from a perspective which respects the integrity of the total system with which it is concerned. Far too many cost-benefit analyses have been too narrow, too limited, and too closed—especially in forestry. We have based our forest policies on economic analyses that seldom reach beyond forest boundaries or the primary users of tangible forest products, with perhaps a somewhat more extended look at recreation. In most cases our concept of "demand" is an anachronism, not only because we simply project the requirements of an ecologically destructive past into a fast changing future with inadequate quantification of environmental costs, but because our roster of costs and benefits is lacking in breadth.

Paper Production

Take paper production as an example. Present policy is based on the probable demand for future forest growth derived from projected paper requirements—requirements that have taken into consideration neither the massive pollution costs borne by the environment in the production of paper nor the fact that almost 50 percent of the solid waste-treatment costs of major metropolitan areas arise from the disposal of paper products. Nor have we dealt realistically with the effects of substantial government subsidies in the form of public land management and protection. This is an example of why we must extend our horizons beyond the immediate limits of the forest and its industries if our economic analyses are to be trusted as the bases of environmentally sound forest policy.

In the case of paper, we need to know what it will mean to internalize pollution costs in terms of the price of paper and resulting consumption schedules. The answer, which we can already guess in terms of direction if not magnitude, cannot, however, be understood simply in terms of the impact on one industry or one sector of the economy. What effect will rising costs of making new paper have on encouraging the recycling of used paper? We now recycle less than 20 percent of the 60 million tons of paper and fiberboard consumed in the United States annually.[8] With present technology we could double that figure. What impact would increased recycling have on relieving the tremendous pressures of solid waste disposal in metropolitan centers where paper and wood fiber constitute over half the problem? What would it mean to New York City, for example, which is forced to seek space in natural preserves and high-value development sites to bury wood fibers which an innovative industry should be buying? What would it mean to reduce the consumption of new wood fiber by over a billion cubic feet a year in terms of land-use conflicts we must mediate as foresters? Would increased paper prices also reduce per capita consumption, further reducing new wood utilization?

In sum, I suspect we would see a new forest policy emerge from the integration of our cost-benefit analyses with those of others closely linked to our policy decisions. In the case of paper it would mean adopting policies requiring industry to internalize all costs of pollution and forest management in order to make recycling more competitive. It would require policies reaching into urban areas to encourage the diversion of solid-waste disposal funds from land fills and incineration to systems of collection that would make fiber economically available and most usable for reuse. Financial support for research and technological changes in the paper industry would have to be an integral part of such policies.

Lumber

We also need to shape policies that will make projected requirements for lumber realistic estimates of true economic demand, rather than mere responses to environmentally unsound policies in other sectors. It is time we stopped responding to a lumber shortage that *is not* our making. Policies that blindly advocate increased yields and harvests are an irrational response. The shortage of housing in this country is more than a forest management problem. *It is* an ill-conceived urban renewal program destroying housing that could be reclaimed: transportation policies encouraging freeway construction and urban sprawl; inadequate regional planning and land-use zoning which discourage cluster development, prefabricated module housing, and multiple-family dwellings; an archaic construction industry, protected by outdated building codes which resists modern construction methods and the use of innovative wood fiber materials; a ''bailing-wire'' lumber industry leaving 2 billion cubic feet of wood fiber in the woods every year;[8] and finally, it is an inflationary economy resulting from explicit governmental policies making defense and space escapades our foremost national priorities. We clearly need bold new forest policies based on expanded and integrated economic analyses that would give us decision models reflecting the realities of the world in which we live.

Recreation

In the area of recreation we must have policies that recognize the need to broaden our concern for urban recreation development, which is closely linked to transportation policy, regional planning, and a host of critical social problems. We must also take a new look at our hardheaded defense of timber production on certain public lands where there are mounting recreational and park demands. At the very time we are fighting bloody battles to keep public forests in full wood production, a number of the largest wood-using corporations in this nation are moving aggressively into the leisure home industry, more than willing to convert timber lands into poorly planned, environmentally destructive recreational developments.[9]

Response and Integration

Integrated forest policy will not only require new approaches to old problems, but will also demand research on a broad range of new questions, such as the forest's role in local climate modification, air pollution reduction, noise abatement, municipal waste neutralization, and a host of social values yet unexplored.

The point is simply this: we have everything to gain and little to lose by responding to the clearest theme of the environmental movement. By opening up the economic bases of our forest policies and integrating that policy with allied professions we will have a more responsive, more rational, more "professional" profession. And we will spend a lot less time mounting elaborate defenses for economically and ecologically naive decisions. What we will, and should, lose is an irrational and indefensible emotional attachment to the idea that *multiple use* is a viable guide in determining sound forest policy.

Multiple use has served a vital role in shaping forest policy during the transition from the custodial period of the first half of the century to the beginning of intensive land management in the 1950's. And there is little doubt that the pattern of management in the future will continue to be multi-purpose use of forest resources. Multiple use may be the *result* of environmentally integrated forest policy; it cannot be the guide. Multiple use is simply not a functional concept. It is an admirable proverb. With its focus on mediating conflict between essentially economic definitions of resources, multiple use either collapses on the issue of assigning values to intangibles, or is forced to accept an inherent bias toward clearly short-run monetary values. This is a flaw in the concept, not in its users, as is too often claimed. Furthermore, the answer is not the replacement of multiple use by the even narrower concept of urban zoning as often suggested.[10] In ecological terms, single-use zoning is even less sensitive to reality than multiple use.

I am not suggesting we merely abandon multiple use, but rather that we integrate it into a broader and more comprehensive systems approach that will make it possible to achieve the balances that were sought and advanced by the Multiple-Use Act of 1960. The mandate stated in that historic act must now be interpreted in light of the broader and more comprehensive mandate of the National Environment Policy Act signed into law the first day of this decade.

Administration

This further suggests the need for a review of the structure and stance of organizations through which forest policy is mediated and implemented.

If we are to seek an integrated policy for the environment, we cannot continue to demand the degree of organizational independence now enjoyed by most forestry agencies. There is no way we can continue to defend the administration of public resources by so many single-use agencies in practically every department of government and still claim to support integrated environmental policy. As a case in point, the issue of whether the U.S. Forest Service is moved to the Department of Interior or left in the Department of Agriculture is no longer a worthwhile debate on historical grounds. I suspect Pinchot took his new agency to Agriculture because it was advantageous to do so in that time and political milieu. I am sure that he would have moved it back to Interior if such a move held the promise of building a new department with broad land management influence.

We must begin to build new bridges to related professions, both organizationally and in less formal ways, if we are to respond to the clear imperatives of the environmental movement. If forestry is to remain a distinctive profession it must seek integration; not independence. We must, in Frome's words, "seek broader views." Our distinctiveness and influence will be lost if we merely seek protection through licensing, organizational separation, and guild-like barriers.

Perhaps we have already turned the corner, at least in terms of cooperative efforts. In April 1970, the Society of American Foresters joined with Friends of the Earth and Zero Population Growth in testifying before the House Appropriations Subcommittee on Interior and Related Agencies to seek an increased Forest Service budget for research on biological pest control and wood fiber recycling. The American Institute of Biological Sciences interdisciplinary meeting last summer was another approach to unity. But these are merely hints of what must occur if our profession is to be responsive to a mounting wave of environmental awareness and sensitivity in America.

In closing, I should like to quote from the youth chairman of the U.S. National Commission of UNESCO at the Commission meeting in San Francisco last November:

> We'll support the individual who does the job. Understand, I represent or belong to no organization or special interest, only the human race. Our request is simple; we seek life, to know our grandchildren will survive.[3]

REFERENCES

1. Bennis, Warren. 1970. How to survive in a revolution (reprint).
2. Boyd, William B. 1968. Pacing the hour. *In* PACE (June). Page 64.
3. Johnson, Huey D. 1970. No deposit—no return. Menlo Park, California: Addison-Wesley Publishing Co. Page 285.
4. Cousins, Norman. 1970. Needed: A new dream. Saturday Review (June 20). Page 18.
5. Fischer, John. 1970. Easy chair. Harpers (April).
6. Trumbell, Van. 1970. Washington outlook. American Forests 76:7:6.
7. Spurr, Stephen H. 1970. Developing a natural resources management policy. *In* No deposit—no return. Page 105.
8. Lassen, L. E., and Dwight Hair. 1970. Potential gains in wood supplies through improved technology. Jour. of Forestry 68:404–407.
9. Unknown. 1970. Lumber firms join leisure home boom. The New Englander (February). Pages 18–21.
10. Sterling, E. M. 1970. The myth of multiple use. American Forests 76:6:27.

7

PERSONALITY, MOTIVATION, AND EDUCATION NEEDED IN PROFESSIONAL FORESTRY

Irving I. Holland
Ronald I. Beazley

First published in *Journal of Forestry* 69(7):418–423. July 1971.

BIOGRAPHY

Mr. Holland is Professor of Forest Economics and Acting Head of the Forestry Department, University of Illinois, Urbana. His career as a social scientist in forestry began at the University of California at Berkeley, where he earned degrees in forestry, in range management, and in agricultural economics. His experience has included 15 years with the Forest Service Branch of Research at both western and eastern locations, a briefer tour of duty at Iowa State University, and consulting assignments abroad and at home. A man of wide interests in his profession and its social setting, he has teamed with his present co-author, a person of appropriately similar concerns, on a number of scholarly projects, ranging from journal articles to a book.

Mr. Beazley is Professor, Department of Geography, and Senior Graduate Faculty Member, Department of Economics, Southern Illinois University, Carbondale. A native of Canada, he commenced his education in forestry and economics at the University of New Brunswick and continued, working for a degree or in a post-doctoral capacity, at Yale, Purdue, Oxford, and Harvard. His professional assignments, primarily in research and teaching, have been with the University of Minnesota as well as Southern Illinois and with numerous foreign and domestic governmental agencies. He has taken a special interest in the problems of the developing nations and in applying quantitative analytical techniques to such problems. He is a past Chairman of the Division of Economics and Policy in the Society of American Foresters.

EDITORS' SUMMARY

Personality and motivation-analysis tests given to university students in Illinois reveal that foresters are in certain respects less well equipped than the average for the coming responsible positions in resource management. Improved forestry-school programs of student recruitment and screening and of curriculum development are required.

QUESTIONS TO CONSIDER

(1) Messrs. Holland and Beazley emphasize the distinction between a *job* and a *role*. What is the distinction? Can you think of specific examples? To what degree is the professional forestry curriculum aimed at the demands of the forest resource manager's role? If there is a deficiency, what can be done to remedy it?

(2) In your observation, why do persons enter the forestry profession? How accurate is the information upon which they base their decision? If their information about prospective demands upon forest resource managers were more accurate, what changes would you expect in the list of personal attributes on page 65?

(3) Where lies the comparative advantage of the person described on page 65? That is, in what sort of occupation is he apt to be most successful by his own standards?

(4) How might the forestry profession go about a self analysis which would lead to a program of recruitment, education, and performance of job and role well suited to social requirements for resource management?

* * *

The capability of professionally trained foresters to successfully manage forest and related land resources within rapidly changing patterns of demands for forest products—extra-market, as well as market—is being questioned with increasing frequency. Particularly in recent years, much of the questioning has come from within the profession itself. It is certain that changes must be made if the profession is to maintain a position of influence in resource-use planning and policy and decision making. The problem is in knowing what changes are needed and how these, once determined, are to be made.

* * *

JOB VS. ROLE IN FORESTRY

It is certain that present-day foresters are being asked to do more than comply with the specifications of a job description. The job description is used to convey to prospective employees and educators the requirements associated with a particular position. It usually covers duties, responsibilities, the necessary education and training required for qualification, place of employment, and level of pay. At least at the lower levels, job descriptions stress the technical requirements of the job. It is toward fulfillment of job-description requirements that much educational attention in forestry has been directed.

If the professional forester has lost influence in the resources management decision-making process, the fault does not lie with his inability to carry out the technical requirements of his job. Instead, it can be argued that he has failed in his ability to fulfill a *role*. A role can be defined as an expected pattern of behavior associated with a particular position. To successfully fill a role requires much more than technical competence. And the role that resource managers are being asked to play (and surely will be asked to play in the future) is becoming more complex. At practically all levels, the forester is now being asked to consider people and social problems, as well as trees, in the performance of his job. He is being required to make technical resource-use decisions within a complex matrix of economics, politics, and administrative dictates and still be sensitive to the aspirations, needs, capacities, mistakes, and inconsistencies of an increasingly environment-conscious and demanding public. Is the profession well supplied with foresters adequately equipped to play these roles? Are we attracting and educating enough young people who can eventually play them better?

The success with which an individual plays his role depends upon several factors, including the requirements of the position, the behavior expected by others, how the individual thinks he should behave in the job, and the individual's actual behavior. The actual role is related to personality and motivations.

A serious shortcoming of job descriptions is that they do not specify much about personality requirements or expected behavior under expected conditions of employment. Until realistic role descriptions supplement job descriptions, there is little chance that more of the right kinds of people will be attracted into professional forestry or that educators will be able to modify curricula to improve the education of professional foresters.

FUTURE ROLES FOR PROFESSIONAL FORESTERS

Henry Vaux[1] asks the question: "In 1980, will a resource manager who has been educated primarily in the role of forests as agents of production still be able to do a satisfactory job?" Put another way, "Will the role of the resource manager in decision making be different in 1980 from what it is today?" All the evidence indicates that this role *will* be different.

Industrial forestry will demand an increasing number of foresters who, according to Cornelius,[1] can plan, direct and schedule programs, and manage people.

The role of the public forester is also changing. Beale[1] believes that because the role of the state forest administrator has changed from custodian and production manager to one of administrator, planner, and salesman, a larger number of foresters must become people-oriented. Winkworth[1] sees two distinct roles for state forest resource managers. One role requires foresters who are motivated to the performance of technical forestry jobs; the other will require foresters with added capabilities in administration and management. In his opinion only about 10 percent of public foresters are presently equipped to play the broader role. Greeley[1] states that increasingly, public foresters must be able to work in the spotlight, take criticism, settle for little commendation most of the time, and still do the resource-management job.

What these people are saying is that we must plan to provide more professional resource managers who can successfully play the roles of managers and administrators of public and private forest resources within an increasingly complex set of economic, social, and political constraints in addition to performing technical forestry jobs. The following sections of this paper consider the prospects for successfully pursuing this course in view of the capabilities of the students attracted to forestry and the structure of forestry curricula.

CAPABILITIES AND PERSONALITIES OF FORESTERS

Forestry students generally have been attracted to technically oriented curricula, not to curricula that are social science– or business-oriented. By the same token young people with latent managerial sales or administrative ability have not been attracted to forestry in very large numbers. Of course, the profession can point to a number of top businessmen and excellent administrators in both public and private forestry. These men are the exceptions one always expects to find; they obviously possess the necessary aptitudes that have made them leaders in their particular professional spheres. The supply of forestry researchers and academicians has, in general, also been drawn from the much larger pool of technically oriented people.

Several years ago, the authors became concerned with the general problem of forestry student aptitudes, personalities, and motivations. We sought answers to such questions as:

1. What sort of an individual is the forestry student?
2. Are the personalities and motivations of forestry students different from those of other college students?
3. If there are differences, what are the implications for better professional performance and/or curricula changes which might lead to better performance?
4. How can the self-selection of students in forestry be improved?

Our hypotheses were that forestry students do differ from other student groups with respect to personality and motivations and that in this regard they are, although probably not unique, at least unusual. To test these hypotheses, a study involving students in the Departments of Forestry of Southern Illinois University (Carbondale) and the University of Illinois (Urbana-Champaign) was initiated in 1962.

A total of 209 junior and senior forestry students were given two different tests. One was . . . [a] Personality Factor Test. . . . The other was . . . [a] Motivation Analysis Test. . . .

* * *

In general [the] findings, in regard to primary personality characteristics, indicate that the forestry student, in comparison with the . . . average [of all male students], is more reserved in manner, more intelligent, more emotionally stable, less demanding, more enthusiastic, more conscientious, more tough-minded, more forthright, somewhat less assured, more conservative, and more self-reliant.

* * *

A general interpretation of [the] motivational characteristics suggests that the forestry student, compared to the average male college student, is more concerned with material activities, expresses hostilities naturally, is willing to meet the world "on its own terms," and tends toward independence and autonomy. Furthermore, forestry students appear to lack an egocentric life style, to have a low need for career status and economic competition, to be low in anxiety, and generally to be satisfied and contented.

In summary, considering the data for *both* personality and motivational characteristics, the typical Illinois forestry student could be described as follows:

1. He is intelligent, resourceful, well-balanced with a relatively high degree of emotional stability; he is not easily aroused and tends to be somewhat submissive and accepting.
2. He is capable of independent decision making and autonomous action.
3. He is dependable and conscientious.
4. He is oriented toward material rather than abstract activities and prefers a nonegocentric life style.
5. He displays a tendency toward personal satisfaction and contentment.
6. He has a relatively low need for career status and lacks intense drive for economic competition.
7. He can be rather easily satisfied with respect to a job but highly involved in his chosen occupation.
8. He is conservative and undemanding.
9. He is only average on leadership, but high on creativity.
10. Although he can work well with others, he is reserved and nonaggressive and not inclined to initiate involvement with people generally.

FORESTERS' CAPABILITIES AND SUCCESSFUL ROLE PLAYING

The above profile does not suggest a person very well suited to attainment of leadership in the management of either public or private forest resources where people, politics, and economics are heavily involved. The complaint that too few foresters reach the top in major forest-products companies appears to be well founded, if, indeed, the typical forestry student lacks intense drive for economic competition and displays a low need for job status. Furthermore, foresters who prefer to avoid involvement with people and are reserved and somewhat introverted do not play the role of salesmen with much success. And the fact that the average forester prefers to concern himself with material activities and concepts, rather than abstract ones, helps explain his preference for subjects other than economics, business, or social sciences in general. It is little wonder that other more vocal, aggressive, outgoing, and articulate people in other professions and walks of life are taking on the role of resources-use decision making, once an important function of technically trained foresters in both business and public administration.

Where do foresters really excel? Foresters make excellent technicians and researchers. They do well where intelligence, dedication, hard work, and resourcefulness are needed. Because the forester is not particularly group-dependent, he is capable of autonomous action. Because he is job-oriented, forthright, undemanding, and on the conservative side, he tends easily to develop an *esprit de corps,* and to identify with an organization. There is little question that the successful development of the U.S. Forest Service, for example, as an effective, decentralized public forest-resources administering agency has been due in large measure to the kind of people it attracted. The kinds of problems the Forest Service has had to solve over the years, for the most part, have been the ones foresters with a technical bent were well qualified and eager to tackle. The same can be said for the forest industries.

The forestry profession itself mirrors the ideas and the philosophies of the people that compose it. If the profession conjures up only the older image of what forestry once was, it will likely continue to draw primarily those young people who, by virtue of their personality and motivational make-up, are naturally attracted to it. How can we effect the changes needed for the successful pursuit of professional forestry in a changing world if we continue to attract the kind of students who are adapted to handling technical resource-management problems but not so well qualified to play the increasingly complex role of the resource manager or administrator?

EDUCATION AND TRAINING OF THE PROFESSIONAL FORESTER

It seems reasonable to conclude that several kinds of action are called for, involving the forestry schools, but also the agencies and industries that employ foresters.

Employers, both public and private, must determine as clearly as possible not only the technical qualifications of the jobs they wish to fill, but also the nontechnical qualifications that will be required of the applicant if he is ever to play a larger role within the organization. The requirements must be communicated to the schools and to young people who are thinking about training for professional forestry careers. A clear distinction must be made between jobs which can be adequately filled by technically oriented foresters, and those which require a "socio-economic" orientation. Forestry employers will err seriously if they think that they can somehow develop all the expertise they are going to need without much closer role specification together with careful screening of applicants. At the same time it will become increasingly unfair to lead the average forestry student into believing that technical training only and a narrow forestry orientation will necessarily suffice for the many more complex resource-management jobs expected in the future.

The schools must endeavor to learn much more about students who come to study forestry and make it possible for those who appear to have the necessary capabilities or inclinations to better prepare themselves as forestry professionals beyond the acquisition of technical competence.

Forestry schools will quite likely need to test their students fairly early, using an acceptable set of test instruments, to determine personality and motivation patterns as well as student interests and aspirations. Proper use of very early and more widely available screening and counseling in advance of initiation of formal training might beneficially direct some students to two-year technical rather than professional forestry schools.

There are also some implications here for student counseling. Indeed, effective counseling becomes a crucial part of the student's college experience. Where forestry faculty expertise in this area is lacking (and it is quite apt to be), then the counseling must be done by persons specifically trained to do this kind of job. The ultimate objective here is to distinguish between the student who is technically and materially oriented and interested in carrying out management directives on the ground or specializing in some area of forestry research, and the student who has the ingredients for success in forest business and/or public forest-resources administration and management.

Whether the forestry schools will be able to provide the proper "mix" of different kinds of foresters required is another question. So long as the professional schools do not screen applicants ahead of enrollment, but accept all students who meet university requirements, there can be little direct influence on the kind of student who enters forestry training. However, if the profession continues to de-emphasize the need for technical foresters in favor of foresters with broader skills and capabilities, the market for professional foresters will eventually determine the mix.

Just what arrangement a particular school should develop in order to introduce needed flexibility into its forestry curricula depends upon the kinds of foresters it believes itself best equipped to train to meet future needs. Conceivably, a number of arrangements are possible and appropriate. However, whatever system is developed, it probably should, as a minimum, qualify students for federal and/or state civil service employment and make possible certification by the Society of American Foresters for those students who wish to qualify for membership in that organization. Additionally, the curriculum should make possible some preparation for graduate work by the undergraduate in advance of his entrance into graduate study for those who wish to take more training at the M.F., M.S., or Ph.D. levels.

Beyond these minimum requirements, the curriculum must permit the student to exercise a well counseled choice of education appropriate for him. This may very well turn out to include most of the subjects offered in a traditional course in forestry where technical and production aspects of the profession receive major emphasis. On the other hand, it may include fewer technical forestry courses and more courses such as accounting, communications, business and public administration, regional science, personnel management, consumer behavior, political science, sociology, wildlife, watershed, or range management, forest recreation, or some combination of courses like these.

Catalogue descriptions of forestry curricula will need to be updated to reflect the increasing opportunities for fully qualified professional foresters if we are to attract a larger share of students who by virtue of their aptitudes, personalities, and interests can be better and more fully educated to meet the challenges resource managers and administrators face. This is a highly important source of advertising at the high-school level. Counselors as well as prospective students depend upon college catalogue curricula descriptions for guidance in choice of career. Furthermore, both industrial and public employers of professional foresters, as well as the Society of American Foresters, should be conducting well organized informational programs directed to the universities and to the public generally in a deliberate effort to change people's notions of what a forester really is, and what he can be expected to contribute to the solution of tomorrow's resource-use problems.

REFERENCES

1. Society of American Foresters. 1970. The Roanoke Symposium. Proc., National Symposium on Undergraduate Forestry Education. February 12–13, 1969. 108 pages.

8

THE FORESTER AS A MEMBER OF A MANAGEMENT TEAM

James G. Yoho

First published in *Organization Management in Forestry;* pages 116–129. 18th Annual Forestry Symposium, Louisiana State University Press, Baton Rouge. 1969.

BIOGRAPHY

Mr. Yoho, an academician-turned-industrialist, is Assistant Director of Woodlands for International Paper Company, with headquarters in New York City. Before joining this firm in 1969, he had pursued a teaching, research, and consulting career over some 25 years. His principal academic positions were in forestry and forestry economics at Stephen F. Austin College and at Iowa State, Duke, and Mississippi State universities. At Duke for 12 years, he developed and led a cooperative program for the graduate education of industry employees in forestry and business management. Affiliated, in the course of his work, with many government-, foundation-, and profession-sponsored projects, at home and overseas, in land-resource development and management, he has written widely of his experiences and findings. His own professional life aptly embodies the goals and achievements which he commends to foresters in the writing reprinted here.

EDITORS' SUMMARY

The average forester neither seeks nor attains an outstanding managerial career. His background and his professional education are to blame. The latter can be strengthened in numerous ways, as by putting more stress on social sciences, humanities, and the arts of communication. Foresters need not all be line professionals: Many can become staff analysts.

QUESTIONS TO CONSIDER

(1) Mr. Yoho, in common with many other thoughtful professional foresters, advocates including more social science—such as political science, managerial science, economics, and sociology—in the forestry curriculum. Yet a large fraction of forestry students take poorly

to such courses. What is the answer? Is it well to force the issue anyhow? change the social-science courses? change the students by recruiting those who are social-science oriented? change the concept of forestry, so that it need not involve "participation on the management team"?

(2) How can people best develop their talents for communication, which Mr. Yoho stresses as so important?

(3) Considering the time required to recruit and educate a professional person and then to give him the work experience that will qualify him to take substantial responsibilities, professional planning should evidently be oriented a decade or more into the future. Do you think that Mr. Yoho's suggestions are so oriented?

(4) Why should not the professional forester be simply a specialist in describing (predicting) resource alternatives, leaving the decision making, persuading, and executing (page 74) to professional administrators?

* * *

How can the forester become a more effective participator on the management team? This question in turn suggests the need for a definition or measure of effective participation. It seems to me that an evaluation of effective participation must be tied to the ability of the manager to mesh his judgment with that of the other members of the team so as to increase the decision-making efficiency of the whole management unit in furthering the interests of the organization while, at the same time, facilitating the career advancement of the individual participator. It definitely should not simply focus on the ability of the aspiring manager to sell the nonforester members of the managerial team on more forestry. Instead, effective participation calls for both give and take which, in turn, demands both appreciation and understanding of the points of view of the other members of the management team.

* * *

FORESTRY PROFESSIONALISM, EDUCATION, AND THE MANAGERIAL CAREER

So few foresters have been successful in breaking into the management team or in becoming leaders of that team, particularly at the top levels of the managerial hierarchy. And this is a matter which apparently bothers many of the leaders in the forestry profession, too, judging from the anguish one usually hears expressed at professional forestry meetings over the lack of influence professional foresters seem to have in almost all manner of natural-resource policy decisions.

* * *

Improving the historically poor probabilities of moving foresters into the managerial ranks and increasing their chances of success once they get there must focus on two broad areas—selection and training. The attraction and selection process which precedes the entry of young people into our schools of forestry has been discussed at great length among forestry school executives and in the literature. Accordingly, we are justified in passing over it very lightly at this point and alluding to it superficially at other points in this paper. It should suffice to remind you, therefore, that forestry appeals to young people who seek to lose themselves in the beauty and marvels of nature in the great out-of-doors and who prefer to look to the back country for adventure. Similarly, forestry tends to repel the young personality who yearns for the opportunity to battle his way to the top of a government bureau or to develop or sell his plan for profitable industrial expansion in a walnut-paneled corporate board room. This negative attraction that forestry exhibits for managerial aspirants has often been ascribed to the image of the forestry profession, an image usually described as poorly or lowly. And in this image-conscious age, it would be hard to quarrel with such an explanation or to doubt the remedial impact that could arise from wholly successful efforts at image building like those which have often been suggested within the forestry profession.

The image-building effort required to attract young men with greater managerial potential into forestry offers all of the profession, and particularly the academics, an almost insolvable problem. The difficulty is one of getting off dead center. If more foresters were to find their way into responsible managerial positions, they could serve as an attractant for young people who have the drive and ambition required to reach the top. What is the best bet for achieving a significant breakthrough in this vicious circle? The answer has to be the schools. The prospects and possibilities of the schools' rising to this challenge are explored in the next several paragraphs.

This brings us to the matter of training. And when we think of training in forestry we immediately think of the academic environment afforded by a large university, with all of its attendant encouragement to seek the truth. But this I fear is somewhat of an illusion because foresters, like the majority of those graduated by an institution of higher learning today, are trained in a program directed by the professional school, which is invariably staffed by professionals. This provides the opportunity needed for the basic tenets of the forestry profession to influence the instruction offered in the forestry program. Consider, for example, the conservation ethic—a fairly descriptive term we might appropriately use to describe a bundle of the forestry tenets which profoundly influences our profession and which in at least a subliminal fashion permeates the young forester's entire educational experience. Eventually, what ordinarily emerges from our forestry schools is a product whose goals are closely allied with the traditional aims of the conservationists and many miles apart from those of the business school graduate of the same university.

Essentially what I have been trying to say here is that formal forestry training begins with a raw product who has chosen forestry because he has abilities, inclinations, and ambitions almost exactly opposite those held by young people with strong managerial aspirations. He next encounters the professional school with a faculty steeped in traditional forestry thinking that does little to motivate him toward higher managerial aspirations or to demand better preparation for eventual managerial careers for all students. Oh, yes, you will find lip service among forestry-school faculties to correct these shortcomings. Also, they often express considerable lament over the fact that so few forestry graduates have been given the managerial responsibilities which they are sure those graduates deserve. But you will find that in all of the tinkering that has been done with forestry curricula over the past ten or fifteen years, additions to the requirements which could contribute toward producing a forester better trained for a career in management have averaged only two or three courses. Incidentally, among the courses added outside of the professional school, one often finds that the requirements for economics have been doubled, i.e., raised from one course to two courses. Nearly everyone, it seems, recognizes that economics is indispensable to the decision-making, managerial area. But in order to accomplish this, the traditional, biologically oriented sequence has gone virtually unmolested.

Now all of us have often heard and heeded the old adage that a little knowledge can be a dangerous thing. But it would seem that the curriculum engineers in forestry have been undisturbed by such conventional wisdom. Apparently it is easy for them to understand that a long sequence of courses is needed on the biological side of the picture in order to give the graduate a depth of understanding and hence a true working knowledge of the subject. However, when it comes to the managerial or decision-making side of forestry, it is assumed that just a sprinkling of exposure will do the job. Of course, I'll have to admit that the 100-percent increase in the requirements in general economics represents a great leap forward; but I would be very reluctant to hand over control of my economic assets—and certainly not a million dollars' worth of timber land—to a person with such a level of economic literacy.

* * *

It would be possible to cite many rather specific factors associated with the professional education of foresters which have tended to retard the advancement of forestry graduates into the managerial ranks; but perhaps none is more unfortunate than the misuse of terminology through which we have long misled ourselves. Consider, for instance, the instruction and research which has gone on under the title of forest management. The content of courses so labeled and the research so designated have seldom borne any relationship with either ancient or modern managerial science. Instead, its essential concern has been with the vegetative manipulation or biological control of the forest; and this may indeed form quite an indispensable part of the total body of technical forestry knowledge. I am confident, however, that exposure to such a body of knowledge did little to enable the forestry graduate to communicate better with other members of the management team. In fact, it probably lured him into assuming he knew something he did not and thereby led to further undermining of whatever confidence he may have initially commanded at face value among real managers.

* * *

If we go back further than twenty years, similar allegations can be made about the instruction and research which the forestry schools titled forest economics. In those days forest economics usually consisted of little more than the compiling of a great array of statistics describing the forest resource situation. A true economic analysis of the situation was totally lacking. Other specialty areas within forestry could be mentioned in the same light.

A final major point regarding the education of foresters or professionals of all types for managerial positions is one that has been made many times by leading executives in business and government but it still warrants repeating. It has to do with the necessity of developing an individual with broader interests, appreciation, and understanding than those ordinarily produced in the four-year, highly technical undergraduate program. This calls for a significant exposure in the total educational process beyond the business and economics courses which I have already cited as essential additions to the basic technical requirements for the undergraduate degree. It calls for some work in the other social sciences and the humanities so as to better understand the derivation of most of the wants of society and the processes by which its values are formulated as well.

* * *

THE ROUTE TO MEMBERSHIP ON THE MANAGEMENT TEAM

Now I should like to zero in a little more directly on my assigned topic, and in this section I should like to focus on the grooming of the forester for an impending entree to the management team from within the organization. In the next section I concentrate on the functioning of the forester as a member of the team, but the words of introduction in the next paragraph apply equally to both sections.

I should like to point out that in visualizing the development of the manager, I have rejected the thesis William Whythe set forth in his book, *The Organization Man,* and accepted in its stead Walter Guzzardi's notions as outlined in his *Fortune* magazine series and repeated later in greater detail in his book *The Young Executives* (1965). That is to say, I have assumed that it is not the aim of modern management development to mold the man so that he conforms completely to the point of view of the organization which employs him. Modern business and government have matured to a point of sophistication to match the educational qualifications of the men who have risen to positions of leadership. Those who are members of today's top management team feel no compulsion to conform, let alone an urge to do so. The very concept of the management team and its presence at most echelons of management tend to substantiate the view that criticism and innovation are sought and not subjugated. As Peter Drucker pointed out in his introduction to Guzzardi's book:

> Practically all of the [members of the management team] see the very essence of their job in making their own top management change itself and the company. An amazingly large number attribute their success to their opposition to higher authority or their defiance of company tradition. And they do not have to conform. . . .

Indeed, reading this book, I could not help feeling that the myth of "the organization man" and of the "corporate octopus" reflects nothing more than the fears of the older generation. . . . And not one [of the new management team] is worried over what big business will do to him. They all ask, "What can big business do for me?" To them the large organization offers freedom of action, opportunities for achievement, and room for effective self-expression.

In the past, most foresters who have successfully risen to positions of influence as respected members of the management team have done so through line jobs within the operations arm of the organization. In addition to being technically competent foresters, they proved to be effective managers of people and their success came through getting a job done at what appeared to be a reasonable cost. The mission of the organization was fairly clear to these men; thus most of their management activity was of the problem-solving type rather than real objective management. Similarly, most of the early staff specialists attained their assignments from line positions judged on essentially the same criteria.

But these old and conventional avenues of advancement appear to be giving way to a new pattern. The new pattern which is emerging seems to be attributable to two basic stimuli. One is the ever-growing complexity and sophistication of business and government. The other is the increasing availability, on a small-scale basis, of foresters who are aquainted with the decision-making disciplines which I have been referring to collectively as managerial science.

The first of these changes is exemplified in business by the increasing amount of analysis that is devoted to what was once routine, or often, almost intuitive, decision making. This greater care is not, I assure you, a passing fad that will only appeal to a few companies and then penetrate only a few facets of the business. Eventually the impact of this analytical revolution will be felt by all areas within every company simply because of the increasing relative cost of making the wrong decision, often through failing to weigh all of the alternatives. More effective competition from companies who make use of these advanced procedures will leave the balance no choice except to follow suit.

A parallel of this analytical revolution is also taking place in government. Here there is growing pressure to apply the cost-effectiveness techniques, first developed in the Defense Department under McNamara, to the activities of all agencies as the only means toward more efficient government in an era of increasingly complex federal budgets. This technique now often goes by the name "program budgeting" and it has already reached the forestry agencies.

The new breed of forester who is capable of performing the kinds of analyses I have just described is typically a man with an undergraduate degree in forestry and a couple of years of graduate study in some aspect of managerial science. Already they find themselves in such great demand by forestry agencies in both government and industry that they are being catapulted into staff positions in the higher levels of management. Since these young men frequently lack much of the cherished practical experience, there has been some inertia in the traditional forestry organizations which has kept the pay scale for such people commensurate with their experience rather than their responsibilities. Unfortunately, such traditionalism on the part of the forestry organizations has resulted in the loss of a few of these uniquely trained young men to fields completely unrelated to forestry, simply on the basis of their training as analysts.

Another problem in respect to managerial advancement associated with this situation looms on the horizon in the form of a staff-line conflict. Given the traditional line avenue for advancement in most forestry organizations, the bright, young, ambitious staff analysts often find themselves by-passed by less well-trained people who are more experienced commanders. Perhaps it is true, as Anthony Jay argues in his book, *Management and Machiavelli* (1967), that the only suitable training for commanding is to command; thus there may be a necessity for top managerial candidates to work their way up from the smaller units where they actually have the opportunity to command. On the other hand, commanding calls for decision making and hence in this increasingly complex world one must somewhere acquire an understanding of the principles of scientific decision making. That is to say, one needs to acquire an understanding of deductive and inductive reasoning, of the analytical tools and the technical knowledge for prediction, and of the workings of the total economic system so as to know the consequences of a given action, both inside and outside of the firm.

Since most graduates become line foresters, this picture of such a man being given increasingly severe competition for advancement from young staff analysts may appear a bit discouraging. But this should not be your reaction if you are thinking about the long-term managerial prospects for the professional forester in general. This follows because the avenue is open for any forester to acquire the basic training required for the staff analyst position. And once trained, this hybrid forester-economic analyst will be in much better position to advance to levels of greater authority than the ordinary forestry graduate, or the general M.B.A. graduate who is handicapped by a lack of understanding of the technical side of forestry. With such golden opportunities in the offing, one is forced to wonder why so few young foresters are rising to the challenge.

SOME PRECEPTS FOR EFFECTIVE MANAGEMENT PARTICIPATION

Now that I have explored ways by which the forester can prepare himself for a managerial career and have also suggested some means by which he might launch himself on such a career, I want to pay attention to a few points dealing with his behavior as one of the top management team. Let me begin this very brief discussion of this broad phase of the topic by listing the essential functions of both the individual team member and the team as a whole. It would be helpful if these functions were kept in mind as the discussion proceeds. They are, in about the sequence in which they must be faced, to analyze, innovate, plan, persuade, and execute.

What problems does the forester face in attempting to perform these functions effectively? In addressing myself to this question, I have not envisioned the ideally trained forester discussed earlier. Instead, I have in mind the typical forester of today who has risen by virtue of ambition, ability, and hard work to a responsible managerial post.

There seems to be almost unanimous agreement that the problem of communication is the greatest difficulty faced by a forester in such a situation. And this communication difficulty appears to be a two-way one. The forester has trouble getting his points across to the other team members and he also does not appear to have an easy time in understanding the views of the other managerial people with whom he must deal.

The explanations for this are numerous. They include the forester's poorly developed ability to speak and write effectively, his inability to reason logically and analytically, his excessive reliance on technical jargon and professional tenets, and finally the fact that he usually does not know his economic ABC's. Conversely, he often fails to understand other team members because he is not tuned to their points of view. From this side of the picture his most glaring deficiencies are attributable to his failure to keep his eyes on the business and financial aspects of the operation. This, in turn, results in his failure to relate his activities to the total objective of the firm as the others see it.

* * *

As far as being understood by other members of a typical management team, the forester would do well to recognize the likelihood that he is the one with the foreign background. The representatives from manufacturing, engineering, research, and sales have backgrounds with a great deal in common and likewise they tend to operate under similar conditions within the organization. The forester's natural resource, his labor, and his tax and production time situations differ considerably from those with which the others are involved. This means that the forester may have to work a little harder at solving his communication problems than other members of the team.

9

THE HEREDITY AND ENVIRONMENT OF PROFESSIONAL FORESTRY EDUCATION

William A. Duerr

First published in *Journal of Forestry* 70(1):21–24. January 1972.

BIOGRAPHY

Mr. Duerr's biographical sketch is given at Item 4, page 30.

EDITORS' SUMMARY

Education is forestry's professional environment; recruitment, its heredity. Recruitment can be made effective through educational excellence; through programs stressing the systems view, modernization, intellectual freedom, and enlarged recruitment efforts themselves.

QUESTIONS TO CONSIDER

(1) Do you think that "meritocracy" is desirable: that society should be led by those of highest intelligence, albeit this intelligence is inheritable? What alternatives are there to meritocracy?

(2) If a profession's recruits are analogous to the genes of an individual, then why be concerned about affecting the former any more than we are about manipulating the latter?

(3) To put the second question in different terms, why would it not be socially desirable to follow a policy of laissez-faire with respect to occupational selection? Will not the most

capable individuals (indeed, all individuals) find their way into the work where they are most rewarded—i.e., most wanted? If our answer is no, then are we not saying that the principle of comparative advantage is invalid?

(4) Are professions analogous to individuals in that the social well-being is best served if each strives for success in its own terms, subject only to the constraints imposed by the culture?

(5) Where undergraduate classes in forestry schools are already unmanageably large, is it not folly to put effort into recruitment?

* * *

The educational process begins for each individual at conception and ends only upon death, in a literal or figurative sense. It has a bearing upon intelligence, at least in its early stages. But on the whole, and certainly for professional education, the process bears primarily upon knowledge and upon the less heritable skills, habits of thought, and attitudes of mind.

Education is not something which one receives from others, but rather, something which one gives to oneself, with more or less help from outside. Thus education and school are not invariably related; school people often take themselves too seriously. The term "self-educated person," surely inspiring, is also redundant. Indeed, there are all sorts of distinctions which turn out at times to be confusing rather than helpful: that between formal schooling and informal education, between education and training, between education and experience. . . .

At any rate, I think of the educational process as all that complex environment in which the individual does what he can and what he will to exploit his genetic potentialities. Among the useful questions which he may emphasize in the educational process are "Why?" and "What are the alternatives?" Among the useful personal attributes which he may cultivate within the process are a devotion to intellectual freedom, to objective truth, and to the interests of other persons.

Further, one needs to propose a concept of *profession* and *professional education*. I think of a profession as a culture—or, if you like, a subculture. A profession has its own language, social structure, means for making a living, aesthetic activities, and philosophy of life, and these aspects of its culture are sufficiently distinguishable and useful in the eyes of its members and of society to warrant its separate recognition. This culture, like any other, is a social instrument, whose function is to promote social survival and well-being.

If one views a profession as a culture, then the idea of professional education is simple. The process is that of enculturation. Those who possess the culture and those who are being introduced to it are brought together. On the one hand, the latter learn the language, social structure, means for making a living, aesthetic activities, and philosophy of life which are the profession's heritage. On the other hand, opportunities may be offered for evolution—even, sometimes, for revolution—at the instigation of the learners, or perhaps of the learned.

Professional forestry education, then, is the process of transmitting and receiving forestry culture, holding it in custody, and keeping it alive by amending it. In sentimental terms, the learner sits at the knee of the learned. In cynical terms, he gets a brainwashing. In analytical terms, he is asked to conform and at the same time is expected to innovate.

Incidentally, by "learner and learned" I don't mean "student and teacher" in the ordinary sense. I see all the participants in a profession as interacting in the learning process. In the forestry school part of it, the teacher may be learning more than most of his students—because, if he's a good teacher, he learns so very much; and if he's a poor teacher, *they* learn so very little.

INTELLIGENCE AND INHERITANCE

The September 1971 issue of *The Atlantic* features a thought-provoking article which reviews and interprets findings of psychology concerning intelligence. Without evaluating the argument, let us trace it and see how it relates to professional forestry education. The author develops the point that intelligence is largely hereditary and reasons that to the degree success, earnings, prestige, and social standing are a function of intelligence (and apparently they are to a large degree), then social standing has high heritability. He concludes that the heritability of intelligence and thus of social status is pushing our society toward an hereditary "meritocracy," a layered caste system hinged on mental ability. Furthermore, all our efforts toward social mobility and equality, designed to reduce such layering, are self-defeating because every improvement in the lives of the less privileged merely makes the environment for intelligence more uniform and thus enhances the influence of heredity upon personal success or failure.

Returning to professional forestry education, let us assume, without supplying a definition, that the goal of our profession is success. The degree of our success is a function of our skill as professional persons, which in turn is the consequence of our intelligence and our enculturation. That is, our skill is part of our identity. Our identity is determined in the process for recruiting us into the profession. Our recruitment depends upon the profession's image—its degree of success. And now we have come full circle: Success means recruitment means personnel means performance means success. Have I left out education? By no means. The circle itself *is* the educational process, for which each new recruit begins with professional conception and ends only with retirement, and which for the profession is unending so long as our culture survives.

I begin to see the subject of intelligence and its inheritance entering our professional process at two points.

INHERITANCE AND RECRUITMENT

First, inheritance, for the profession, takes place in its recruitment. That is to say, the analogue of the genes which man brings with him into life is the recruits which the profession brings into *its* life. The genes of the individual are largely unchanging. The genes of the family are only slowly changing, and so it is with the genes of the profession, represented in the names of its members. And just as man's intelligence is seen as largely a function of his genes, so the counterpart of intelligence in the case of the profession is largely a function of its membership. This counterpart, which I shall term the profession's *special genius,* is a compound of our degree of intelligence, extroversion, articulateness, aggressiveness, sensitivity, and other such qualities. Now if the profession's success is a product of its special genius and genius is hereditary, then the key to the success of our profession is its recruitment. Here is a familiar conclusion, reached by a somewhat unfamiliar path.

In taking that path, we come to an arresting realization. If man's intelligence is an inheritance, then it also helps determine what he passes along to succeeding generations: Level of intelligence changes in the line as slowly as it does *because* of the degree of its heritability. So with a profession's special genius: Not only is success, through genius, a function of recruitment, but recruitment is a function of success, and consequently the degree of a profession's success, being hereditary, tends to be *not* susceptible of rapid change. Is this true of forestry? Under what circumstances may mutation occur in the line of a profession?

The second point at which the subject of intelligence and its inheritance enters the professional process becomes evident as we turn our attention for a moment away from heredity toward environment. For the profession, this is a turn away from recruitment toward education: taking what you get as a raw material and doing the best you can with it. But who is this *you* referred to? He is primarily a person who has been recruited into the profession, whether directly or vicariously. Recruits become not only learners, but also the learned. They not only personify the profession's heredity but are also mainly responsible for the environment which embodies its educational process.

Am I saying that for the profession's success, recruitment is everything and subsequent education nothing? That would be like saying that heredity is everything in human performance and environment nothing. The position I want to take is to express greater concern for the former than for the latter—but at the same time to recognize the vital bearing of education upon recruitment, taking "education" in the very broad sense which I intend. Although it may be true that one can't spoil a first-class student even in a third-class educational system, the fact remains that circumstances militate against such a student's entering such a system to begin with. And here again, as we gain social mobility and environmental equality, the chances of getting the good student into the poor system diminish. The function of excellence in professional education is not to *produce outstanding members*, but to *attract recruits who are already outstanding*.

Let me, then, express a few thoughts on this subject: effective recruitment into the forestry profession by means of excellence in professional forestry education.

EDUCATIONAL SCOPE

First, it is helpful to adopt the broad view of our educational process. It is not just something which takes place during the academic year in forestry school, but something which is identical with the whole life of the profession. For each one of us, professional forestry education extends continuously from the time of our recruitment to the end of our career. We are the learners and the learned. Our classrooms are the office, the field, the laboratory, the city park, the supervisor's headquarters. I think such a view of the professional education process is helpful because it makes you and me responsible for it and not just the other fellow, because it is congenial to the idea of maintaining the individual's professional vitality and currency throughout his career, and because it is the only concept of education which is fully consistent with my major point: that excellent recruitment is both the end and the means for excellent education.

Second, I believe it is helpful to rethink continually the subject matter in our educational process—that is, to rethink our definition of forestry, and to make such rethinking a conscious and explicit part of our professional obligation. Every one of us has his ideas about how the forestry curriculum, as it were, should be changed. Now and then we are able to agree reasonably well on major points. What I would stress is that forestry has such abundant potential for challenging outstanding persons! It did so yesterday, and it can be fitted to do so tomorrow.

I have heard the relation between forestry and the environmental movement discussed, and personally I think that our profession has some distance to go to achieve a reasonable degree of empathy, not to mention leadership, in the environmental push. I do think that we are in a position today, in view of our intense interest in ecology—indeed, our rooting in ecology—to help environmentalism on both the action and the research fronts. We can stress that ecology is seeing things in context—seeing the system—and that the system upon which public concerns today are centered is primarily social, and then, it is biological and physical.

In further reference to social orientation, I personally think that forestry can better meet tomorrow's challenge by giving a larger share of attention to consumers of forestry goods and services and a smaller share to our traditional clients, the producers. Most of these consumers are poor people, and we can well devote a larger share of interest to them and a smaller share, again, to the traditional objects of our concern, the middle and upper income classes. These ideas are merely mine. I voice them to illustrate what I mean by opportunities for change such as we can debate and try to resolve, both in the forestry schools and in the other, larger realms of professional education.

But we may as well recognize that if our profession proposes to take a larger interest in recreation, amenity values, water, wildlife, range, social impacts, and other nontimber matters, then we are working for change on a broad front. We are changing not only the idea of forestry. We are changing also the idea of *forest,* or whatever one is to call the land on which the profession stands. We are changing the boundaries of the whole system of which we are part.

To be sure, changes are hard to accept, especially in a professional culture which is fortified against uncertainty to the degree that ours is, by rules of conduct and tenets of faith. However, explicit recognition that so many of our ideas are cultural in character can improve our chances of effectuating constructive change.

The realization that forestry, over the years, has become less monolithic, less well defined, more fragmented has, I know, been disturbing to the profession. In reference to Society of American Foresters national meetings of the recent past, some of us have deplored the fact that we came together only to split at once into little separate groups. It was this anxiety, perhaps, which led us to the present more consolidated format for our national meeting. Yet we may discover that the little separate groups are not a weakness, but the prime strength of forestry. The professional field may prove to be like an artichoke, such that if one peels back all the little separate leaves, he hasn't much left. Here again, I voice these thoughts, not to make a point about what forestry is, but to make the point that we can never rest on our current concept of it.

EDUCATIONAL FREEDOM

So much, then, for the idea of viewing education broadly and the idea of rethinking forestry. A third idea about effective recruitment through educational excellence is that it will be helpful to cultivate those aspects of our professional education which provide intellectual freedom for the individual: freedom to learn, make mistakes, take responsibility, innovate, express himself. The point here is not primarily that such freedom is a nourishing environment for education, although this is true enough. The primary point concerns recruitment into the educational process: Every instance, in the schools or beyond, where freedom is upheld and innovation encouraged within the bounds of social integrity wins another capable recruit into the profession. On the other hand, every unnecessary rule, every act of censorship, every assignment for a forester which could have been carried out by a subprofessional loses outstanding recruits.

Professional collaboration with outsiders is an exercise of freedom, in contrast to the running of a closed corporation. Experimentation and innovation in teaching method are symptomatic of freedom—and here again I stress the teaching we receive after the school years. It would be interesting to know how much of the issue of "relevance," an issue widely recognized today, is a function of teaching method rather than content. The act of listening to students, of the learned listening to the learner, of the latter listening to the former, of both listening to the general public and not simply "educating" the public—all are symptoms of freedom. A center of freedom is the researcher or the research group. To what extent may the profession achieve self-improvement by revising its research structure? Can greater freedom be attained, particularly in the policy-sensitive social-science areas of forestry research, by separating such research from the control of the public administrative agencies, which have a tradition to uphold and a "party line" to follow? To what extent, I wonder, may our profession's desperately serious deficiency of knowledge and even interest in the social-science aspects of our work be traceable to deficiencies of freedom in our research? Here are fistfuls of issues for our profession to air, debate, and try to resolve.

RECRUITMENT PROGRAMS

Fourth and last, I believe it would be helpful to explore the possibilities for improving our recruitment program—that is to say, our deliberate projection of an image, as distinguished from our inadvertent projections. Forestry is a bundle of exciting opportunities to work with people and contribute to their lives. It is a fascinating aggregation of social concerns and politics and engineering and biology. It finds itself in the midst of some of the hotly argued causes of the day, both domestic and international. No publicist could ask for richer material.

Yet the man in the street, if he gives any thought at all, still inclines to think of the forester in the traditional terms: a man with a horse; a fire lookout; perhaps a friend of sick trees; perhaps even a tree butcher—thoughts over which is imposed, incongruous or not, the image of Smokey Bear. Bears, collie dogs, hootie owls, and the like are good attention getters and effective allies in conservation publicity. But before we create them and turn them loose, we need to investigate their probable impact upon our professional environment—i.e., upon our professional education, notably its recruitment phase—just as we demand to know the effects of a drug or other chemical before its release. The forestry profession needs a stronger identification, not with animals, but with human beings. And here I am referring, not to Johnny Horizon, but to the man in the street: the member of the urbanized society of which we wish to be a part.

Today's young, intelligent extrovert is serious minded, interested in policy and politics, deeply concerned over the unjust and the selfish behavior in the world around him, eager to find and pursue a human cause. Surely, to manage for peoples' well-being one of our basic renewable resources *is* such a cause. Among our potential recruits are new evangelists, new forest scientists, new leaders in resource management, new forest and product engineers, new innovators in the artistic and human side of forestry. How many of these persons we are able to captivate will largely determine the mould and significance of our profession in the urbanized society of tomorrow.

10

BLACK FORESTERS NEEDED, A PROFESSIONAL CONCERN

Brian R. Payne
Donald R. Theoe

First published in *Journal of Forestry* 69(5):295–298. May 1971.

BIOGRAPHY

Mr. Payne is forest economist and project leader of the Northeastern Forest Experiment Station's urban environmental forestry research unit in Amherst, Massachusetts. His academic degrees, in forestry and in agricultural economics, were earned at the University of California and at Duke University. He began his Forest Service career at the Pacific Northwest Forest and Range Experiment Station in Portland, Oregon, moving thereafter to join the Southern Forest Experiment Station. Here he set up headquarters at Tuskegee Institute in Alabama and helped organize the first preforestry program at a predominantly black U.S. university, an accomplishment for which he was given a Superior Service award by the Secretary of Agriculture. Out of his Tuskegee experience came some of the thoughts reprinted below.

Mr. Theoe holds the headquarters-staff position of Professional Programs Director in the Society of American Foresters and is stationed in Washington, D.C. In this position, his major concerns include professional recruitment, education, and employment, some facets of which are the subject of the present article. Educated at the University of Washington, he has resource-management experience with the State of Oregon and with the Forest Service, and teaching experience in Everett Community College, State of Washington, and in the University of Washington. An immediate forerunner to his joining the SAF staff in 1969 was a nationwide contest for recently graduated foresters, in which his essay on the assigned subject, "The Young Forester Speaks Out on Forestry Education," was chosen for publication by the Society.

EDITORS' SUMMARY

The forestry profession has attracted very few members of minority groups. Negroes, for example, either are not entering universities with forestry programs or are uninterested in forestry. This situation represents a loss to the profession. Recruitment of Negroes must be stepped up, as through the Tuskegee program.

QUESTIONS TO CONSIDER

(1) The population of the United States can be subdivided into a large number of classes based on personal characteristics over which the individual has minimal control—for example, sex, race, inherited traits, national origin, religious background, and economic and other social circumstances during early life. What would be the salient features and methods of a study designed to (a) identify the significant classes of the population and (b) estimate the proportion of each class that would be represented in a forestry profession which could best serve the wants of society (the requirements of other professions and occupations being considered)?

(2) In the optimally constituted forestry profession envisaged in the first question, would there, do you think, be a larger percentage of women than at present? a larger percentage of Negroes of West African origin? a larger percentage of men of Russian Jewish descent and socially humble background (one of the outstanding national leaders in forestry, Raphael Zon, was in this category)?

(3) What special gifts can Negroes bring to forestry, aside from the general benefits of more equal representation? What special gifts can professional membership bring to Negroes?

(4) Why has the issue of women's professional representation become so prominent in recent years?

* * *

Despite our diversity of background, education, and experience, we [foresters] don't represent all segments of the society we serve. How many Negro foresters did you see the last time you attended a professional meeting, at your last class reunion, or the last time you met other foresters in a work situation? If you saw even one, your experience is exceptional.

At the present time, there are probably no more than five Negro members of the Society of American Foresters. The Forest Service employs only two or three Negroes having a degree in forestry. Probably no more than a dozen black Americans have ever graduated from a forestry program in this country, a remarkable figure compared to the 40,000 or more foresters turned out so far.

Is this situation likely to change? Will there be any Negroes qualified as foresters? Not many, and not very soon. A survey conducted by Tuskegee Institute found only 18 black undergraduates in forestry schools during the 1970 spring term. A similar survey by the Society of American Foresters disclosed virtually the same figures for the 1969–1970 school year, and further revealed that one-third of the black students were freshmen. Thus, of approximately 11,000 undergraduate students in forestry, less than two-tenths of one percent are Negroes. Yet Negroes make up 11 percent of the general U.S. population.

For graduate students, the situation is nearly the same. The Tuskegee survey found six Negroes among the nearly 1,800 U.S. students pursuing advanced degrees in forestry, or about three-tenths of one percent.

The SAF survey further revealed the number of current students from other minority groups, as well as the number who have graduated from a technician or professional program. Underrepresentation of minority groups is not restricted to Negroes. However, this article is limited to the problem of increasing black representation in our profession.

A PROBLEM SITUATION

Is the lack of black foresters and black forestry students a problem? Do we want or need—indeed, should there be—more Negroes in forestry? We believe the answer to each of these questions is Yes.

The underrepresentation of the Negro race in the forestry profession limits its ability to serve society. We need the best talent we can get in our profession, but we have been passing up an important source. Foresters are traditionally in the forefront of concern for wise use of natural resources, but as a profession we have failed to tap some major human resources.

There is no question that black Americans can be effective in providing professional service. In certain circumstances their contributions could be particularly effective. For example, much of the forest work force and a significant portion of farm forest ownership is black. An increasing number of forest recreationists are black, particularly in areas near our major cities. The growing emphasis on urban forestry and in providing forest-related activities for inner-city dwellers will bring us increasingly in contact with a largely black clientele. As long as we remain a wholly white profession it will be difficult to convince these black Americans either of our sincerity or our ability to serve them.

Further, we should encourage Negroes to enter forestry because they badly need a greater opportunity to use their talents in productive, remunerative occupations.

We should be anxious to share with them the excitement, the challenges, and the rewards of forestry. We must fulfill our moral duty to remove any barriers to full participation by Negroes in the profession.

CAUSES OF THE PROBLEM

Why are there not more Negroes in forestry? One factor is certainly a lack of knowledge by black young people of what the profession is all about. The "lookout tower" impression of the forester is shared by students of all races. A "pulpwood-cutter" concept of the forester is particularly strong among blacks in the South. The idea of a highly qualified, professional manager of natural resources is hard to communicate to any student without personal contact or example. Because there are so few Negro foresters, a black student has no role models with whom to identify. He therefore assumes the profession is closed to him.

Another factor is that forestry suffers from an anti-agricultural bias among blacks. This bias is expressed by the statement of a black student, "We've been trying to get out of the woods for so long, why should we want a job that's going to keep us there?"

Students of many groups ascribe higher status to medicine, law, and engineering than to forestry, but observation suggests that this view is particularly prevalent among blacks.

The lack of Negroes in forestry is undoubtedly attributable in part to the past segregation of facilities for higher education, especially in the South. In spite of the anti-agricultural bias it is reasonable to think that rural blacks, almost all of whom live in the South, would have greater knowledge of forestry and perhaps would assign it higher status than would urban blacks. But until the 1950s, there were virtually no Negro students at southern universities having forestry schools. Although these universities have been desegregated for some time and have made strenuous recruiting efforts, they have enrolled few blacks.

Fear and racial pride have each worked to keep Negroes from entering the predominantly white universities in the South. Northern and Western forestry schools have been open to blacks for a considerable time, but they are in areas where the Negro populations are either small or concentrated in cities—and, in the inner city, forestry has both low visibility and low prestige. Therefore, in the part of the country where black students might be most interested in forestry, they are not entering universities with forestry programs. And where black students are in universities with forestry programs, they are not interested in forestry.

The attitudes of Negroes toward forestry and forestry schools are the basis for the reasons given above to explain the lack of black foresters. But we believe the attitude of white society (which, of course, includes most professional foresters) toward Negroes is at least as important a cause of the present racial imbalance in the forestry profession. That attitude can be described as white racism.

White racism, or the institutionalized subordination of blacks in our society, has been documented in reports such as those by the Kerner Commission[1] and Anthony Downs.[2] Many of the findings apply to our professional segment of society.

Even though we intend no racial discrimination, our way of doing things just doesn't make it as easy for blacks as for whites to enter forestry. Sure, there are good and logical reasons for our ways of doing things, but the effect is to exclude blacks. And that *is* institutional racism.

For example, how many of us have spoken to classes at primarily black elementary and secondary schools? How many have served as forestry merit badge counselors for predominantly black Boy Scout troops? In filling jobs, how many have made special efforts to seek out blacks, in the realization that our normal channels of vacancy announcements reach only whites? How many have attempted to help black colleges develop courses in forestry or have publicly expressed concern that our profession includes so few blacks?

POSSIBLE SOLUTIONS: THE TUSKEGEE PROGRAM

* * *

Some new and vigorous efforts are being made. One of them is underway at Tuskegee Institute, a predominantly black university in southeastern Alabama. Founded in 1881 by Booker T. Washington and the home and laboratory of George Washington Carver for over 40 years, Tuskegee Institute has been a leader in teaching, research, and community service among black colleges in the South and throughout the world.

In 1968, Tuskegee established a preforestry curriculum with the help of the University of California at Berkeley, the University of Michigan, and the Forest Service. The cooperating universities agreed to accept transfer students as juniors in forestry after they had finished the preforestry curriculum at Tuskegee, and to provide them with substantial financial aid during their final two years of study. California helped to set up the preforestry curriculum. Michigan, aided by a federally financed exchange program, is helping to teach an introductory course.

Also in 1968, the Forest Service assigned a research forester to Tuskegee Institute to assist in formulating a program of cooperative research, to initiate and participate in the development of a plan for managing forest land owned by the school, to administer the preforestry curriculum, and to assist in teaching introductory courses in forestry and natural resources management.

The attempt at Tuskegee to bring forestry to black students has been quite successful. As of the fall of 1970, 11 students are majoring in preforestry. Two students have completed the program and have transferred, one to California and the other to Michigan. In addition, Iowa State University has offered firm support to the program, and several other universities have shown strong interest in cooperating.

Approximately 40 undergraduates in other major fields of study have been exposed to forestry through the introductory courses. Seminars in current forestry research have reached faculty and graduate students in a wide range of disciplines. Several students plan to work toward advanced degrees in forestry after completing the Bachelor degree in related fields at Tuskegee.

The location of a Forest Service employee at Tuskegee has enabled nearly two dozen students to find summer jobs, primarily in the national forests. Several graduate students are candidates for employment in forest research.

One scientist, who is a Negro, has been added to the Forest Service research team at Tuskegee to study outdoor recreational opportunities, needs, and benefits of low-income, rural, black people.

The participation of Tuskegee faculty members and graduate students in research has added to the quality of the forestry program. Three faculty members and six graduate students have been studying soil nutrient availability fertilization of longleaf pine seedlings, and physical and chemical treatments of tree seed to overcome dormancy. These studies have received financial support from the Forest Service.

The management plan for the Institute's forest land has been completed with the material aid of the Southeastern Area, State and Private Forestry of the Forest Service, and of the Alabama Forestry Commission. The plan recommends reforestation and other rehabilitation so that the tract can be used for research, instruction, demonstration, and sustained output of forest values, including timber.

Considering the obstacles discussed earlier, the location and recruitment of students interested in forestry have been surprisingly easy. Black high school students are very responsive to the idea that there is a forestry program at a black university. They are also impressed to learn that a forester may earn as much as $8,000 per year as a starting salary. This information helps counteract the pulpwood-cutter image of the forester. In addition, the Lassie television series and Smokey Bear are familiar to many black students, and have exposed them in a limited way to the variety and importance of a forester's work. When the student is convinced he can be a forester and is told that scholarships are available, he can give forestry an equal chance in considering career alternatives. But personal contact and attention are required to aid the student in reaching this point.

CONCLUSIONS

What is the future of forestry at Tuskegee? Because of its size (3,000 students) and its status as a private university having a limited endowment, Tuskegee is unlikely to attempt development of a traditional, 4-year curriculum. However, it is building an environmental sciences program that can incorporate the present pre-forestry courses. Perhaps some professional work in forestry will be added. But the main source of professional training will have to come through cooperation with other universities at both the graduate and undergraduate levels.

The program at Tuskegee is only a small beginning, but perhaps it can serve as a model for cooperative efforts elsewhere. There have been discussions between existing forestry schools in several states and neighboring predominantly Negro institutions regarding cooperative forestry offerings. The recent addition of a U.S.D.A. representative to the staff of each of the "1890" colleges may provide additional opportunities for encouraging blacks to enter forestry.

One thing is certain. Avenues for bringing forestry to black students must be strengthened and broadened. The impetus for such efforts is unlikely to come from the Negro segment of American society. It must come instead from concerned foresters.

REFERENCES

1. National Advisory Committee on Civil Disorders. 1968, Report of. Washington, D.C.: U.S. Government Printing Office. 426 pages.
2. Downs, Anthony. 1970. Racism in America and how to combat it. U.S. Commission on Civil Rights. Clearinghouse Publication. Urban Series 1. 43 pages.

11

THE OBJECTIVES OF FOREST POLICIES

Albert C. Worrell

First published in *Principles of Forest Policy;* pages 11–12. McGraw-Hill Book Company, New York. 1970.

BIOGRAPHY

Mr. Worrell is Edwin W. Davis Professor of Forest Policy in the School of Forestry and Environmental Studies of Yale University. His continuing interest and his major professional contributions have been in the political economy of renewable natural resources. After some 12 years of forestry work with state, federal, and private organizations and further years of faculty service at the University of Georgia, he returned to his alma mater, the University of Michigan, for a Ph.D. in forestry economics. He has served in Chile as forestry economist with the United Nations Economic Commission for Latin America and in Germany as guest professor at the University of Freiburg. His textbooks, *Economics of American Forestry* and *Principles of Forest Policy,* are standards in those fields.

EDITORS' SUMMARY

Forestry policies form hierarchies, in which each item represents a means for reaching the end embodied in the next item. Since such hierarchies function as guides toward the ultimate goals of society (or some other entity), it is desirable that these goals be made as explicit as possible.

QUESTIONS TO CONSIDER

(1) Mr. Worrell sees his work as the *principles* of forest policy. The distinction intended is evidently with the practice or actuality of policy. Can you explain the difference and give some helpful illustrations?

(2) The purpose of policies, according to Mr. Worrell, is to help assure that social goals will be pursued. Do we have social goals for forestry in the United States? Where does one go to find what these goals are? What, briefly, are the U.S. social goals for forestry?

(3) What is the process by which goals are established? Where does a consideration of means enter the process? What is the difference between goals and means?

(4) Does one "reason from ends to means" (page 90), or is the reasoning ordinarily performed in the other direction?

(5) Mr. Worrell speaks (page 90) of the "hierarchy of policies." What is a hierarchy? Construct an illustrative hierarchy of forest policies (chain of ends and means), extending the chain as far in each direction as you can.

* * *

The purpose of a society in establishing policies about what its members do is to try to assure that their actions will contribute as much as possible toward some ends which the society deems desirable. Before the society can intelligently establish such policies, it must have a clear idea of the ends it wishes to attain. It also must understand the possible effects on these ends of the actions which it seeks to control through the policies.

A policy is thus a means to some end or ends, and its effectiveness can only be judged in terms of those ends. However, many of the ends toward which policies are aimed are desirable only because they in turn become means toward the achievement of other ends. For example, a society may have an objective of adequate wood supplies for all consumer needs. It, therefore, establishes a policy of increasing the growth of timber in its forests. One way of doing this is to reduce the occurrence and severity of wild forest fires. A second objective thus arises of keeping forest-fire damage to a minimum. In order to achieve this, a policy may be established of suppressing immediately all wildfires that are discovered. But if this policy is to be effective, it is necessary that the fires be discovered. So a further objective exists of discovering all wildfires as soon as possible. As a means of achieving this, the state forestry organization may put into effect a policy of manning all lookout towers on every day when the burning index is above some stipulated figure.

This example shows that in many cases there will be a hierarchy of policies as a result of the chain of ends and means. For society, the objective is to have an adequate supply of wood; for the state tower watchmen it is to discover all wildfires. The two are linked together, however, and policies about manning towers are essential if broader policies about growing timber are to be made effective. It is risky to say that one of these policies is more important than the other. But in order to measure the value or effectiveness of any given policy, it may be necessary to look beyond its own immediate ends.

The objectives of forest policies are often not clearly stated and probably in many cases are not even clearly known. We do not always go through a logical process of reasoning from ends to means in the development of forest policies. This cannot be entirely overcome for reasons which will show up as we consider the policy-formation process. But confusion about objectives leads to many difficulties, and it is desirable that the objectives of forest policies and of other policies which affect or conflict with them be made as explicit as possible.

12

POLICY FORMULATION IN VIEW OF SOME PROBLEMS WE FACE

Albert C. Worrell

First published in *The Future, Forest Policy and 1970 National Timber Review;* pages 3–7. Society of American Foresters, Columbia River Section, Portland Chapter, Part No. 7. June 1971.

BIOGRAPHY

Mr. Worrell's biographical sketch is given at Item 11, page 89.

EDITORS' SUMMARY

Policies are predictions, and their formation requires predictions. So it is with timber consumption (production) goals. These goals are best conceived as forecasts of our needs, or requirements, for wood—assuming, not that wealth is distributed as in the past, but that broad social goals are met respecting such vital matters as wealth distribution and population growth.

QUESTIONS TO CONSIDER

(1) Why are policies "always concerned with the future" (page 92)? That is to say, what is wrong with developing policies for the past?

(2) Is economics unique or unusual among sciences in respect to its involvement in forecasting? What other branches of forestry can you name which are also heavily involved in forecasting?

(3) Is "every economic forecast a projection of what will not happen" (page 92)? Can you cite examples of the opposite—that is, cases in which people will "go busily to work" in an effort to ensure that a forecast will materialize? If you cannot find examples, what does this

mean: that accurate or useful forecasting is impossible? Why, then, do we find the major share of human thought being devoted to the future?

(4) Citing the difficulty of setting forestry production goals by analyzing supply and demand, Mr. Worrell speculates (page 93) about goals that will enable people "to live at a satisfactory and comfortable level." Can you describe the process of calculating such goals?

(5) What procedures does economic theory give us for analyzing not only the efficiency of resource combinations but also their equity in terms of income distribution? What procedures does social science give us for analyzing the social acceptability of resource combinations? their administrative manageability?

* * *

One important characteristic of . . . policy formation . . . is that it is always concerned with the future. When people accept and follow a policy, they are always determining in advance the course of their future actions. When they make policy decisions, they are always thinking about future objectives and charting courses of action which they hope will achieve them. Now policies can be changed and it is desirable that they be flexible. But a society that does not stick to any set courses of action but changes its mind with the whims of the moment does not really have any policies.

In order to develop policies that will be better than extemporaneous decisions, people must have a convincing picture of the future in which they will be acting. They must anticipate future conditions and estimate the effects their proposed policies will produce under those conditions. This is why forecasts are so important to policy. There has been a running argument about forecasting in forestry for many years, much of it concerned with the Forest Service practice of forecasting "requirements" rather than consumption. It seems to me that this has been a footless argument and I would like to explain why.

Economists and weather forecasters are frequently compared but the only real characteristic they have in common is that both try to say something useful about the future. The meteorologist attempts to forecast the kind of weather that will be produced at specific future times by natural physical phenomena and conditions over which he has no control. He does not expect the people who read his forecast to try to do anything to change the predicted weather pattern, for he knows there is virtually nothing they can do.

The economist, by contrast, forecasts events or situations which people can do something about. Very often, the purpose of his forecast is to stimulate them to do something. John Zivnuska once criticized a Forest Service forecast as "a projection of what will not happen." Now, if you stop and think about it, every economic forecast is a projection of what will not happen. For as soon as it is published, people go busily to work to keep it from materializing. Forecast a prolonged period of depressed lumber prices, and what happens? Marginal sawmills shut down, expansion plans are dropped, people try to get out of the business—and future supply heads downward. Builders start substituting lumber for other materials, people begin the playroom or vacation home they had thought they couldn't afford, government moves to promote exports—and future demand heads upward. Prices don't decline as was forecast and may even go up. It is never this neat, of course, but in general this is what happens if people believe the forecast.

Trying to forecast actual consumption at any distant future date is a rather futile undertaking unless in some way account can be taken of the kinds of responsive actions just described. But even then, all such a forecast could indicate is the future levels of consumption, production, and prices that would result from interactions among existing market forces and any changes in those forces which may occur spontaneously between now and the forecast date. If this were a world of perfectly competitive markets, we might accept the idea that the outcome of such a free working of the market system would be the best of all possible future situations. Under those conditions, an economic forecast—like a weather forecast—would merely forewarn us of a future situation which we could or should not do anything to change.

But it is obvious that the economic world which foresters deal with is a far cry from perfect competition. The great number of governmental interferences and restraints is evidence enough of that. But even without those and the large private concentrations of economic strength, the nature of stumpage as a market commodity is such that it violates practically all the necessary assumptions behind the perfect competition model. There is little reason to assume that the future balance between supply and demand which such an imperfect market will produce would be the most desirable one.

Under such conditions, it seems more meaningful to try to forecast the quantities of forest products which future populations will need if they are to live at a satisfactory and comfortable level. Now, scarcity is going to continue to be an economic fact of future life and people will only be able to have forest benefits by sacrificing some amount of other things that they also want. So forecasting ''requirements'' will always be a difficult and uncertain undertaking. But it has the great advantage that requirements forecasts produce production goals which are conceptually valid. The future is certain to bring massive governmental activities in the natural resources area. Housing is an example which will have tremendous import for forestry. Our present economic system has failed miserably to provide adequate housing for all our people. What we clearly need to guide future productive activities is an estimate of reasonable future housing requirements and not a forecast of how many dwellings our inefficient construction sector would build if it continued unchanged into the future.

Economic analyses in forestry almost always seem to assume either that the present distribution of wealth is satisfactory or that nothing can be done about it. Many people who want to do something about inequitable distribution feel the proper approach is through programs aimed specifically at that problem and not through adjustments in other programs. But I have trouble accepting that idea. It sounds too much like the widely held view among economists that their job is merely to assemble facts and not to promote policy changes, no matter how convinced they are that the changes are needed. It also resembles the idea sometimes expressed that the forester's job is to manage forests efficiently to produce whatever he has been told is needed. Judging by the current flood of criticism, I suspect that many members of our profession may have been doing just that.

I believe that an important objective of managing our resources can and should be to improve the distribution of income, with income defined in the broad sense of human satisfaction and not money alone. When approached from this viewpoint, many current and proposed policies do not look too good. The Sierra Club, for example, shows up as an elitist organization. Much of what it advocates would benefit only a small minority in the upper income and social strata.

While the Club's proposals would not necessarily harm disadvantaged people, it is advocating very little that would directly benefit them. Recreation research has demonstrated a close correlation between people's level of education and income and their participation in activities such as backpacking and wilderness camping. A similar correlation with skiing and horseback riding is obvious. In view of this, we might ask whether our present wildland recreational developments are merely servicing the status quo or are actually trying to make outdoor recreation available to those who have not been able to enjoy it in the past.

I don't want to overdo this because there are some straws in the wind like the urban forestry programs. But I think an analysis would disclose that most forestry programs are aimed at providing current benefits or assuring future benefits for those segments of society which already enjoy them. And this brings us back to the idea of forecasting requirements. It is not enough to plan ahead for future production of adequate supplies of forest products to cover the kinds of consumption that are going on today. We must be planning to produce whatever will be needed to distribute these forest benefits more equitably in the future.

Since our forest resources are scarce, this planning is going to have to be done rationally if we are to produce the greatest value in future benefits. But here we encounter another serious problem: we have at present no satisfactory means of determining comparative values for all of the goods and services produced by forests. Many people have pointed this out in recent years but no one has succeeded in doing much about it. To complicate things even further, I feel that we do not even have a satisfactory way of determining comparable values for marketable products like timber over such long periods of time as are involved in forest management planning. Analyses based on compound interest formulas produce such absurd conclusions about the rotations which are biologically necessary to grow many timber species that they have either been ignored or else fudged to produce acceptable results by generations of European and American foresters.

There is probably no forestry area that warrants more intensive current study than this question of how to determine *useful* comparative values for all kinds of forest benefits. There also is probably no other area in which the needed research will be more difficult or frustrating. A consortium of German universities is just starting a long-range study to develop techniques for evaluating what they call the "infrastructure" effects of forests. We too should be making a greater effort here in the United States to help find solutions to this central problem of rational forest use.

And finally we come to what I feel is the most important problem of all. Even if we should succeed in doing all the things that I have talked about, we would still be only fighting a losing rear-guard action unless the inexorable growth of the world's population is somehow brought under control. What point will there be in accurately determining the relative social values of recreation, scenery, and other forest benefits if eventually every possible acre will have to be devoted exclusively to producing food, wood, and other essential commodities? What good will it do to temporarily achieve a more equitable distribution if inevitably everyone is going to be reduced to an absolute minimum? What will we accomplish by attaining a maximum efficiency in production if even that will not provide enough in the long run to go around?

Our whole forestry undertaking will be as futile as the program of the Montana hunters to carry in hay for the starving Gallatin Elk Herd. Wildlife managers know that winter starvation was not the problem; the herd was just too large for the carrying capacity of its range.

Man also lives in a finite world and our wildlife friends would say he is beginning to press hard on its carrying capacity. Perhaps this is why the biologists have been the ones to take the lead in promoting Zero Population Growth and other programs now trying to bring man's excessive reproductive activity under control.

But population is a forestry problem too because if the number of people continues to grow, the resource that we prize so highly will inevitably be destroyed. Gifford Pinchot could not have foreseen that "the greatest number" might eventually become so large that "the greatest good" the forestry profession could produce for them would be totally inadequate and impossible to sustain "in the long run." We cannot afford to let that day arrive. Foresters must start to adapt their planning to the idea of a stabilized rather than a growing population. Even more important, professional foresters as an organized and socially minded group must join actively and resourcefully in the most significant campaign of our day: to bring human population under control.

13

COMMENTS ON FOREST POLICY FORMULATION

George H. Weyerhaeuser

First published in *The Future, Forest Policy, and 1970 National Timber Review;* pages 5–8. Society of American Foresters, Columbia River Section, Portland Chapter. 8 pages. 1971.

BIOGRAPHY

Mr. Weyerhaeuser is President and chief executive officer of Weyerhaeuser Company, with headquarters in Tacoma, Washington. A great-grandson of one of the Company's founders, he became an employee in 1949, after being graduated from Yale University in industrial administration. From then until 1966, when he took his present position, he worked in such varied activities as timber production, pulp-mill and sawmill operation, administration, branch management, and corporate management. In the Weyerhaeuser tradition, he has taken a large interest in forest resource policies, as evidenced not only in his firm's performance, but also in his advisory work with the Stanford Research Institute, the Yale School of Forestry, and the Forest Industries Committee on Timber Valuation and Taxation.

EDITORS' SUMMARY

A good approach to setting policy goals for forest land management is to describe target forests, which will produce the services consumers want at prices they will pay. Two groups of consumers must be studied for this purpose: the consumers of today and those of 50 years hence.

QUESTIONS TO CONSIDER

(l) Do you agree that it is especially true of natural resources that their function is to serve the needs of people?

(2) How may we characterize the rule cited by Mr. Weyerhaeuser (page 97), "When in

doubt, grow wood"? What is the necessity for a rule in this case? What alternative rules might sensibly be proposed?

(3) Mr. Weyerhaeuser describes the "target forest" approach to setting a private corporation's production goals in forestry. Is this same approach suited to setting social production goals? Explain how one might develop such goals by means of the target-forest idea. What use would be made of "demand projections" in this case?

(4) When Mr. Weyerhaeuser states (page 98) that the professional forest manager is not *the* decision maker, what decisions is he referring to? Are there, nevertheless, decisions for the forest manager to make? Are there decisions which are made neither by him nor by anyone else in the corporate organization? Who, then, is the forest *manager* in the sense of *decision maker?*

(5) In his statement just cited, is Mr. Weyerhaeuser at odds with Mr. Yoho (Item 8)?

* * *

It is clear that forests, as a basic resource, have only one ultimate function—to serve the needs of people. This is true of all businesses and governments but especially so of a natural resource. It is not always as clear, however, in today's society, which voices are really representative of the people's long-term interests as opposed to those that may be talking loudly about resources without knowledge of or sincere concern about the real balanced interests of people in those resources.

In the midst of all these voices it becomes increasingly important for the professionals and the decision makers—public and private—to look harder at the policy-making procedure. This is true in many arenas. It is especially true in forest management because a lot of people must live with today's forest-policy decisions for several decades.

Forest management is a complex art, not just because of the long time spans involved, but also because a wide range of skills enters into effective forest management.

In a period of accelerating change in knowledge about forest growth, harvest systems, genetics and wants of people, the long time spans of forest growth make it imperative that forest needs thirty to seventy years in the future be made more visible. Trends must be examined and broad assumptions used with flexibility built into programs wherever practical. We must continue to meet current wood needs.

In our company we have one basic rule, "When in doubt, grow wood." But this cannot be done blindly. We accelerated intensive forest management in our operations in 1966 using all available research knowledge and technology. Our goal was to use forest capital to grow wood in the way that would best meet people's needs for wood at prices they would be able and willing to pay.

We then asked our forestry research people to be sure we understood the growth capacity of each of our properties. There are several answers to this question. One can seed, plant, fertilize, thin or use combinations of these practices. For each level of intensive management, there is a reasonably predictable balanced forest in the future when harvest and growth are balanced with a reasonably stable combination of age classes. After evaluating the economics of the alternatives, a plan for each major property was selected and identified as the "Target Forest." There are usually several interim harvest and management plans that can eventually achieve the selected "Target Forest." One must be chosen that achieves a logical balance between short-

and long-range considerations. If we continue to learn and take advantage of the opportunities for change, we will probably never have a total forest property which meets the then-existing conception of the "Target Forest." The opportunities and the "Target Forest" objective will continue to improve as men do the job and learn from doing. At any given time, the things we are doing to cause wood to grow should be measured against our standard of using forest capital to grow wood in the way that will best meet people's needs for wood at prices they will be willing to pay. Of course, the problem is a little more complicated than this because people want some other things from a forest complex. They want recreation, water, air, and scenery from all or parts of the forest.

All of these wants enter into a perfect description of a "Target Forest." But the test is the same. Managers of forests must invest people's money wisely to meet people's needs at costs they will be willing to pay.

It is of no consequence that the tourist or the fisherman may not buy a ticket as he enters forest land the way he does when he buys a two-by-four or a tablet of writing paper. The consumer ultimately pays the bill for any extra costs incurred in making possible his access to an enjoyable forest experience. This principle applies to both public and private forest land. It applies under any economic system.

In the end, the people pay—or the forest venture collapses; the forest managers and the decision makers, public or private, can continue to collect their wages as servants of the people only if they make reasonably accurate estimates of what prices the people will pay and gear their forest policies accordingly.

In this environment, the responsibilities of the forest managers and the decision makers are clear. The professional forest manager is not the decision maker.

It is the function of the forest manager to make visible (1) the facts, (2) the management alternatives, and (3) his professional recommendations.

It is then the function of the appointed decision makers (be they legislators or corporation managements) to (1) look at the history of what people have wanted, (2) listen to and sort out people's current expression of wants, (3) examine trends and those future factors that are becoming visible, and (4) against this knowledge, to select, from among the alternatives offered by the forest manager, that course of action which is judged to be the best way of investing people's savings to meet people's needs for wood or other forest benefits within the biological limits of the resource and forest environment.

The process of listening to and sorting out people's current expressions of wants is the one that puts today's decision makers to a real test. No man can hear and understand all of the voices being raised today. This is especially difficult when the decision maker realizes that the decibel output is not a measure of sincerity and concern for "all" people's investments and "all" people's wants.

Perhaps we who are involved with forest policy would do well to listen for two voices. These are representative people whom scores of millions of men and women resemble even though their specific characteristics vary.

One of these people is, let us say, a man old enough to have a family and young enough to have a good range of recreational interest. He may be in his late thirties in 1971. He has a modest savings account, some portion of which he or his banker is willing to invest in a business venture with a modest risk. He hopes to live in a better home in a few years if the cost is reasonable. He reads newspapers and magazines and his family occasionally likes to fish, hunt, ski, hike, picnic, drive in pleasant scenery, and, perhaps twice in a lifetime, will visit a wilderness area. This family likes to swim. They drink water and want it to be available and clean 365 days per year. They breathe air and want it clear of foreign substances. So much for the first voice we are seeking.

The other man has almost precisely the same description. The only change is that, instead of being in his late thirties in 1971, he is in his late thirties in the year 2021, fifty years from now. He will be around when many of the trees planted today are harvested. One problem is that his voice is not very audible today.

Nonetheless, these are the two people today's forest managers and decision makers must satisfy. They are the judges. They don't wear black robes and render wordy decisions. But they do make the decision in the end. They will lend, will invest, or withhold their savings, buy or pass up our products and they select legislators who will eventually do these two men's will in shaping tax systems, tax rates, and regulatory statutes governing resources like forest lands. They are all-powerful. Now that we have identified them, we must find out what they want.

One of them is occasionally represented by one of today's pressure groups. It is often hard to identify the voice that sincerely represents him. But he is always in the marketplace, picking and choosing. He is always represented in the capital markets and in legislatures.

Aside from what people say these two men want, what else do we know of their wants?

They both like to use some wood products *and* they like forests for recreation. The nation must furnish both.

The 2021 man has one special characteristic. He has several billion more neighbors on earth than does our contemporary judge. Hopefully, he and these neighbors will have an even higher standard of living than their ancestors who are here today. That will take natural resources. Wood is a renewable resource. Accordingly, it can and should be an important basic raw material. Wood will be replaced for some of today's uses, but because it is renewable, it will be converted into many new products.

Therefore, our 2021 judge wants large supplies of wood on hand. He will use a lot of short-fiber hardwood, but the greater strength and versatility of long-fiber conifer wood will make its supply very important to him.

Since he wants long-fiber wood in large quantities in 2021, this creates a little conflict with the wants of our 1971 judge. Mr. 1971 wants wood today, but he wants as much good scenic hiking, fishing, and touring as he can have.

When the forest manager describes the management alternatives within a timber property, there are apt to be very different wood-growth potentials per acre in different parts of the property—and between properties.

The world actually has a limited amount of land which will grow long-fiber wood fast enough and close enough to transportation facilities to be able to meet the cost limits of our two judges. A significant proportion of this land is in the United States. Mr. 1971 would like to use some of this land for wilderness, parks, power development, utility lines, and highways. That means the land left for growing commercial wood must be used intensively to meet the needs of Mr. 2021 and his billions of neighbors on Earth.

The answer is clear for the decision makers today. If the land will grow large quantities of wood which can be delivered at low cost and adequate low growth rate lands have been set aside for pure recreation, then the good commercial growing land—whether publicly or privately owned—must be made to grow wood as intensively as the forest manager knows how at a reasonable cost.

There are grounds for expecting sound results. Mr. 1971 expects us to do an efficient job of resource management. He is a reasonable man who will support sound forest management decisions when he is adequately informed. To merit his support we must listen for his views, obtain forward-looking professional forestry inputs and alternatives, base decisions on the public's visible and predictable choices for capital allocations, update our target forest descriptions, gear current forest management activities to those targets, and show Mr. 1971 what we are doing and why.

14

PLANNING RESEARCH FOR RESOURCE DECISIONS

C. H. Stoltenberg
K. D. Ware
R. J. Marty
R. D. Wray
J. D. Wellons

First published in *Planning Research for Resource Decisions;* pages 5–7. Iowa State University Press, Ames, Iowa. 183 pages. 1970.

BIOGRAPHY

Mr. Stoltenberg is Dean of the School of Forestry and Director of the Forest Research Laboratory, Oregon State University, Corvallis. A graduate in forestry and in economics from the Universities of California and Minnesota, he served for some years, in various regions, with the Forest Service's Branch of Research. It was then that he developed the concern for research planning and administration reflected in the book from which the present reading is taken. His academic career included faculty positions at Minnesota and Duke universities and the headship in forestry at Iowa State University.

Mr. Ware is project leader of the Forest Ecosystems Decisions Institute maintained by the Forest Service at Athens, Georgia, as a branch of the Southeastern Forest Experiment Station. Jointly, he is Professor of Forestry in the State University. His central professional interest is in bringing modern quantitative, analytical procedures into forest resource management. In pursuit of this interest, he has studied forestry and biometrics at West Virginia and Yale universities, taught statistics and forest measurements at Iowa State University, performed survey techniques research in the Forest Service, and served widely as lecturer and consultant.

Mr. Marty is Professor of Forestry and Resource Development in Michigan State University. Before taking this position in 1967, he was following a career in the Branch of Research of the Forest Service, both at the regional forest experiment stations and in Washington, D.C. His specialization, in the economics and administration of natural resources, is supported by degree studies at Michigan State, Duke, Harvard, and Yale universities and is evidenced by his extensive writings. He has served as a consultant to state and federal government agencies, evaluating their natural-resource programs in economic terms.

Mr. Wray is Chief of Information Services in the North Central Forest Experiment Station, Saint Paul. A forester by education, he began his Forest Service career, nearly three decades ago, on the nationwide *Forest Survey.* It was as a resource analyst on that project that he demonstrated his talents as a writer and editor. He was made Chief of Information Services in the Central States Forest Experiment Station and commenced a professional regime which included the training of Forest Service employees in scientific writing and the production of his own magazine articles and books.

Mr. Wellons is Associate Professor in the School of Forestry at Oregon State University, where his professional responsibilities are focused upon the field of wood chemistry. Graduated in forestry in 1960, he studied at Duke University for advanced degrees in wood science and chemistry. His research and teaching service has included tours of duty with the Research Triangle Institute in North Carolina and on the forestry faculty at Iowa State University. He is an active participant in the affairs of the Forest Products Research Society, the American Chemical Society, and related professional organizations.

EDITORS' SUMMARY

Three alternative purposes of natural-resources research are (1) to develop new options in resource management, (2) to answer questions of fact that arise during management, and (3) to provide research workers with information essential to their studies. Thus all research may be viewed as problem-oriented.

QUESTIONS TO CONSIDER

(1) What is a "question of fact," the answering of which is said to be a major purpose of resource research? What other kinds of questions are there, and how do they differ from questions of fact?

(2) What is "applied research"? "basic research"?

(3) How is it possible to weigh the value of a research proposal, as every research administrator and budgeting or granting agency must do in making choices? How is it possible to weigh the value of research results?

(4) By what means can the research planner (page 103) anticipate problems accurately, so as to identify those which will be critical 5 to 20 years hence? Choose a forestry subject-matter area especially interesting to you, and try your hand at developing a brief "problem analysis" which will spot the research questions that will call loudly for answers a decade or so hence. What are your criteria for identifying such questions?

(5) What, today, are the three most urgent unanswered questions in social-science aspects of forestry in the United States?

* * *

Natural-resource science and scientists are valuable to society for the same reasons that natural resources and their managers are valuable—they can help satisfy

people's wants. And this is the main reason that society supports their research.[1] Just as the resource manager's contribution is measured by how effectively he helps others to satisfy their wants efficiently, the productive value of resource scientists must ultimately be determined by how much their efforts increase the efficiency of the resource manager.

This, then, provides the basis for a tentative statement about the purposes of natural-resources research. One major purpose is to develop new alternatives for the resource manager. These may be new practices, tools, and concepts, or new products and services. A second common purpose of resource research is to answer questions of fact that arise during management. And inasmuch as resource management is viewed as the process of solving clients' resource problems, these answers would be the information needed to solve these problems. That is, resource research provides the information needed to define or compare alternative means for achieving a resource user's or owner's objectives. A third purpose is to answer questions of fact that arise during resource research since it is only after some of these basic questions have been satisfactorily answered that the first two objectives can be achieved most efficiently.

Many resource researchers do not directly provide information for the resource manager. Some provide information to solve other researchers' problems. Resource research may be viewed as a continuous spectrum of scientists with the resource manager at one end of the spectrum. The manager is principally concerned with his client's problem, that of the resource owner or user. But when he lacks the information needed to help evaluate the client's alternative resource practices, he may experiment with several of those practices. Next on the spectrum are the scientists who are attempting to answer the managers' immediate questions of fact, such as resource-use trends, prices, and technologies. Following them are the developmental researchers who create new alternatives for the manager, particularly alternatives that will help solve problems not only of tomorrow but of years to come.

Further along this research spectrum are the scientists who serve a clientele of other researchers. These scientists, usually from the basic disciplines, provide the facts and relationships needed by other scientists who conduct the more "applied" developmental research. The relationship between applied and basic research is central to the history and philosophy of science and technology. Kranzberg[2] discusses this briefly but comprehensively.

Although the ultimate client for all research activities is the resource-manager's client, the immediate clientele of a researcher may be managers or other researchers. But every productive researcher has a clientele to whom he provides research results. This clientele uses his results to solve their resource-management, or research problems. When the client is another researcher, he in turn is able to conduct his research more efficiently and then provide his clientele with improved answers to help solve their resource-management research or practice problems.

As the distance on the spectrum increases between the manager and any researcher, successful problem anticipation, and hence planning, become more difficult. Although a scientist focuses primarily on his immediate client, his most basic research is really intended to help solve a resource-management problem. The research planner needs to anticipate that problem accurately. Thus researchers in the basic resource sciences are helping to solve important problems not expected to be critical for 5, 10, or even 20 years. This scientist may have great difficulty in anticipating the correct problem, but he is very valuable to society when he succeeds.

The results of some scientists' research will be useful in the solution of several clients' problems. This is particularly true for basic research. For example, a tree physiologist, a biochemist, and a wood anatomist might be teamed to study how trees form wood. Their results could help wood chemists develop pulping processes. But the same results could help organic chemists define the structure of lignin and likewise help forest biologists control tree growth. In turn, silviculture researchers can develop appropriate fertilization systems to satisfy the timber managers' needs.

Clearly, this research team may question the need for being concerned with which particular client's problem they are working on. In such instances, overemphasis on an ultimate problem (that of the resource manager's client) can result in confusion. In these situations, problem orientation can be achieved by having the silviculture researcher understand the biological-data needs in the probable decision-making framework of the timber manager and his potential clients; having the forest biologist understand the cultural needs in the decision-making framework of the silviculturist; having the physiologist understand the data and physical-relationship concerns of the biology researcher; and having each in turn respond to these needs.

Some researchers may feel that basic research is inherently nonproblem oriented. But we and many other researchers believe that a problem orientation both helps the researcher develop his research plans and enhances the value of even the most basic resource research.[3]

REFERENCES

1. Heady, E. O. 1961. Public purpose in agricultural research and education. J. Farm Econ. 43(3):566–581. Dubos, R. J. 1967. Scientists alone can't do the job. Saturday Review (Dec. 2, 1967):68–71.
2. Kranzberg, M. 1967. The unity of science-technology. Am. Scientist 55(1):48–66. Kranzberg, M. 1968. The disunity of science-technology. Am. Scientist 56(1):21–34.
3. Balch, R. E. 1964. The future in forest entomology. J. of Forestry 62(1):11–18. Quinn, J. B., and R. M. Cavanaugh. 1964. Fundamental research can be planned. Harvard Bus. Rev. 42(1):111–124. Spieth, W. 1964. Basic research on a problem. In Letters to the Editor, Science 143:197. Hoover, S. R. 1965. Research and purpose. In Letters to the Editor, Science 147:1523.

15

NEW TOOLS FOR PLANNING AND DECISION MAKING

O. F. Hall

First published in *Journal of Forestry* 65(7):467–472. July 1967.

BIOGRAPHY

Mr. Hall, Head of the Department of Forestry and Forest Products at Virginia Polytechnic Institute and State University and previously an administrator in forestry and in wider resource fields at the University of New Hampshire, is a pioneer and leader in the application of management science and operations research to forestry. During his years on the forestry faculty at the University of Minnesota and at Purdue University, his research set targets for managers, and his students advanced into influential staff positions, notably in southern industrial forestry. At one time the Editor for Forest Management of the *Journal of Forestry* and long a forest-management consultant to firms in the pulp and paper industry, he is well known for his development of forest-management decision methods, especially for complex problems best approached with a computer.

EDITORS' SUMMARY

Applications of management science to forest land resources emphasize two basic concepts: that of the organization as a system and that of the investment center. Classes of models promising for forestry include simulation, mathematical programming, and discounted cash flow. Forestry requires its own, tailor-made models, typically developed for the computer by a research-management team.

QUESTIONS TO CONSIDER

(1) What is the distinction between planning and decision making?

(2) What are the distinguishing characteristics of a *system* in the sense used by Mr. Hall?

Are these characteristics possessed alike by, for example, the economic system, the ecosystem, the forestry system, the forest system, and the organization system?

(3) If system A contains both the variables within B and those within C, B and C are termed *subsystems* of A. Cite examples from social sciences in forestry.

(4) How does an "investment center" differ from a "profit center"? How do these differ from a "cost center"? Why is it more generally more appropriate for a firm to create investment centers from horizontally related enterprises than from vertically related ones? Why is it appropriate for a firm to establish different guiding rates of return for its different investment centers?

(5) List the variables that you would include in a simulation of timber-stand development over time, and sketch the functional relationships among variables.

(6) Evaluate the argument that operations-research techniques are applicable to quantifiable questions of intermediate size (weight, scope) but not to others. You will need to supply a definition of *intermediate* as opposed to *large* or *small.*

* * *

Tools of the manager just as important as detailed analytical methodologies or information systems are the conceptions the manager holds of the environment in which he operates and of his functions and goals in that environment. The past decade has seen two definite concepts evolve and become widely recognized as valuable tools of practicing management. These are (1) the concept of an organizational system, and (2) the concept of investment centers in an industrial organization. Recognition of them is essential to effective use of the more specific techniques.

Organizations as Systems

Organizations, regardless of their size or social location—whether industrial or governmental—have common characteristics. The study of these characteristics has produced a body of knowledge which has proved of great value to many practicing managers. The systems approach has led to steadily more definitive descriptions of organizations even to the extent of describing certain aspects of them in charts, logical networks, and even groups of mathematical formulas.

From this approach the overall job of management has come to be regarded as the development, maintenance, and improvement of the operating system under its administration to meet the multiple goals for which it exists. The system for a large industry is a complex of people, equipment, finances, natural resources, and institutional constraints. Space does not permit further elaboration of this viewpoint, but it is at sharp divergence from the earlier viewpoint of the executive as the person in authority who renders seemingly arbitrary judgments about the disposal of people and resources placed under him. The new conception of the role of the manager has led not only to a broader scale for analyses of the efficiency of production processes, but also to new ways in which the manager exercises his authority to get cooperative effort from the people with whom he is associated.

With regard to forest management, the new concept means a drastic change. No longer is concern solely with manipulation of land tracts, soil productivity, distribution of tree sizes and volumes, measurement methods, and maximizing production from the ecological community which is the forest. Now the managers must develop entire organizations of people, land, trees, markets, and financial resources into smoothly functioning units, defining their goals and fitting them into the larger economic and social environment upon which they depend and which depends upon them. This new concept of management requires an outward as well as an inward view, and emphasizes coordinative control as well as maximizing productive efficiency.

Investment Centers

There has been a strong trend toward establishing within companies divisions which are treated as individual *profit* centers, each responsible for yielding a profit above its own expenses. More recently these centers have been evaluated on the return on investment (ROI) they generate, and have been termed *investment* centers. A recent survey by Muriel and Anthony[1] of 1302 leading companies in the country showed that over 80 percent are using the decentralized responsibility approach, the great majority with investment centers. Their survey shows that the wood-using industries have moved in this direction even more strongly, as Table 15-1 shows.

Table 15–1 Results of Study by Muriel and Anthony Showing over 80 Percent of Companies Polled Using Decentralized Responsibility Approach

INDUSTRY	COMPANIES POLLED	COMPANIES REPLYING	NOT USING PROFIT CENTER	USING PROFIT CENTER BUT NOT INVESTMENT CENTER	USING INVESTMENT CENTER
	Number	*Number*	*Percent*	*Percent*	*Percent*
All industry	1648	1302	18	22	60
Pulp and paper	73	62	8	16	76
Wood products	52	40	5	35	60

Further analysis of their returns showed that the wood-using industries were similar to other industries in the accounting practices used in calculating investment base and in whether or not they require the same ROI for all centers (about half do and half do not).

The extent to which industrial timberlands are operated as investment centers within their firms is not known, but the question exists of how such a management-control system should be applied to woodlands, and many timberland excutives face it. To freely operate as such a center, the division should have control over its expenses and its revenues. Timberlands may not qualify under the revenue portion of this restriction, since transfer price, the price credited to timberlands for wood transferred to the mill investment center, may be arbitrary. If the price used is that for wood from the open market, delivered to the mill or other designated loading points, it may be too low, reflecting only direct logging costs, and not the costs of land management for a continued supply of wood. If timberlands are treated as an investment center, the best solution may be a cost-based transfer price, including land management costs of building up timber production capacity, but the rate of build-up remains a question to be resolved.

Muriel and Anthony have pointed out some misevaluations commonly found in investment-center control systems, and a number are especially appropriate to timberlands. They point out that when net book value is used as the investment base in calculating ROI, as it is for 73 percent of the firms surveyed, depreciation reduces the base, and rate of return rises even with a constant income. Since land is not depreciated, and maintenance of timber land productivity requires continued investment, this center may suffer in comparison with centers using depreciating assets. Many concerns have realized that the use of investment centers tends to favor short-term thinking and those activities which expense expenditures rather than capitalize them. All these misevaluations stem from the fact that when a ratio is used as a criterion, it can be increased in either of two ways: increasing the income in the numerator, which is the goal sought, or decreasing the valuation in the denominator, which may be desirable, but may simply mask a trend toward long-run decrease in profits. Muriel and Anthony conclude that, where investment centers are used, (1) there is a need for an internal evaluation system separate from the accounting system for external reporting, (2) judgments based on ROI must avoid the danger of overemphasis on that figure and must temper it with other criteria, and (3) the use of a single target ROI for all centers may be inappropriate.

SOME SPECIFIC TESTS

Simulation

The first technique to be considered, and perhaps the one with the greatest potential for immediately usable answers to operating questions, is simulation. Simulation is the construction of a mathematical model, usually a series of equations translated into a computer program, that approximates the important operating features of the system being managed. By running through the series of calculations repeatedly one can simulate the passage of time and obtain some idea of the condition of the system after any elapsed period under any assumed conditions. In this way one can have a "pilot plant" for testing alternative policies in operating systems that are too large or expensive to be run experimentally: Various policies may be tried, and the results of the simulation taken into account by the manager as he makes the decision on which one to implement.

Almost any system that can be described verbally can be simulated in this way. Of course, the more quantitative data there are on the operating characteristics of a system, the closer the simulation will be to reality. Equally as important as finding the state of the system at any future time, is the understanding of the interconnecting relationships among factors of the system gained from constructing the simulation model. Many times the requirement to define precisely relationships and to get particular data where none has previously existed are sufficiently helpful to justify the expense of the simulation, even if it were never run to completion. Sometimes, when precise knowledge of relationships is not obtainable, approximations can be entered in the simulation, permitting it to function. Later sensitivity testing often shows that the approximated factors are not critical in obtaining useful information from the simulation. In this way such modeling has the advantage of permitting usage of all knowledge, even though imperfect, in making decisions.

Simulation has been said to be most useful in problems of logistics, a good example of which is control of woodyard inventory at a pulp mill. This is a problem to which the technique has been applied successfully.[2,3,4] Models have been constructed and used incorporating mill-consumption rates; dealers, producers, and their labor crews; and delays in ordering, producing, and delivering wood. Weather influences are included. By simulating the passage of time, it is possible to determine how frequently, and under what conditions, the mill inventory will be exhausted. Repeated trials can help to determine the inventory goal to minimize this possibility. Gessford[3] has demonstrated by this means how it is possible to organize a large group of woodyards so that intershipment of wood can reduce the inventory that must be carried in each, and thereby the average cost of wood storage.

Simulation has been demonstrated as probably the best immediate means of realistically scheduling cutting and replanting operations on a large managed forest.[5] By repeatedly running the model with a variety of cutting policies and schedules, one can arrive at the schedule which comes closest to meeting restrictions of continuous production, maximum growth, minimum wood cost, limits of budget, etc. The technique has been demonstrated for even-aged forests by the FOPS program at the University of Georgia,[6] by Gould and O'Regan,[7,8] and for uneven-aged forests by Pope.[9]

The difficulty of developing and running simulation models on the computer has been greatly reduced by a number of general simulation languages such as:

DYNAMO, built for the industrial dynamics modeling concept by Forester;[10]
SIMSCRIPT, developed by Markowitz, Hausner, and Karr at Rand Corporation;[11]
GPSS, developed by Gordon and Efron, available through IBM.[12]

Once the conventions of the language are learned, a matter of a few hours, one can concentrate on writing the equations that represent the systems he wishes to simulate, with little concern for the difficulties of implementing it on the computer.

Once time-sharing computers become generally available, the possibility of rapid formulation and rerunning of simulations for real-time help in decisions becomes feasible. Under those circumstances a language such as OPS-3[13] will be useful, although at the present time it is operational only on the time-sharing system at Massachusetts Institute of Technology for which it was designed.

Simulation models may give accurate indications of future events, *provided* all the conditions built into the model can be accurately predicted. Uncertainties in these preconditions, however, may also be incorporated into the models by probability distributions and random number generators, to give a realistic range in outcomes that may be encountered, and thereby aid the manager in coping with uncertainty. This technique has been applied by Hertz[14] in the evaluation of risk in capital investment.

Despite its advantages, simulation is not a panacea for all business decisions. A management team may be tempted to build an all-encompassing model of a business that can simulate any possible decision question. Such models may take many months for construction, elimination of errors, and obtaining essential data, so that the organization may undergo changes that make the simulator obsolete before it is completed. Furthermore it must be remembered that models, no matter how complex, do not provide decisions; they only test the outcomes of decisions, which must originally come from the managers and their understanding of the operating environment. Finally, finding the optimum is usually impossible. The simulator can test, from the infinite combinations of factors, only a relative few which past simulations and the experience of the manager indicate may lead to improvement. This may fully justify use of the technique.

Mathematical Programming

A second technique, or really a class of techniques, of proven usefulness in management decision is mathematical programming. The most highly developed single method is linear programming, from which have sprung, in reaction to its severe restrictions, variations called integer, quadratic, convex, and dynamic programming. The basic method of mathematical programming is to try to maximize or minimize the value of some mathematical expression subject to a number of restricting equalities or inequalities, all of which must be solved simultaneously for the values of the variables found in all expressions. These variables are amounts of factors, some of them controllable, about which the managers must make decisions. The procedure (called an algorithm) for solving this set of simultaneous equations by the methods of matrix algebra may be quite complex, and usually requires a computer for problems of sufficient size to be realistic.

Scheduling of cutting, planting, and other silvicultural operations seems to be the major area of potential application of these techniques in the management of forest properties. Determination of which tracts of land to allocate to cutting and regeneration in each year can be financially crucial, since these decisions determine the rate of cash flow from invested capital, and the rate of further investment.

The last six years have seen a large number of publications that phrase forest operations scheduling in terms of the linear programming problem.[15,16,17,18,19,20,21] With the exception of Curtis's[16] work, none of these models are known to be in use regularly, but they have given valuable experience in stating forest variables and restrictions in suitable terms. Development of models for real situations continues. Several industrial concerns are using models to associate known product potentialities of standing timber with anticipated short-term demand, in order to shift cutting to meet this demand most efficiently.

Research is proceeding under R. J. McConnen at the Pacific Southwest Experiment Station of the U.S. Forest Service on a land management system which incorporates linear programming for prescribing short-term cutting and other activities on forested watersheds.[22] Numerous partially conflicting goals are incorporated in the single model, thereby providing a means of approaching multiple-use problems. The system includes computer-produced map overlays to aid managerial analysis.[23]

These efforts to use mathematical programming in real scheduling problems have brought into focus several unresolved difficulties stemming partially from the nature of the linear-programming model. The first difficulty results because the model treats variables as continuous—that is, they may take on any fractional values. Translated into real management, this means that the model schedules a number of acres to be treated or a volume of timber to be cut, but it still leaves the problem of allocation to particular tracts of land undetermined. This situation may be advantageous in allowing latitude for managerial judgment, but the economics of logging, planting, and other treatments, and the requirements of continuous record keeping, usually prescribe that land be permanently subdivided into blocks, each treated as a unit in any one year. Treatment of fractional blocks as prescribed by LP solutions is often impractical, and simple rounding of acreages to whole blocks may invalidate the entire optimization for which the model was formulated. Efforts to schedule real blocks of land, rather than homogeneous but disconnected acres, require modification toward integer programming, where techniques are still developmental. It is possible that the methods worked out by Pierce[24] for paper trimming will point toward solutions.

A second difficulty arises from the large size of realistic scheduling problems, so that either they cannot fit existing computer programs, or they take so much running time that solution is too costly. One solution may be the decomposition technique, which breaks a large matrix into almost independent submatrices, which are then solved subject to a single master matrix. An effort has been made to apply decomposition to a large forest scheduling problem. While the study was very helpful, it did not achieve the great economy of computer time that was hoped for it. Other recent studies seem to indicate similar long running times with decomposed models.

Another limitation on linear programming is that it imposes a very definite time horizon. The model permits planning only for a restricted number of periods. In addition it prohibits for even-aged forests consideration of the second rotation when acres, once treated, will be treated again, perhaps in different sequence. These limitations can be partially circumvented by restrictions requiring certain conditions, such as a minimum growing stock or acreage forested, to be retained at the end of the planning period. Still, complete freedom to bring in variations in rotation and thinning cycle is not available. Pioneering efforts to use dynamic programming to circumvent these difficulties have been made, but much further research is needed.

Discounted Cash Flow

One technique, in which industry at large is only now catching up with the forest manager, is discounted cash flow for evaluating capital investments. There has been much controversy about it in the business press, but there is now general acceptance that the internal rate of return (yield, ROI) that equates discounted investments with discounted returns is a valid criterion for comparing alternative investment programs. Accountants are coming to accept the approach. There remain a number of theoretical questions, and certain cash flow patterns especially encountered in the petroleum industry favor the present value approach. Nevertheless for the great bulk of long-term business investment decisions internal rate of return is the most readily usable criterion.

One obstacle in the way of adopting this tool has been the difficulty of calculating it, but with special tables or with the digital computer this difficulty is largely overcome. A general program for calculating it has been demonstrated,[25] and Row in 1963[26] and Lundgren, Schweitzer, and Wambach recently at the Lake States Forest Experiment Station have developed special forest-oriented programs for the calculation. General programs are also available through the program libraries maintained by computer manufacturers.

Fedkiw[27] described the use of rate of return for ranking alternative forestry activities, and other research men have demonstrated its use in a similar way. Coan[28] described its use for evaluating purchases of timberland, and several pulp and paper companies use it in this way. Ratliff[29] has reported its use by the Crown Zellerbach Corporation for comparing proposals for research and development investment. Trials for justifying woodland investments have not been altogether happy for industrial forest managers, because the rates obtained frequently come out lower than alternative investment opportunities in other parts of the corporation. However, now that the measure has attained coroporate respectability, it is probable that timberland executives will find ways to use it that yield results more comparable with other corporate divisions, possibly including projected increases in land and timber values, and savings from higher yields, better quality control, lower inventory costs, by-product incomes, and independence of high-cost wood sources.

Real-Time Management Information Systems

Until recently, getting results from computer centers has involved delays of hours or sometimes days, because efficient use of the expensive machines has required programs be run in batches. Now the time-shared computer permits a number of input units to feed varying types of programs into a large central computer, with the computer apparently handling a number of programs simultaneously. With this innovation comes the possibility of real-time computer usage, that is, such rapid response from the computer to managerial inquiries that use of the computer fits into (or actually speeds up) the normal time schedule of the organization. Present technology makes it possible for the input/output units to be located miles or hundreds of miles from the computer, and for the central computer to store large "data banks" of current business information for immediate access.

These systems, with a potential of aiding business decisions by greater speed, rapid analyses regardless of complexity, and instant access to large stores of information, are being installed by large businesses and other organizations throughout the country[30] for handling problems in finance, personnel, logistics, marketing, research and development, and strategic planning. One of the most advanced of such systems, being installed by Mead Corporation, is described on pages 391 to 408 of Prince's book.[31] Several other large wood-using industries are installing the computers and communications to make such systems possible. While there is yet little evidence of tying woodlands directly into these systems, there are a number of types of woodland data to which real-time accessibility would be of great value in decisions by woodlands managers, corporate planners, wood procurement directors, accountants, legal advisors, and mill operating managers:

1. Current inventory of standing timber, frequently updated for cutting and growth. The various types of continuous forest inventories could provide this information.
2. Current inventory of cut wood at yards, landings, and in transit.
3. Financial records on woodlands expenditures—budgeted and actual.
4. Current average costs.
5. Road system—status, investment, and plans.
6. Logging and transportation equipment, operating and maintenance costs, depreciation.

7. Land descriptions. Title information.
8. Maps, base maps and current status of planned operations, possibly using something like MIADS.[23]

Practical Applicability

A practical person may ask, "Are these so-called tools really useful, or are they just mathematical playthings of academic theoreticians who never got their hands dirty in a real operating industry?" The best answer to the question would be to cite the many applications of the methods in real industrial settings. Some of that has been done above, but there are many more examples which are not listed because of lack of space, and industrial security. Nevertheless, there *is* ample evidence in trade magazines, in the size of operations research staffs of many large industries, and in large sums spent on management consulting firms, that this general approach to management decision making has great application. The petroleum, mining, and power industries are heavy users, as are transportation, steel, and many types of manufacturing. Anyone interested in the applications in the pulp and paper industry alone should obtain CA Report No. 9 of TAPPI.[32] Considering the wider area of natural resource management, from which forest land management cannot divorce itself, applications of linear programming[33] and simulation[34] have been demonstrated for management of water-resource systems.

SOME GENERALIZATIONS

Trying to apply these new tools to timberlands, and observing the efforts of others doing so, lead to several tentative conclusions, which seem to be borne out by the progress of operations research in other industries.

The first is that rapid solution of new problems by adopting models developed by other industries and simply "plugging in" new variables will not usually be successful. Again and again experience by operations researchers has shown that each problem is just enough different to require some special features in the model or solution algorithm. Furthermore, as solutions to larger and more complex problems are attempted it is found that more than one technique and more than one model are often needed for the solution. Woodland problems have been shown to be just as difficult as those in other areas, with their own special aspects that require special approaches. Hufschmidt and Fiering[34] discovered the same thing in attempting to build a generalized simulation model of water-resource systems. This conclusion would seem to justify increased research directed at the application of these techniques to timberland management rather than hoping that the research of others will take care of timberland needs.

The second conclusion is that the computer is an essential assistant to application of most of the new tools and that the most efficient and creative solutions to problems come when the analyst using the tools has ability and experience with computer programming.[34,35] On large projects the technical analyst may have a programming specialist working for him, but the analyst needs to have some mastery of the techniques of computer programming himself. This need for cross-disciplinary training goes even further, with the probability that a forester with some training in the management techniques will arrive at more realistic solutions than the operations research technician without training in forestry. These conclusions seem to have implications for the kind of training that should be given the men who will be the timberland executives of the future.

A third and final conclusion is that successful implementation of solutions to woodlands problems gained by these new tools requires close rapport and frequent feedback between the line managers and the analysts working on the problem. All parties must be prepared for the first-proposed solution possibly to be very difficult to implement, and for the solution finally accepted to be the result of considerable give and take on both sides, including unsuccessful trials, delays for assembly of new data, and experimentation with different approaches. Only in this way can managers be forearmed against apparent disillusionment and the cutting short of implementation before full benefits can be achieved.

REFERENCES

1. Mauriel, J. J., and R. N. Anthony. 1966. Misevaluation of investment center performance. Harvard Business Review 44(2):98–105.
2. Hewson, T. A. 1960. Simulation of pulpwood inventory dynamics in the operation of an integrated pulp and paper mill. TAPPI 43(6):518–527.
3. Gessford, J. E. 1962. Computer simulation yields evaluation of procurement policies. Management Technology 2(2):91–100.
4. Hamilton, H. R. 1964. Attacking a paper industry problem by simulation. TAPPI 47(11):678–683.
5. Hall, O. F. 1962. Computers in forest management. Tech. Paper 62-TP-101. American Pulpwood Association.
6. Clutter, J. L., and J. H. Bamping. 1965. Computer simulation of an industrial forestry enterprise. Proc., 1966, Society of American Foresters. Pages 180–185.
7. Gould, E. M., and W. G. O'Regan. 1965. Simulation—a step toward better forest planning. Harvard Forest Paper No. 13. 86 pages.
8. O'Regan, W. G., L. Arventis, and E. M. Gould. 1966. Systems, simulation, and forest management. Proc., 1965, Society of American Foresters. Pages 194–198.
9. Pope, R. B. 1965. Processing inventory data. 56th annual meeting, Western Forestry Conservation Association, Vancouver, B.C. December 8, 1965.
10. Pugh, A. L. 1963. Dynamo users manual. Cambridge, Massachusetts: MIT Press.
11. Markowitz, H., B. Hausner, and H. Karr. 1963. SIMSCRIPT: a simulation programming language. Englewood Cliffs, N.J.: Prentice-Hall, Inc.
12. Anonymous. 1963. General purpose systems simulator II. Reference Manual B20-6346, IBM DP Division.
13. Greenberger, M., M. M. Jones, J. H. Morris, and D. N. Nass. 1963. On-line computation and simulation: the OPS-3 system. Cambridge, Massachusetts: MIT Press. 126 pages.
14. Hertz, D. B. 1964. Risk analysis in capital investments. Harvard Business Review 42(1):95–106.
15. Coutu, A. J., and B. W. Ellertsen. 1960. Farm forestry planning thru linear programming. Tennessee Valley Authority Rep. No. 236–60.
16. Curtis, F. H. 1962. Linear programming the management of a forest property. Jour. of Forestry 60:611–616.
17. Donnelly, R. H., R. W. Gardner, and H. R. Hamilton. 1963. Integrating woodlands activities by mathematical programming. Battelle Memorial Institute (for American Pulpwood Association).
18. Kidd, W. E., Jr., E. F. Thompson, and P. H. Hoepner. 1966. Forest regulation by linear programming—a case study. Jour. of Forestry 64(9):611–613.
19. Kinne, S. B., III. 1966. Maximizing returns of a farm-forest enterprise using dynamic linear programming. Unpublished thesis, Purdue Univ.
20. Loucks, D. P. 1964. Development of an optimal program for sustained yield management. Jour. of Forestry 62:485–490.
21. McCurdy, D. R. 1961. A linear programming simulation of an Indiana woodland. Unpublished thesis, Purdue Univ.
22. Navon, D. I. 1966. Computer-oriented systems for wildland management. Jour. of Forestry 65:473–479.
23. Amidon, E. L. 1964. A computer-oriented system for assembling and displaying land management information. Res. Paper PSW-17, U.S. Forest Service. 34 pages.

24. Pierce, J. F. 1964. Some large-scale scheduling problems in the paper industry. Englewood Cliffs, N.J.: Prentice-Hall, Inc. 255 pages.
25. Hall, O. F. 1962. Evaluating complex investments in forestry and other long-term enterprises using a digital computer. Res. Bul. 752. Agric. Exp. Sta., Purdue Univ. 11 pages.
26. Row, C. 1963. Determining forest investment rates of return by electronic computer. Res. Paper SO-6. U.S. Forest Service. 13 pages.
27. Fedkiw, J. 1960. Capital budgeting for acquisition and development of timberlands. *In* Financial management of large forest ownerships, Bul. 66, School of Forestry, Yale Univ. 45 pages.
28. Coan, N. A. 1962. "ROI" cash flow is guide to help answer basic woodland management questions. Pulp and Paper 36(9):12.
29. Ratliff, F. C. 1966. The discounted cash flow method as a tool in research and development. Paper presented at TAPPI meeting, Boston, Massachusetts (November).
30. Greenberger, M. 1966. The uses of computers in organizations. Scientific American 215(3):193–202.
31. Prince, T. R. 1966. Information systems for management planning and control. Homewood, Illinois: Richard D. Irwin, Inc. 416 pages.
32. Pierce, J. F. 1966. A KWIC index and bibliography of operations research and related topics in the pulp and paper industry. CA Report No. 9, TAPPI, New York.
33. Maass, A., M. Hufschmidt, R. Dorfman, H. Thomas, S. Marglin, and G. Fair. 1962. Design of water resource systems. Harvard Univ. Press.
34. Hufschmidt, M. M., and M. B. Fiering. 1966. Simulation techniques for design of water resource systems. Harvard Univ. Press. 212 pages.
35. Ledley, R. S. 1965. Use of computers in biology and medicine. New York: McGraw-Hill Book Co. 965 pages.

16

ECONOMIC MODEL BUILDING AND COMPUTERS IN FORESTRY RESEARCH

Daniel E. Chappelle

First published in *Journal of Forestry* 64(5):329–333. May 1966.

BIOGRAPHY

Mr. Chappelle is Professor of Resource Development and Forestry at Michigan State University. Resource Economics is his special field—particularly its quantitative side. His writing emphasizes forestry microeconomics, resource planning, and the application of computing equipment to resource decision-making systems. A native Virginian, he studied forestry and economics at Colorado State and Duke universities and at the State University of New York. He served with the Forest Service, in its Branch of Research, both in the Southeast and in the Pacific Northwest. His academic experience, prior to his present assignment, was gained at the State University of New York.

EDITORS' SUMMARY

Models are simplified representations of systems. Abstract models may be classed as static, comparative-static, or dynamic—also as verbal, graphic, or mathematical, and the last-named as deterministic or statistical. Model building begins with data from the real world and ends with tests of predictive ability in this world. Computers are valuable in model building and use.

QUESTIONS TO CONSIDER

(1) What is meant by the "real world," of which models are said to be a representation? What other sorts of world are there?

(2) Models, according to Mr. Chappelle (page 117) must be less complex than the world, but still "sufficiently complete." Construct a model for determining the optimal complexity and completeness of models.

(3) What is the "scientific method"? Is it applicable to social science? What role is played by a hypothesis? Is this the role played by a model—i.e., is a model identical to a hypothesis?

(4) The argument that economic theory is derived from "real world" data and tested against such data (page 119) may be countered by reference to theory obviously based on unreasonable assumptions, such as that of the "economic man." What is a helpful rejoinder to the counter-argument?

(5) Both Mr. Chappelle and Mr. Hall have a good deal to say about the use of computers in research and in management. How do you explain and characterize the place of computers in forestry—now and prospectively?

* * *

About the only thing that all scientists have in common is a working knowledge of scientific method. The success obtained by the use of scientific method is its most convincing advertisement.

Central to scientific method is the concept of the model. A model is a representation of a system and is used to explain the behavior of some aspect of the system. It is an expression of the individual researcher's view of the system based upon experience, a knowledge of past work, and data. A system may be defined as an integrated network of interacting elements, receiving certain inputs and producing certain outputs with the objective of optimizing some function of these inputs and outputs, given certain constraints. A production system, for example, requires inputs in the form of agents of production and transforms these into products of some type.

To be useful, the model must be less complex than reality itself and at the same time be sufficiently complete to approximate those aspects of the real world in which we are interested.[1] One might say that the model must capture the "essence of reality" to be useful.

All scientists are model builders, either implicitly or, hopefully, explicitly. A clear presentation of his model is one of the major obligations of a scientist when research results are presented. It is in the statement of the model that the researcher's view of the real world is developed and all of his simplifying assumptions are explicitly recognized. Finally, it is the scientist's presentation of the model that enables other scientists to build on his work.

It should be pointed out that a model is not identical to a theory. A model becomes a theory only when it passes the test as an adequate representation of some real world system. A model can be right or wrong only on logical grounds, whereas a theory is right or wrong on grounds of predictability of the behavior of a real world system.[2]

Models have played and will continue to play a major role in the development of economic theory. As Kaplan[3] has noted, models save us from self-deception by forcing our ideas out into the open at an early stage of inquiry.

TYPES OF MODELS

Models may be characterized in many ways. One important way is by mode of expression. That is, whether the model is expressed verbally, graphically, or mathematically. The trend in economics is to mean a "mathematical model" whenever the term "model" is used. However, if the model is well thought out, it makes little difference how it is expressed. The formulation most appropriate should be preferred. On the other hand, complex models nearly always call for a mathematical form, because only in this way is it possible to show all of the interrelationships.

It should be recognized that the use of models does not represent a new way of looking at problems. The most common models are of the verbal type and are formulated intuitively. Verbal models have played an important role in economic research; particularly, in exploratory research.[4] The current emphasis on mathematical and statistical models is a result of the many difficulties economists have encountered in dealing with verbal models. I believe all of us recognize many of the difficulties that may develop, e.g., equivocation of language (inclusive and exclusive disjunction), so I will not spend time discussing them.

Mathematical models may be either deterministic or statistical. Deterministic models are those which imply prediction without random elements. In contrast, statistical models are those in which the relations are stated in terms of probability distributions. In statistical models, the recognition of variance is an essential feature, whereas in deterministic models the variance is assumed to be zero. In economics, statistical models are termed econometric models.

Economic models may also be divided by their scope of subject. That is, they may represent either microeconomic or macroeconomic systems. A microeconomic system is one which operates at the level of the consumer, firm, and industry. In contrast, a macroeconomic system is one that operates at the level of the economy as a whole.

We can distinguish between static, comparative static, and dynamic models. A static model is one that represents a system at one point in time and for a given set of conditions. Static models deal with stationary systems, in which no attention need be paid the effect of passage of time. A comparative static model depicts a system in a similar manner but does so before and after some change which is attributable to the passage of time. In contrast a dynamic model attempts to represent a system at any given instant and involves the explanation of how change over time occurs in the system.

The most common type used in forestry economics seems to be the comparative static model. Dynamic models are often held up to us as an ultimate aim, but their development has not measured up to expectations because of both theoretical and computational difficulties. It appears that forestry economists have made little use of truly dynamic models. In contrast, in general economics, dynamic models appear to be reaching a fairly high state of sophistication, especially in the area of growth models. Dynamic models may be further classified into those which are continuous dynamic and those which are discrete dynamic. Variables in continuous dynamic models are thought of as changing continuously through time and require the use of differential equations. In contrast, variables in discrete dynamic or sequence models are thought of as changing in a stepwise fashion and require the use of difference equations.

The above classification of models is not comprehensive, but it does lay the groundwork for a discussion of the process of model building.

MODEL BUILDING PROCESS

The formulation of a model involves the use of assumptions to simplify the analysis of a complex situation, which is termed a system.

The model building process is illustrated by the diagram[5] in Figure 16–1. In this diagram, the dark arrows indicate primary flow, whereas the lighter arrows indicate possible feedbacks.

This diagram illustrates the fact that economic theory is checked by facts at both ends of the model building operation. This operation must begin with reasonable assumptions derived from real world data and end by surviving exacting tests which utilize real world data.

The ultimate test of any model is its predictive accuracy. The prediction of the system's behavior, and not necessarily the actual future action, is our test criterion. One might say that a key step in the progress of a field of knowledge toward scientific maturity is the building of models which permit successful prediction—i.e., permit formulation of theories.[6]

The building of both deterministic and econometric models begins with experience, data, and past work, which provide the researcher with a basis for the formulation of a symbolic model. Once the symbolic model is developed, however, a split in the process occurs. In the case of deterministic models, the researcher merely performs some symbolic manipulation and then proceeds into the prediction phase. Deterministic mathematical models and verbal models are tested against reality mainly on grounds of consistency and reasonableness. Only partial information about the real world is brought forth in their support—and this often intuitively.[7] In contrast, the process of building econometric models proceeds from the symbolic model to sample data which provide estimates of the model parameters. Predictions are then

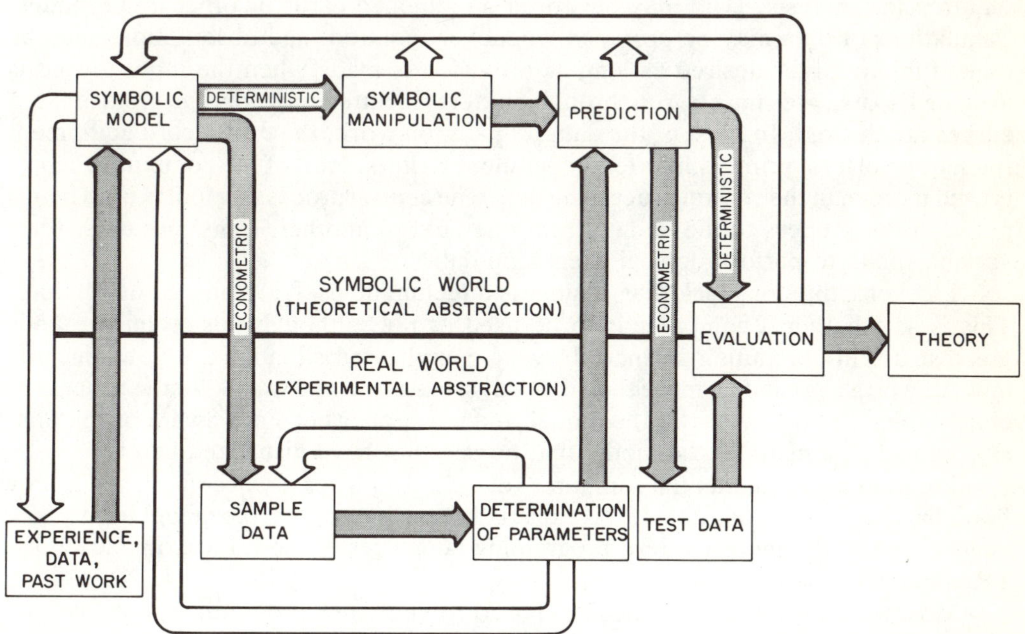

Figure 16–1. The model building process.

calculated on the basis of the derived parameters. Econometric models are designed to make systematic use of statistical data in assessing their adequacy.[8] That is, test data from the real world are used to test the adequacy of the econometric model. In both cases, evaluation does occur in the model building process. If the model is found to be an adequate representation of the system, then we may say we have formulated a theory. If not, then we must start over again. As noted in the diagram, we may find that we must backtrack at various junctures in the process (denoted by lighter arrows). For example, it may be found after parameters are derived for an econometric model that they do not make sense in terms of the subject matter. In such a case, it would be necessary to either reformulate the model or collect more data.

Although I have stressed the difference between deterministic and econometric models, the functions of these two types of models overlap and it is not particularly useful to attempt to define boundaries very sharply.

ROLE OF COMPUTERS IN RESEARCH

Data processing systems may assist the researcher at many points in the model building process and in subsequent use of the model. For example, data processing systems can assist the researcher with the enormous job of keeping track of past work in any field of inquiry. Automated information retrieval systems have been developed and are being improved constantly, e.g., the National Institutes of Health medical library system. Although there are many problems to be overcome in this field, e.g., classification difficulties, it is likely that all sciences will eventually be involved with this aspect of modern scientific investigation.

Also, the computer may assist in organizing and transforming data to a form required for analysis. Data may be stored on punched cards or other input modes. Tabulations of data may be provided whenever required, and observations may be ranked in any order desired as long as provision is made when the input format is designed. Data cards may be run through plotters if scatter diagrams and plots of fitted curves are desired. In most of the data preparation work, the unit record equipment (punches, sorters, printers, plotters, document writers, etc.) of the computing center is used more than the central processing unit. The central processing unit is used in this phase of the process to move data from one field to another, transform data, form variables, and to perform logical checks on data.

Flow charting, or block diagraming, is a technique useful to any model builder. This is so whether a computer is to be used or not, although this technique finds greatest use in computing science. In very complicated situations, a flow chart is, indeed, worth a great deal more than the proverbial thousand words. In the process of constructing the flow chart of the model, the researcher becomes aware of relationships among elements of the model that he may not have visualized before.

Programing the model for computer solution is most easily accomplished after a flow chart has been constructed. It is especially easy to program the model from a flow chart if one of the more modern programing languages is used, for example, FORTRAN or ALGOL.

Most scientists who have been exposed to computer programing have learned the FORTRAN language. Although FORTRAN is relatively simple as far as programing languages go, its use still demands extremely careful work. Even the smallest programing error can foul up the solutions, if the program runs at all. FORTRAN compilers can detect most of the common programing errors but there are many errors that can get by without detection, the most important being errors in logic. All programs must be checked by hand calculations with test data, and it is generally advisable to build into the program checks on the input data, especially consistency checks. When iterative techniques are used, it is important that both truncation and round-off errors be tested to insure desired precision.

Once the model has been programed, the researcher will wish to experiment with it so as to be fully aware of its limitations and deficiencies. A deck of punch cards or a reel of tape is a convenient way in which to store one's model, even if it is not a very complex one. Also, if the program is written in one of the problem-oriented programing languages, such as FORTRAN, its listing is in itself a convenient expression of the researcher's model. If the program is on punch cards, it may be easily modified whenever the model is and listings made whenever desired. Programing languages such as FORTRAN and ALGOL are used by people in computing science as a direct mode of communication, just as a mathematical statement of the model has long been used in other sciences. It is certainly a very exact way of conveying the model on which any study is based.

Statistical analysis of data, to provide for testing of variables in the model and estimation of the parameters of the model, may be accomplished most quickly and accurately with the help of a computer. Most of the standard statistical techniques (particularly the parametric methods) have been programed for digital computers. These computer programs have been checked out to varying degrees and are available to users in computing center libraries. Availability of computing facilities has made it possible for the researcher to process data from large samples in a reasonable amount of time. In addition, data links permit time-sharing of large computers by small-scale users some distance from the computing center.

The next step, prediction, may be most conveniently accomplished with the help of a computer. In fact, when very complicated models are used, the computer may be essential in this phase.

After predictions have been made, comparisons with test data from the real world are facilitated by the computer. In this way, the model may be tested for soundness by comparing predicted time series generated by the model with the observed behavior of the system. The computer program of the model may be easily modified to provide residuals (predicted outcome minus actual outcome). In fact, the residuals may be calculated and plotted by machine.

Once the researcher is satisfied with the model, his estimates of the parameters, and the computer program of the model, he might wish to use it to simulate the system it represents for a wide range of conditions. In this way, it is possible to experiment with the model and determine the implications of various assumptions made in model development. The effect on decision variables when one factor varies and all others are held constant is possible once the model has been developed and programed. Sensitivity of the results to changes in parameters may be ascertained. Inconsistencies in logic may show up at this stage of research if they are not discovered earlier. It may be found desirable to revise and refine the model to make it a closer representation of the real world system. As a matter of fact, the model may serve as a general framework for constant expansion and improvement. As mentioned before, symbolic and problem-oriented programing languages permit frequent and easy modification of model programs.

There are a number of advantages which flow from the use of a computer in model building. As mentioned above, if nothing else, the computerization of a model is likely to illuminate aspects of the model to the economist to a greater extent than before. For one thing, a researcher who uses the computer as an integral tool of research is unlikely to fail to be aware of simplifying assumptions he has made in model construction. Also, use of the computer enables the researcher to examine aspects of problems which could not otherwise be examined. For example, problems of such great complexity that solution is not possible except by iteration (e.g., certain optimization problems) could not be touched by scientists until computers were available.

Another advantage in using data processing equipment in research is that the forest economist can escape a large part of the dull and routine jobs and drudgery commonly associated with empirical research.

An advantage which should be emphasized is the increased accuracy in results which is possible if a computer is used in the analysis. This is an especially important feature if the model is very complex and includes numerous feedbacks. In such a case, values computed in one sector of the model enter as inputs to another sector.

Of course, there are a few disadvantages involved in using a computer. One is that in working with computers the researcher may become so interested in the programing and the equipment that he begins to spend most of his time in computation and not in forestry economics. The degree of involvement in computing science which is optimal can be debated and likely varies a great deal among individual researchers.

Also, there is a tendency to make snap decisions regarding the future steps of analysis when one is using the computer in a study. This comes about because calculations may be made rapidly when a computer is used. Some computer programs are designed to minimize this tendency.

The point I have tried to stress in this discussion is that the computer can be a fundamental tool of the forestry economist. Its role in the research process is best understood as an integral part of model building.

REFERENCES

1. Miller, D.W., and M. K. Starr. 1960. Executive decisions and operations research. Englewood Cliffs, N.J.: Prentice-Hall, Inc. 446 pages.
2. Coombs, C. H., H. Raiffa, and R. M. Thrall. 1954. Some views on mathematical models and measurement theory. *In* Decision processes. New York: John Wiley and Sons, Inc. Pages 19–37.
3. Kaplan, A. 1964. The conduct of inquiry, methodology for behavioral science. San Francisco: Chandler Publishing Co. 428 pages.
4. Bross, D. F. 1953. Design for decision. New York: Macmillan Co. 276 pages.
5. Bross. Work cited. Coombs et al. Work cited.
6. Bross. Work cited.
7. Beach, E. F. 1957. Economic models. New York: John Wiley and Sons, Inc. 227 pages.
8. Same reference.

Part Two

FORESTRY'S PRODUCTIVE AGENTS

17

THE FOREST OWNER— A PERSON

Robert F. Keniston

First published in *Journal of Forestry* 60(4):249–254. April 1962.

BIOGRAPHY

Mr. Keniston (*1910–1971*) was, the last 25 years of his life, a member of the forest management faculty at Oregon State University. There he taught, among other subjects, forestry economics and valuation. His appreciation for the land-owner was fostered in the Yale School of Forestry, where he wrote his doctoral dissertation in 1962 on the forest management decisions of small-tract owners. A Bachelor of Arts in economics (Nebraska) at the age of 18, he continued his academic career with a B.S. in forestry and an M.S. in forestry economics from the University of California. Work with the Forest Service in varied regions and with Counter-Intelligence in Japan during World War II preceded his appointment at Corvallis. His extracurricular affairs included advising the Oregon State Tax Commission and officiating in the Economics and Policy Division of the Society of American Foresters.

EDITORS' SUMMARY

Once oriented primarily to the forest, research on the ownership of small tracts has, during the past decade or so, given increasing attention to the owner. The usefulness of such research can be further increased by stressing the owner's personal characteristics and aims and the ways in which he wants to be assisted in his management.

QUESTIONS TO CONSIDER

(1) What do you conceive to be the purpose or purposes of the research advocated by Mr. Keniston? To be sure, the immediate purpose is to learn more about forest-resource owners; but *why* should we want to learn more about them? Deal with the question as a chain of ends and means à la Worrell (Item 11), naming the successive links up to the point at which the chain crosses the boundary of the forestry system.

(2) In the first question, if we trace the chain in the opposite direction—i.e., toward the means for doing the research and the means for those means—what do we learn about the chain's structure and its relation to the forestry system? For example, does the chain have branches? How many times does it cross the system boundary?

(3) Suppose that you wish to study the timber-production plans of a group of forest owners so as to learn how much timber they will be likely to harvest under various circumstances. That is, you wish to construct a supply function, a model of the owners' supply responses. What questions will you ask them?

(4) How do you explain the fact (page 133) that farm extension has been oriented primarily toward farmers' goals, while forestry extension has tended to work for social or "forestry" goals?

* * *

The basic reason why ownership studies to date have not been wholly satisfactory, and why public programs directed at improving the management of small private forests have not had great success, was summed up in 1950 by Stoddard:[1]

> It is said that our present efforts are insufficient, more foresters are needed in the field to contact more owners; more film, slides, and pamphlets are required, etc. No one ever asks why more of the same dosage is prescribed when the patient has not shown much visible improvement. Perhaps it is time to do a little research on the doctor, on the medicine, and on the patient.

It was the intent in this study to do "a little research on the doctor, on the medicine, and on the patient," or at least on the "patient." It was strongly suspected by many research foresters, before this study was undertaken, that what was needed was more attention to the forest owner and less, proportionately, to the property owned. This view was expressed by James[2] in 1953:

> . . . to learn about people, primarily the forest-land owners, who are responsible for timber management—to learn who they are, how they differ from or resemble each other in their timber management, and why they are following their present practices. This is the more neglected field for research.

Most of the forest ownership studies undertaken since 1948, except those related to the *Forest Survey* or the *Timber Resources Review,* have paid more attention to the owner than did previous studies. Even in *Timber Resources for America's Future,*[3] Josephson and McGuire discuss some of the characteristics and difficulties of the small-forest owner. So far, however, no forest ownership study has been primarily concerned with the owner as a person. It was necessary to go into the literature of agricultural extension and related studies in rural sociology and social psychology to find work emphasizing the human qualities of the landowner.

* * *

OBJECTIVES AND APPROACHES OF STUDIES

The general objective of most ownership studies during the period 1942–1961 has been to determine the current ownership of forest land, the management status of such land, and to relate the management status in some way to selected characteristics of the ownership unit or of the owner. This information was intended, in turn, to provide part of the basis for national or local resource reports or policies or programs. Some studies have sought to explain the observed variation in management status; but others have been content with describing the variation, often in very general terms. At the end of this section an alternative statement of objectives for future studies will be presented. First, however, we will consider the approaches and apparent objectives of some of the studies of the past 20 years. We will also consider psychological studies in agriculture, and articles on agriculture and forest extension.

Management-status studies—inventory type. This group of studies typically has broad objectives, is concerned more with ownership units, forest condition, and management status than with characteristics of the owner. In fact, the only attention given to the owner is usually to classify him first as to the size of his holding, and second, as to broad occupation class or purpose of ownership. The chief studies in this group are those by the United States Forest Service and by The American Forestry Association. The studies are often part of a national or local resource survey or land-use study. This group of studies likewise has the longest history of any category of forest ownership studies.

The procedure in this type of ownership study was concisely described by Horace J. Andrews[4] as ". . . take an inventory and get the facts. List the assets . . . , where they are and what they are worth." Private ownerships are usually divided into large, medium, and small. The large ownerships may be further classified as industrial and non-industrial. The industrial may be divided into forest industries and "other," with the former including pulp-and-paper companies and lumber companies as major subdivisions. The small forest ownerships are divided into farm and non-farm. In the latter category, absentee-owned units may be given special note. Management status and cutting practice are then related to these different owner-categories; and that is usually about as far as the ownership part of this type of study goes.

Special mention should be made of the California studies by Poli and Baker.[5,6] These studies were concerned with land use, and gave a little attention to the owner; but they were mainly concerned with description and classification. The 1957 California study by Dana and Krueger[7] is also a little different from the forest-survey or forest-resource type of study, in that it sought to locate and describe problems. It was, of course, exploratory in nature.

This inventory type of study serves useful and necessary purposes, such as compiling ownership statistics and giving a broad picture of the current status of management as related to general categories of ownership; but it does not contribute much information on the reasons for the variation in management as related to type of owner. Thus it is not a reliable basis for designing programs for inducing owners of "poorly-managed" woodlands to improve their practices. Programs based only on this type of study are likely to fall into the error illustrated by the very worthwhile objective stated in a Forest Service report (citation omitted to protect the writer's relations with his friends): "Each forest acre growing a commercial forest product." The implicit assumption is that every tract of woodland can be made to produce the wood crops it is capable of, as soon as foresters succeed in pointing out convincingly to the owner concerned the error of his ways; then since he presumably accepts the silvicultural ideal or thinks and feels as the "economic man" should, he will "see the light" and practice constructive timber forestry. The optimum allocation of all resources (including human) from an economic or social point of view is evidently disregarded. What is best for forestry is not necessarily best for the individual or for the nation.

Management-status and economic studies—with considerable emphasis on the owner. Since 1942, most of the forest-ownership studies made by state agencies or educational institutions have paid much more attention to the owner than has been done in the inventory-type study. It was not until 1957 that U.S. Forest Service studies became somewhat owner-oriented. There is no implication that studies giving major attention to the owner are superior to those studies giving chief attention to the owner's property; but it seems reasonable to expect that owner-oriented studies would tend to provide more useful information to guide forest conservation programs aimed at education of the forest owner.

Eighteen of the principal studies of this type, in approximately the order of their occurrence, are [those by]:

[Stoddard[8] (Lake States, 1942)
Folweiler[9] (Louisiana, 1943)
Chamberlin et al.[10] (Arkansas, Louisiana, Mississippi, 1945)
Barraclough[11] (New England, 1949)
James et al.[12] (Mississippi, 1951)
TVA[13] (Six states in Tennessee Valley, 1956)
Christensen[14] (New York, 1956)
Yoho[15] (Michigan, 1956)
Mignery[16] (Texas, 1956)
Laybourne[17] (Ohio, 1958)
McDermid et al.[18] (Louisiana, 1959)
Webster and Stoltenberg[19] (New York, 1959)
Sutherland and Tubbs[20] (Wisconsin, 1959)
Perry and Guttenberg[21] (Arkansas, 1959)
Worley[22] (Kentucky, 1960)
Miller and Southern [23] (Texas, 1960)
Anderson[24] (North Carolina, Georgia, 1960)
Martin[25] (Alabama, 1960)]

Although the U.S. Forest Service or the agricultural experiment stations cooperated in several of these studies made before 1957, only one study was conducted primarily by a federal agency (the Tennessee Valley Authority).

Five eastern forest experiment stations of the U.S. Forest Service made or completed during 1958 a series of small-forest ownership studies which were somewhat owner-oriented. The over-all results of these Forest Service studies were reported by T. A. McClay in the February 1961 Journal of Forestry.[26] According to McClay,

> The objectives of these studies were to determine the characteristics of owners of small forest properties; attitudes toward their forest lands and toward various public and private programs to improve forestry practices on private land; and reasons why these landowners adopted or did not adopt specific forestry practices.

Thus it would appear that program evaluation was at least as important an objective as gaining an understanding of the owner in these Forest Service studies.

These eighteen studies, with three or four possible exceptions, seemed more concerned with the welfare of society than with the welfare of the individual forest owner. For example, Barraclough[27] in his introduction states, "The management of these forest lands in such a way as to maximize the net returns to society from this region's [New England's] most extensive natural resource is a challenge to the economist and to the forester." It was notable that nearly every investigator by the time he had completed his work expressed more concern for the point of view of the individual owner than he did at the beginning. For example, Yoho[28] in his "suggestions for further research," says,

> The most need seems to call for a fundamental type of study concerning the motivations of private forest owners. Forest economists in the past have tried to explain owner actions in terms of the theory of the economics of the firm. This may explain the behavior of industrial owners, but with the present means the economist has of measuring the intangible values of the forest, economic theory is limited in explaining the actions of most forest owners. This is particularly true for non-farm owners. Even in the case of farmers whose actions can be explained fairly well in agricultural production by economic theory, forestry appears to be an exceptional enterprise.
>
> The author believes that a psychological study of owners' attitudes and behavior might make the contribution needed to fill this void.

Dr. John D. Black of Harvard, although not a forester, was quite well versed in forestry problems. Barraclough and Gould[29] relate the following:

> After a long discussion of various forest management programs that could be used in the Northeast, a forester asked Dr. Black, "Which would you advise an owner to use?" The answer was, "I never tell an owner what he should do; I like to help an owner lay out his alternatives and evaluate them, and then let him decide what he wants to do."

Perhaps foresters could benefit by this point of view. At any rate Barraclough and Gould used this approach in their budget-analysis work. They state that

> the nine farm analyses show that the owner's objectives and capabilities are often deciding factors that determine what kind of forest management is most desirable. An interest in forest production and a relatively high value placed on future as well as immediate profits are essential if high-intensity management is to appear attractive. In addition to profit from production, building up the value of the farm is often a major objective. Intangible values may also influence forest management.

Yoho[30] emphasizes the fact that ownership of forest land may be for purposes not limited to production of wood or other commercial products. He states,

> . . . forest land as a producer's good has been superseded by forest land as a consumer's good, or that forest land has become a producer's good held for its yield of certain intangibles. . . . Present private forest owners may be maximizing their satisfaction of owning forest land by the production of values other than those expected from the sale of forest products.

Barraclough[31] also emphasizes the importance of the satisfaction of ownership as a reason for land ownership.

Studies in this category typically made objective ratings of management practices or cutting practices, then related these ratings to such owner characteristics as the following (based on the work of James[32] and Yoho[33]): occupation, age, control of timber cutting, attitude toward timber management, concept of timber management, recognition that the management could be improved, explanation of poor timber management, objective of management, reason for ownership of forest land, length of tenure, manner of acquisition, taxes, distance of owner from forest, date of most recent cutting.

Various investigators found that some or all of these characteristics were of value in explaining variations in quality of forest management.

James[34] found that the power of example was important in explaining variations in management.

The TVA[35] found that certain personal characteristics of the owner were significant in explaining variation in quality of management. The successful forest manager apparently is the type of person who is interested in civic affairs. There was a strong correlation between management success and participation in community activities. The better managers customarily sought and used technical assistance in all their affairs—not only in forestry matters. Successful business managers were also good woodland managers.

An economic study in Sweden by Dr. Streyffert, rector of the Royal College of Forestry,[36] provides a good economic analysis which helps explain variation in quality of forest management as related to form of ownership (industrial, farm, etc.) and size of the ownership unit.

In a 1958 paper, Worrell[37] did a realistic job of interpreting some of the motivations and limitations of the typical small forest owner, with particular reference to the South. He emphasized the importance of length (or shortness) of the owner's "planning horizon" as affecting his discounting of the future. He showed how the present condition of the woodland may influence the opportunities open to an individual owner. He suggested that

> the idea of growing a short-rotation crop like pulpwood may motivate [the owner] to manage his woods. . . . Since his planning horizon will continue to extend for about the same length of time into the future, he progressively will be able to visualize benefits from growing larger and larger timber.

This interest in learning about and considering the small owner's point of view was also evidenced by Christensen,[38] who wrote,

> . . . more research at the "grass roots" level is needed. . . . It would be advantageous to investigate the objectives, attitudes, and interests of the owners, the results of which help to define the problems or obstacles confronting the private owners. A knowledge of this, in turn, provides a sounder base for policy development and program formulation.

In his thesis[39] Christensen stated:

> Public education in forest conservation, to be successful, must either be designed around the aims of the individuals at whom it is directed or be prepared to change these aims . . . In either case, people's aims or objectives need to be understood.

The point of view of "nonmanagers"—small-woodland owners who do not follow recommended practices—was studied by Martin,[40] in a departure from the usual small-forest ownership study.

The economic situation of the forest owner is receiving increasing attention as a clue to his willingness or unwillingness (ability or inability) to invest in intensive forest management. Many investigators have found that the larger the acreage of forest land owned, the greater the tendency to adopt recommended forest practices. Webster and Stoltenberg[41] found that the assessed value of the owner's property (within one taxing area) was also positively correlated with adoption of forest practices.

As to the effects of low income on forest conservation, Duerr[42] in 1948 wrote the following:

> . . . studies made by the writer during the past decade in the eastern United States lead him to the conclusion that low income is the heart of the problem on at least a fourth of the small-owner forest land, and is intermittently a key factor on an even greater portion of our forests. . . . In any case, depression deepens and spreads the two primary problems: low income and depleted timber. . . . The small, low-income property is not solely, or even largely, a forest problem. In most cases it is a problem also in sociology, farming, industry, or indeed, a problem of the whole regional economy. . . . So broad a problem calls for research on a broad front—primarily research along the lines of economics and sociology, and including forestry only as one of several fields for consideration.

As to economic circumstances, McMahon[43] states:

> Empirical evidence shows a direct relation between asset level and an interest in and capacity for forest management. . . . Lack of capital, low income, and consequent inability to accumulate capital have been identified as obstacles to intensification of management. . . . When income is insufficient to meet immediate needs, then disinvesting or capital consumption may result. . . .

McMahon, under the sponsorship of the U.S. Forest Service and the University of California, has completed a study designed to uncover the economic factors that determine the manner in which "small owners" manage their properties. Another study of small-forest motivation now nearing completion is being conducted by R. F. Keniston for Oregon State University's Agricultural Experiment Station, with sponsorship by the Yale University School of Forestry. This latter study attempts to determine how various situational, economic, and personal factors influence forest-management decisions.

Yoho,[44] in a paper stressing sound economics and straight thinking in defining the small private forest ownership problem, warns foresters against what Mason Gaffney (a general economist) describes as "sheer sylvan fundamentalism" by quoting Gaffney and Ralph W. Marquis. He quotes Gaffney as follows:

> To some persons, "good forestry" is a primary value transcending cost considerations, costs being based on use of resources in nonforest alternatives which the [sylvan] fundamentalist discounts heavily. When interest cost thwarts otherwise feasible forestry enterprises, it is not forestry that must yield, but interest.

Yoho quotes Marquis as follows:

> It cannot be held wasteful in an economic sense to leave land idle if the costs of cultivation could not be equaled by the value of the tangible and intangible products resulting from cultivation. It is economically wasteful to devote capital, labor, and land to a submarginal enterprise.

Nonforestry studies—dealing with the landowner as a person. To date, apparently no one has undertaken the primarily psychological type of study advocated by most of the foresters cited in the previous section. Such studies are not lacking, however, in the field of agriculture. Much good information of this type has been obtained by various extension specialists, rural sociologists, and social psychologists. A few of these studies will be mentioned here, because the findings applying to farmers as agriculturists can be expected to apply also to farmers as woodland owners.

Probably the best summary on the psychology of introducing new farm practices was written by Beal and Bohlen[45] in 1957. They list the five stages of the adoption process for a new recommended practice as (1) awareness, (2) interest, (3) evaluation, (4) trial, and (5) adoption.

Most investigators found that education, income, size of farm, and social participation were positively associated with the adoption of improved farm practices. These investigators included Bonser, working in Tennessee,[46] Copp in Kansas,[47] Marsh and Coleman in Kentucky,[48] and Wilkening in North Carolina.

Wilkening, as reported by Soth,[49] found "that the tendency to accept new farm methods was associated with willingness to accept new ideas in religion, education, and recreation. In other words, receptivity to change is a general trait."

Group influences were found to be important in inducing or inhibiting change, according to Marsh and Coleman[50] and others.

Penders[51] has an excellent compilation on extension psychology and methods.

The futility of trying to change convictions by sheer logic is pointed out by Lewin and Grabbe:[52] "Arguments proceeding logically from one point to another may drive the individual into a corner. But as a rule he will find some way—if necessary a very illogical way—to retain his beliefs."

Restatement of objectives of forest ownership studies. We are now ready to present the alternative statement of objectives for future studies referred to previously. On December 20, 1957, Carl Stoltenberg stated:

> For purposes of our research we will define good and poor forest management as follows: Good management exists on land in which there are no opportunities for profitable investment of additional funds (or time) in forest management measures. Poor forest management is relative; it depends upon the number of profitable investment opportunities that are not being taken advantage of. The objective of public forest management programs would be to remove obstacles which prevent the private owner from taking advantage of any profitable opportunities. . . . Ownership research is designed to identify obstacles which exist and to aid in specifying ways in which they may be removed.

An article by Stoltenberg[53] develops this investment-opportunity approach to management.

CONCLUSIONS AND RECOMMENDATIONS

* * *

The principal recommendation is that future forest ownership studies should seek to understand the owner, especially his personal characteristics, background, objectives, interests, possibilities, and limitations. This implies relatively less emphasis on the physical characteristics and condition of the forest property, except in those broad studies where determination of over-all condition of and compilation of statistics on the forest resource are the objectives.

The second recommendation is that the primary concern of future studies should be with advancing the welfare of the individual owner, as a means of maximizing the returns to society from the human and natural resources. This means, also, seeking to help the individual in the way that he wants to be helped. It means allowing him to set the goals for the management and use of his property as long as the general welfare is not actively and immediately threatened thereby. This change of emphasis would mean adopting a goal more like that of the Agricultural Extension Service—helping the individual to help himself, not crusading for "conservation" for conservation's sake. Education can cause the individual's goals to rise. This different psychological approach is analogous to the psychology of selling which involves at least three steps: (1) Discover the wants and needs of the buyer. (2) Demonstrate the ability of the product to satisfy those wants. (3) Bring about action (help the buyer to sell himself).

Insofar as foresters want to be salesmen, they should use sales psychology and devote their research to learning more about the "buyer," and not try to sell him a product which will not, in his opinion, fill his wants and needs. Products can sometimes be adapted to better fit the needs.

Specifically, it is recommended that future ownership studies should seek to obtain information on the following items in addition to the traditional ones: (1) management and land-use objectives of the forest owner; (2) personal objectives and outlook of the owner as they affect his "planning horizon"; (3) stage of the forest-practice adoption process[54] where the owner appears to be; (4) what improved forest practices appear to be feasible and advantageous for the owner; (5) personal characteristics and circumstances of the owner that tend to influence his adoption of new ideas (education, income, social participation, assets and liabilities); (6) specific sources, personal and other, of new ideas for the individual owner.

Interviews with county extension agents, county assessors, bankers, and longtime residents may help in obtaining information about items 5 and 6.

It is believed that shifting the emphasis of future studies to the owner as a person in the ways suggested will produce more useful information.

REFERENCES

1. Stoddard, Charles H., Jr. 1950. Needed: a research program in forest owner education. Jour. of Forestry 48:339–341.
2. James, Lee M. 1953. Forest-land ownership and its relation to forest management. *In* Research in the economics of forestry. Washington, D.C.: Charles Pack Forestry Foundation.
3. U.S. Forest Service. 1958. Timber resources for America's future. Forest Resource Rep. No. 14.
4. Andrews, Horace J. 1924. The Michigan land economic survey. Ames Forester 12:36–42.
5. Poli, Adon, and H. L. Baker. 1953. Ownership and use of forest land in the coast range pine subregion of California. U.S. Forest Service, C. F. & R.E.S., Berkeley, California. Tech. Paper No. 2. 64 pages.
6. ———, and ———. 1954. Ownership and use of forest land in the redwood-Douglas-fir subregion of California. U.S. Forest Service, C.F. & R.E.S., Berkeley, California. Tech. Paper No. 7. 76 pages.
7. Dana, Samuel T., and Myron Krueger. 1957. California lands. Report for the American Forestry Association.
8. Stoddard, Charles H., Jr. 1942. Future of private forest land ownership in the northern Lake States. Jour. of Land and Public Utility Economics 18(3):276–283.
9. Folweiler, A. D. 1943. Ownership of forest land in selected parishes in Louisiana and its effect on forest conservation. Unpublished Ph.D. thesis, Univ. of Wisconsin. 307 pages.
10. Chamberlin, H. H., L. A. Sample, and R. W. Hayes. 1945. Private forest land ownership and mangement in the loblolly-shortleaf type in southern Arkansas, northern Louisiana, and central Mississippi. Louisiana Agric. Exp. Sta., Baton Rouge. Bul. 393, 46 pages.

11. Barraclough, Solon L. 1949. Forest land ownership in New England. Mimeo. Ph.D. thesis, Harvard Univ. 269 pages.
12. James, Lee M., W. P. Hoffman, and M. A. Payne. 1951. Private forestland ownership and management in central Mississippi. Mississippi State College, Agric. Exp. Sta. Tech. Bul. No. 33. 38 pages.
13. Tennessee Valley Authority. 1956. Influence of woodland and owner characteristics on forest management. Rep. No. 217–56. 38 pages.
14. Christensen, Wallace W. 1957. A methodology for investigating forest owners' management objectives. Ph.D. thesis, State Univ. of New York, College of Forestry. 184 pages.
15. Yoho, James G. 1956. Private forest land ownership and management in 31 counties of the northern portion of the lower peninsula of Michigan. Ph.D. thesis, Michigan State Univ.
16. Mignery, A. L. 1956. Factors affecting small woodland management in Nacogdoches County, Texas. Jour. of Forestry 54:102–105.
17. Laybourne, William. 1958. Survey of timberland owners. Ohio Forestry Association Inc., Columbus, Ohio. (Mimeo.)
18. McDermid, Robert W., Paul D. Kitt, and Sam Guttenberg. 1959. Ownership factors affecting management of small woodlands in St. Helena parish, Louisiana. Louisiana Agric. Exp. Sta., in cooperation with U.S. Forest Service, Southern Forest Exp. Sta. Louisiana Agric. Exp. Sta. Bul. No. 520.
19. Webster, Henry H., and Carl H. Stoltenberg. What ownership characteristics are useful in predicting response to forestry programs? Land Economics 35(3):292–295.
20. Sutherland, Charles F., Jr., and Carl H. Tubbs. 1959. Influence of ownership on forestry in small woodlands. U.S. Forest Service. Lake States Forest Exp. Sta. Paper No. 77. 21 pages. (Processed.)
21. Perry, Joe D., and Sam Guttenberg. 1959. Southwest Arkansas' small woodland owners. U.S. Forest Service Southern Forest Exp. Sta. New Orleans, Louisiana. Occasional Paper 170.
22. Worley, David P. 1960. The small woodland owner in eastern Kentucky—his attitudes and environment. U.S. Forest Service Central States Forest Exp. Sta., Columbus, Ohio. Tech. Paper 175.
23. Miller, Robert L., and John H. Southern. 1960. Management intent of small timberland owners in east Texas. MP-439. Texas Agric. Exp. Sta. College Station, Texas.
24. Anderson, Walter C. 1960. The small forest landowner and his woodland. U.S. Forest Service Southeastern Forest Exp. Sta., Asheville, N.C. Station Paper 114. 15 pages (Processed.)
25. Martin, Ivan R. 1960. Some characteristics of non-managers of small woodlands. Thesis, Louisiana State Univ.
26. McClay, T. A. 1961. Similarities among owners of small private forest properties in nine eastern localities. Jour. of Forestry 59:88–92.
27. Barraclough, Solon L. Work cited.
28. Yoho, James G. 1956 work cited, page 294.
29. Barraclough, Solon L., and Ernest M. Gould, Jr. 1955. Economic analysis of farm forest operating units. Harvard Forest Bul. No. 26. Petersham, Massachusetts.
30. Yoho, James G. Same reference.
31. Barraclough, Solon L. Work cited.
32. James, Lee M., et al. 1951 work cited.
33. Yoho, James G. 1956 work cited.
34. James, Lee M., et al. 1951 work cited.
35. Tennessee Valley Authority. Work cited.
36. Streyffert, Thorsten. 1957. Influence of ownership and size structure on forest management in Sweden. (In English.) A study of fundamentals. No. 23b Bul. of the Royal School of Forestry, Stockholm, Sweden.
37. Worrell, Albert C. 1958. Short rotation forestry and the small forest owner. (Mimeo.) Paper presented before the annual meeting of the Association of Southern Agric. Workers, Little Rock, Arkansas.
38. Christensen, Wallace W. 1956. Objectives of plantation owners. New York Forester 13(1), 6–9.
39. Christensen, Wallace W. 1957 work cited.
40. Martin, Ivan R. Work cited.
41. Webster, Henry H., and Carl H. Stoltenberg. Work cited.
42. Duerr, William A. 1948. The small, low-income landholding: a problem in forest conservation. Iowa State College Jour. of Science 22:349–361.
43. McMahon, Robert O. 1961. The economic determinant of management intensity of private nonindustrial forest lands. Paper presented at April meeting of Portland, Oregon, Chapter, Society of American Foresters.
44. Yoho, James G. 1959. Have we a valid basis for defining the small private forest ownership problem? (Mimeo.) School of Forestry, Duke Univ.
45. Beal, George M., and Joe M. Bohlen. 1957. The diffusion process. Special Rep. No. 18. Agric. Extension Service, Iowa State Univ., Ames Iowa. 5 pages.
46. Bonser, Howard J. 1957. Rural community clubs in Tennessee—their bearing on adoption of farm and home practices. Dissertation Abstracts. Pages 1139–1140.

47. Copp, James H. 1956. Personal and social factors associated with the adoption of recommended farm practices among cattlemen. Kansas Agric. Exp. Sta., Manhattan. Tech. Bul. 83. 31 pages.
48. Marsh, C. Paul, and A. Lee Coleman. 1955. The relation of farmer characteristics to the adoption of recommended farm practices. Rural Sociology 20:289–296.
49. Soth, Lauren. 1952. How farm people learn new methods. An Agricultural Committee Report. Washington, D.C.: National Planning Association. Planning Pamphlets No. 79. 23 pages.
50. Marsh, C. Paul, and A. Lee Coleman. 1956. Group influences and agricultural innovations: some tentative findings and hypotheses. American Jour. of Sociology 61:588–594.
51. Penders, J. M. A., et al. 1956. Methods and program planning in rural extension (In English). Wageningen, Veenman. 324 pages. Ref. 275.1, page 37. (Issued by) Wageningen, Holland: International Agricultural Study Centre.
52. Lewin and Grabbe. 1945. Conduct, knowledge, and acceptance of new values. Jour. of Social Issues 1:3:53–64.
53. Stoltenberg, Carl H. 1959. An investment-opportunity approach to forestry programming. Jour. of Forestry 57:547–550.
54. Beal, George M., and Joe M. Bohlen. Work cited. Also, Iowa State Univ. Extension Service. 1955. How farm people accept new ideas. Iowa State Univ. Agric. Extension Service Rep. 15. 12 pages. North Central Regional Publication No. 1, Agric. Extension Services.

18

A THEORETICAL MODEL FOR EXAMINING FOREST LAND-USE ALTERNATIVES

Perry R. Hagenstein
Barney Dowdle

First published in *Journal of Forestry* 60(3):187–191. March 1962.

BIOGRAPHY

Mr. Hagenstein is Executive Director of the New England Natural Resources Center, Boston. Schooled in forestry and in economics, he went on to a varied career: as forester for a lumber company, as forestry economist in the Northeastern Forest Experiment Station, as senior policy analyst for the Public Land Law Review Commission, and as Bullard Research Fellow in Harvard University. He is a past chairman of the Division of Economics and Policy in the Society of American Foresters. His special interests include natural resources economics and public land policy.

Mr. Dowdle is Professor of Economics and Forestry, an interdepartmental appointment in the University of Washington. He gained his interest in both forestry and economics early in life, working as an independent logger in the Pacific Northwest. Higher education followed, culminating in a Ph.D. program at Yale University. A 3-year stint with the Forest Service at the Northeastern Forest Experiment Station preceded his joining the faculty at Seattle. His current interests are the economics of public forestry.

EDITORS' SUMMARY

One of the important costs of reserving forest lands for uses other than timber production is the opportunity cost of the timber production forgone. This cost may be analyzed by means of an isoquant model in which land and capital are the two inputs. Such analysis helps us to understand the implications of our legislative and administrative forest-use decisions.

QUESTIONS TO CONSIDER

(1) In Figure 18–2, and on page 140, why does the production-possibilities line (expansion path) coincide with the *Y* axis? And why can the iso-cost lines not be steeper than those described; indeed, with land assumed to be homogeneous, would it not be valueless when not all used, and the left-most iso-cost lines thus vertical?

(2) What forces do you picture as being responsible for the adjustments described on page 140 and following? What periods of time do you see as being required for the adjustments? What supply condition (long-run, short-run, etc.) are the authors referring to when they mention changes in supply?

(3) Noting that point *R* in Figure 18–3 is labeled *T* in Figure 18–4 and that the uppermost curve in 18–4 is the curve in 18–3 without the segment *TA*, redraft the two figures to make them easier to follow.

(4) Why is the opportunity cost of the forest land allocated to alternative use equal only to the timber value forgone by not using *CD* capital (page 142)? Why is the value of capital *used (BC)* not included?

(5) What is the central point of the argument in this article? It is introduced in the fourth paragraph on page 141. Restate it in your own words.

* * *

For the country as a whole, forest land is limited in extent. A firm (or an individual), when it needs to expand its output, has the alternative of extending its ownership of forest land. It can probably buy forest land or land suitable for planting—at some price. But the country as a whole can expand its acreage of forest land only by changing land from other uses to forest land, presumably back to forest land. While the shift back to forest land may be made immediately, the production of forest values from the new forest may involve a lag of many years.

As pointed out above, there are a number of possible uses of forest land *qua* forest land that do not involve removal of the timber. Use of forest land for recreation, wildlife, watershed, and wilderness falls within an area in which foresters have special interest and some knowledge. For the moment, we will lump all these alternatives together and examine them as opposed to other uses of forest land.

Among the spectrum of land values based on different uses, the value of land for timber production is usually near the bottom of the list. This has been the case since colonial times. The colonists considered forests an impediment. They cleared forest land as soon as possible to make way for agriculture. This is still true today: forests are not allowed to stand in the way of suburban housing, highways, industrial developments, or agriculture when it is felt that the potentialities of these alternative land uses are even remotely promising. One can observe that urban sprawl is no less a problem in areas where stumpage is relatively valuable than in areas where it is relatively cheap. One can also observe the magnitude of the difference between the values of forest land and agricultural land, which is probably only one step above forest land.

Within some limits, and disregarding for a moment recreation, wildlife, and the like, we can consider forest land as having no alternative use; that is, the value of forest land for other uses is zero. Forest land in the United States is something of a residual. We produce our timber on lands that have been disqualified as agricultural, industrial, or residential lands. The bulk of our forest land today is in the mountains of the West, the swamps of the North, and along the sandy ridges and overflow bottoms of the South. This land generally has little value for other uses. On the other hand, the best land for the production of timber, and for the production of some of the other values derived from forest land, may have been preempted long ago for other uses.

Before moving on to a description of the model, we should make one final point concerning forest land. In the production of timber, we would expect to find at some point that there are decreasing returns to investments in forest management. If land area were held constant, each succeeding unit of input spent on the land would result in a smaller return than the preceding unit of input. It is possible, when forest lands are first put under management, that increasing returns will accrue to the very first management inputs. On much of the forest land in the United States the level of management probably has already passed this point. Since fire protection and other first measures cover virtually the entire country, we may reasonably assume that most of the remainder will soon reach this point.

A THEORETICAL MODEL

The model that we propose for use here is the traditional two-input, isoquant model of economic theory. This model can be best described with the use of a graph (Figure 18–1). The axes of the graph represent inputs. Output is represented by isoquants showing the various combinations of the two inputs that will give the same output. Visualize the isoquants as representing contours of equal production, with

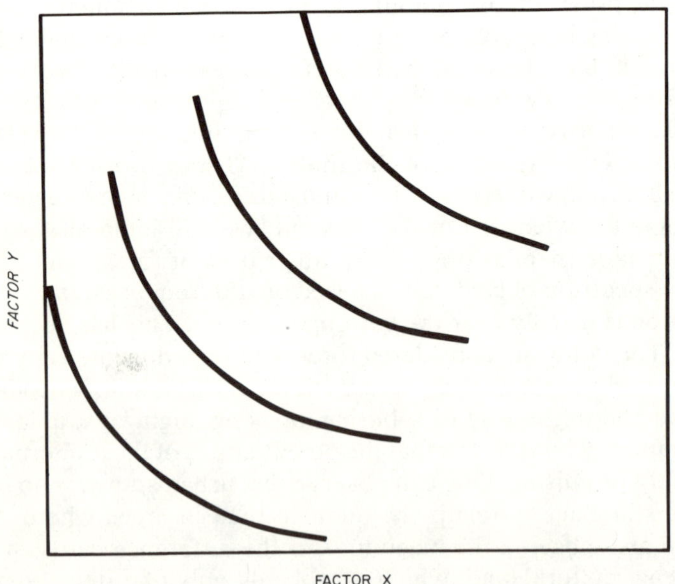

Figure 18–1. The two-input isoquant model.

the level of production increasing with distance from the origin, or across successive contours.

The two inputs that we will consider for forestry in the United States are land and capital. It has already been pointed out that the total area of forest land in the country—the first input—is fixed. We will also assume for the moment that forest land is homogeneous; that is, the production from 2 acres of forest land is exactly double the production from 1 acre if other inputs are held constant. The second input—capital—will be used to describe all forest management activities. Each successive addition to the capital that is applied to forest land can be thought of as an intensification of forest management for the production of timber. These management activities are brought down to a common denominator—money.

The production isoquants in Figure 18–2 represent timber output. The limited total quantity of land is denoted by the top line of the graph.

Several interesting points are to be noted on this graph. First, observe that one can get some timber production with no capital. Furthermore, some minimum land area is required to attain any given output of timber regardless of the extent to which capital is used on this land. This minimum area is determined by the biotic potential of the land. As the biotic potential is approached, the amount of capital needed for a given increase in output becomes very large.

The figures on the isoquants—10, 20, 30, 40, 50—represent the levels of output that can be obtained with the various combinations of land and capital. For example, note that 20 units of output can be obtained either by applying very little capital to a large quantity of land, or by applying a relatively large quantity of capital to a relatively small quantity of land. In other words, land can be substituted for capital. This is true only up to the point at which the biotic potential of the land is reached. After this, the addition of capital will achieve nothing at best, and may possibly even lead to decreasing total returns.

Since the total area of forest land in the country is fixed, it can be seen that capital must be applied to the land to achieve a level of output greater than 20 units.

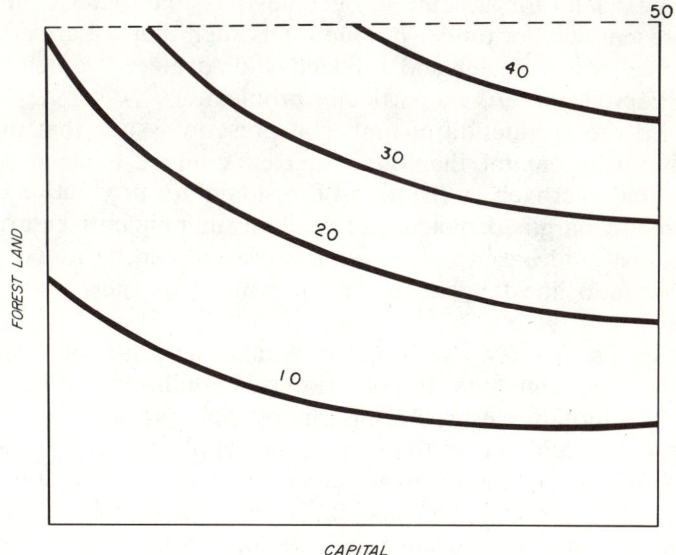

Figure 18–2. Two-input model for timber production.

Any level below this can be, in fact will be, attained without the use of capital. Also, levels of output greater than 50 units would require the use of all, or nearly all, of the forest land.

There is one final point to be made concerning this model before we move on to examine some of its implications. We will always use land instead of capital wherever possible up to the limit on forest land area. The production-possibilities line for Figure 18–2, that is, the line showing optimum combinations of land and capital in the production of timber, will be coincident with the *Y* axis of the graph up to the limit of land area and will then move to the right along the top line of the graph, which represents the total area of forest land.

The capital/land factor/cost lines, which show the value of land relative to capital, will be tangent to the production isoquants at the point of production (along the production-possibilities line). As the amount of capital used increases relative to land, the factor-cost lines become flatter, indicating a greater relative value for forest land *qua* forest land. Thus, the opportunity cost of reserving forest land for other uses becomes greater as more capital is applied to the land.

Notice from this model that any given level of production will in general be obtained most efficiently by using all the productive forest land available and the smallest quantity of capital necessary to reach the desired level of production. This, then, constitutes a definition of efficient forestry for timber production. If the country's needs for timber can be met with small amounts of capital, this is efficient forestry.

DISCUSSION

There are several conclusions that can be reached by using this model. Let us suppose that there exists some alternative use for forest land—for example, agriculture—for which prices and costs are determined in a competitive market. Let us further assume that the demand for agricultural products shifts so that the use of some of the forest land for agriculture becomes possible under competitive conditions. If the present level of timber production is such that the area of forest land is sufficient to meet the new demand for land and to meet the demand for timber without using capital, there is no particular problem.

However, if the production of timber at present is such that there can be no increase without using capital, then there is a clear conflict. In a competitive market, the new use would encroach on the use of forest land for production of timber up to the point where the marginal efficiency of the investment in both enterprises is equal. A value for forest land based on the alternative use can be found graphically by drawing a price-ratio line tangent to the isoquant at the new point of production (Figure 18–3).

After the shift to the new use, which now takes a portion of what was formerly forest land, has been completed, the situation is in equilibrium and forest land again has no value for alternative uses. Because more capital is being used in the production of timber, one would expect to find that the supply curve for timber has shifted upward, resulting in a higher price and a smaller quantity being used (if demand is neither wholly elastic nor wholly inelastic). Because some of the forest land has been shifted into the new use and because the level of production is closer to the biotic potential of the land, the extent to which total timber production can be expanded is lower than previously.

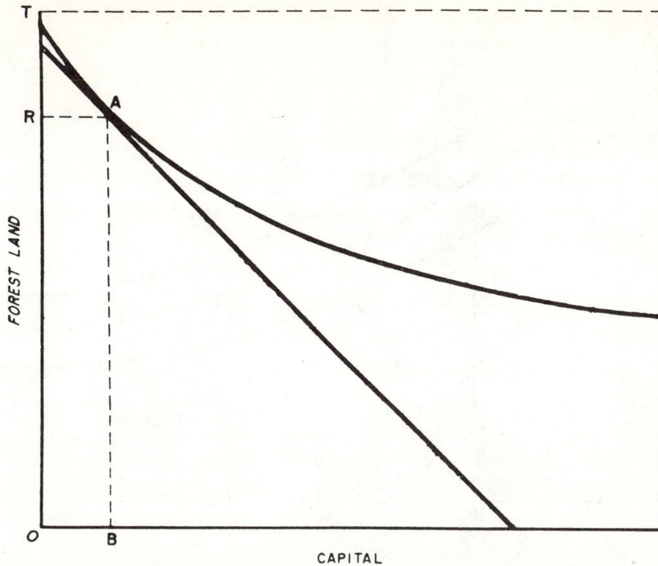

Figure 18–3. Allocation of forest land to two uses.

Now see what would happen if forest land were used for recreation, wilderness, wildlife, or watershed protection, all of which are valuable, but none of which normally has a market value in the usual sense. Assume for the moment that these uses are incompatible with timber production. Even if they are not incompatible, but as long as they compete with timber production, no violence is done to the model, nor are the essential conclusions changed.

The allocation of forest land to these uses is not made in the market place. Instead, it is usually made through legislation or by administrative fiat. As a result, any resemblance to the allocation that would have been made if these uses were allocated in the market place is coincidental. It would indicate that legislation and administrative action are more responsive to economic pressures than it is commonly thought they are.

Suppose allocation of forest land to one of the above uses has been made through legislation and that Figure 18–3 represents the equilibrium position after the act. As in the case of a market allocation, the reservation of forest land for other uses shifts the supply curve for timber, increases the price and decreases the quantity of timber used, and limits the extent to which timber production could possibly be expanded. The price-ratio line in this case can be used to determine graphically the value that has been placed on the alternative land use by legislative action. It is not the same as the market value, except in the case noted above where legislation is unusually responsive to economic pressures.

From the social viewpoint, it might be argued that the cost of forest land for an alternative use is the cost of producing sufficient timber so that the equilibrium position with respect to timber production remains on the same production isoquant. However, we can show that the additional cost is made up of two parts. Figure 18–4 illustrates this.

Suppose the level of timber production is at *A* on the production isoquant, representing 20 units. Assume then that by administrative fiat some forest land is reserved for an alternative use. This is done to placate some body of enthusiasts,

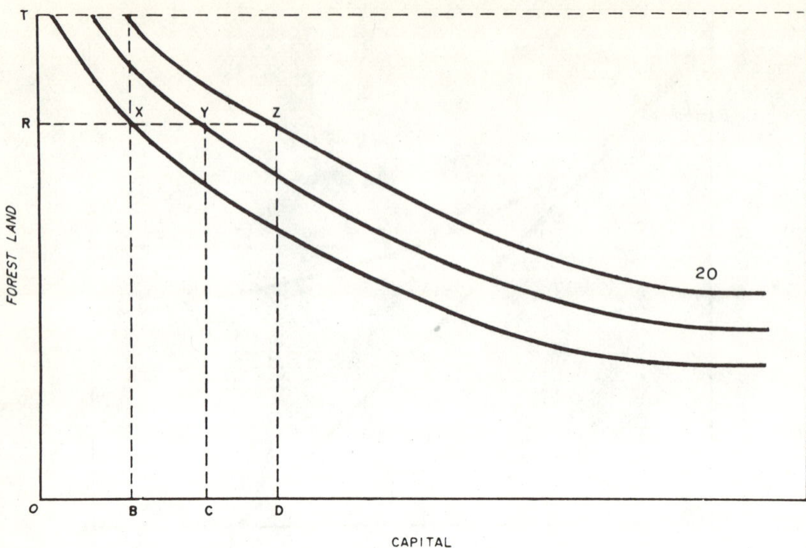

Figure 18–4. The costs of reserving forest land.

either large or small. As a result, suppose the forest land for timber production is reduced from *OT* acres to *OR* acres. After the timber land has been reserved, more capital will be required to increase production over that possible by using the original *OB* units of capital. Note that using *OB* units of capital would result in production at point *X* on an isoquant below the original isoquant. To stay on production isoquant 20 would require an additional *BD* units of capital.

We can argue that, in a free market, forest land would be allocated to the alternative use at the margin, resulting in a solution at some point *Y*, requiring an additional *BC* units of capital. Assuming that the demand for timber is neither wholly inelastic nor wholly elastic, one would expect that the price of timber would rise and the quantity used would fall. In so far as economists (and others) commonly suggest that a free market provides an optimum solution to problems of allocation, this solution is optimum.

However, we have already pointed out that these decisions are not made in a free market. If in equilibrium the level of production is at the optimum point *Y*, it is, as pointed out above, purely a matter of chance. You can see from Figure 18–4 that additional *BC* units of capital would be required to increase production to point *Y*. It would be highly unlikely to find that decisions made through the ballot box or through legislation would specify the use of exactly this quantity of capital in timber production.

The second part of the additional costs required to maintain the original level of timber production is *CD* in Figure 18–4. Clearly, from an economist's standpoint the allocation of this capital to timber production would constitute an inefficient allocation of capital in a free market. The value of the timber production forgone by not using the additional *CD* units of capital is the opportunity cost of the allocation of forest land to the production of nonmarket-determined values.

It may be impossible in practice to identify this opportunity cost. However, the fact that it exists should be recognized and evaluated carefully in allocating forest lands through legislation or administrative fiat. So, from an economist's standpoint again, the benefits obtained from the alternative use should exceed the opportunity cost.

We are not making a case here for free market operation in allocating forest land to these various uses. There is nothing sacrosanct about the free market as an allocating mechanism. In fact, the market place is notably inefficient in allocating resources in cases where benefits cannot be precisely defined or priced. However, in so far as many consider it to be important, the free market allocation should be considered regardless of the actual criteria used in making the allocation.

The fact that forest land is not homogeneous, as assumed in the model, does not weaken the results. Differences in productivity can be readily integrated into the model. Without diagramming the results, one can see that the marginal efficiency of forest management (investment) on the various productivity classes of land should be equated to reach any given total level of timber production most efficiently.

It should also be apparent that alternative uses, the only real cost of which is the opportunity cost of timber production that is lost or the increased cost of maintaining timber production at some given level, should be directed toward the forest land of lowest productivity. Such action, for uses that are not allocated in the market place, will minimize the cost of reserving any given number of acres of forest land for uses other than timber production.

* * *

19

AN ESTIMATE OF CAPITAL NEEDED IN FORESTRY TO MEET PROJECTED TIMBER REQUIREMENTS FOR THE YEAR 2000

Charles H. Stoddard

First published in *Journal of Forestry* 56(7):484–488. July 1958.

BIOGRAPHY

Mr. Stoddard is a consultant in natural resources. He was educated in forestry and economics at the Universities of Michigan and Wisconsin, and at the former held a Charles Lathrop Pack Research Fellowship. He began his professional career in the Forest Service and continued at Resources for the Future, Inc., where he made the analysis reprinted here. Moving to the Interior Department, he rose to the Directorship of the Bureau of Land Management. His varied and imaginative writings include books and bulletins on forest farming, small holdings, forest credit, and land use, topics which he addresses with special conviction by virtue of 40 years' management experience on his Wolf Springs Forest in northern Wisconsin. During a World War II Navy assignment in the Solomon Islands, he discovered a new timber tree species, which botanists then named *Mastixiodendron stoddardii.*

EDITORS' SUMMARY

In this early, illustrative analysis, national total costs and benefits are estimated for a program to raise timber growth to the requirement forecast for the year 2000. Necessary additional investments in forest capital and additional current expenses for protection are expected to return 5 to 6 percent annually in timber value alone, with no allowance for collateral forest benefits.

QUESTIONS TO CONSIDER

(1) If you were to undertake a thoroughgoing analysis of forestry investment in the United States as a basis for social planning, what procedure would you follow? That is to say, what models would you use in order to estimate the consequences of alternative courses of social action respecting forest resources? You may wish to think through only a fraction of this question because it is so large, but do not fail to consider how you would provide for putting forestry in context—i.e., estimating its effects on other segments of the national life, and vice versa.

(2) Mr. Stoddard emphasizes the scantiness of data available to him. As you follow his analysis and think about your own (Question 1), can you identify the principal areas in which a strengthening of data would be desirable?

(3) The Forest Service (page 147 and following) uses the concept of "realizable" timber growth, presumably a quantity socially feasible, and in any case lower than the "biological potential." How would you estimate the realizable growth for the United States? How would the figure relate to the growth levels implied in your solution of Question 1?

(4) What is the contribution of a preliminary, speculative analysis such as Mr. Stoddard's?

* * *

As the United States approaches the end of a long era of liquidation of virgin timber we face the problem of making a transition to a timber-cropping economy. Encouraging progress has been made during the past 20 years in reforestation of cutover lands, protection of forests from fire, insects and disease, and in the advancement of applied forestry sciences. Nevertheless, our decisions to invest money in forestry have lacked an overall economic rationale as a basis for intelligent planning. Since timber growing takes decades instead of seasons, with long intervals between investment inputs and yields, planning entails a degree of complexity not common to most other kinds of financial forecasting.

In recent years, economists have given increasing attention to the question of proper allocation of capital investment in forestry without benefit of quantitative estimates. Some have felt that intensification of forest management should largely be relegated to a public works category to be undertaken in periods of underemployment; while a more recent feeling has developed that the current transition from a forest economy based upon growing stock liquidation to a forest cropping system must inevitably be accompanied by more systematic capital inputs. There remains, however, the problem of determining in quantitative terms the intensity of management and the size of the capital investment economically justified.

The author is fully aware of the many limitations of data in making an estimate of this nature. It must be recognized that choices in making assumptions were unavoidable and that in some instances a good case may be made for different assumptions. Nevertheless, it is felt that the time is past due for the development of a methodology in this very important aspect of forest economics. This paper is offered in the spirit that it will begin to stimulate future discussion, criticism, and further development as a basis for decision making.

ESTIMATES OF FUTURE TIMBER REQUIREMENTS

The Timber Resource Review of the U.S. Forest Service[1] has appraised our current forest situation in more detail than any of the previous efforts. Of immediate interest to this study are the overall national estimates of forest growing stock (capital), of annual rates of growth and cut, and of area primarily devoted to commercial forestry. . . .

In attempting to forecast America's capital requirements, both the President's Materials Policy Commission[2] and the Twentieth Century Fund[3] made estimates of the investments required to increase the average annual growth from 13.4 billion cubic feet to 20, which was estimated to be the volume of timber the nation would need in the year 2000. These estimates, which are based upon the U.S. Forest Service's Forest Reappraisal Report of 1946,[4] may be compared with more recent T.R.R. data which set forth several projected future demand levels for forest products for the years 1975 and 2000. Several years ago the Stanford Research Institute[5] completed an economic forecast of timber consumption, using similar techniques. A sizeable group of economists have concentrated on forecasting techniques with increasing refinements. In each case certain assumptions are made with regard to rates of economic growth based upon population changes and shifts in the production and consumption of the particular products being analyzed. This type of economic forecasting is gaining increasing acceptance as a process for both public and private planning, although occasionally it is still viewed with alarm. The assumptions, the techniques, and the forecast results give rise to critical appraisals and differences of approach of which the Forest Service estimates have had their share.[6]

Although differences exist over the future levels of total timber requirements, all of the forecasts assume that we shall need to grow more timber than at present, even without any rise in current living standards. Even with a continuing decline in our per capita consumption, it has been assumed that population increases will bring added pressure to produce still more wood. The Forest Service has set forth two projections for the year 2000, known as "upper" and "lower-level" projections. These upper-level estimates are based on the assumption that industrial timber products will occupy the same position relative to the consumption of all physical structure materials as they did in 1952. Lower-level estimates assume a decline in relative consumption. For purposes of this study, the basic data and the assumptions underlying the upper-level projections are the starting point from which this analysis is developed. All figures used are from the T.R.R. unless otherwise indicated; and no attempt is made to deal with any but overall figures.

This writer further assumes that because of the long growth cycle of timber, plans and programs required to assure the level of growing stock needed by the year 2000 must be decided and executed within the next ten years. Capital outlays required for intensified forestry to reach this production goal are developed from estimates of the projected requirements; but the policy determinations needed to assure the required investment are not given consideration in this study.

. . . Our 1952 cut of all roundwood products amounted to 10.8 billion cubic feet, divided between softwoods (7.5 billion cubic feet) and hardwoods (3.3 billion cubic feet) while total growth amounted to 14.2 billion cubic feet. In terms of board feet, growth and commodity drain are about in balance. The excess of cubic foot growth over drain is due to the growth occurring in large areas of pole size timber.

The T.R.R. upper-level estimates for the year 2000 indicate a 54-percent overall increase in our cubic-foot volume growth from the present (from 14.2 to 22.1 billion cubic feet) and more than 100-percent increase in board-foot growth (from 47.4 to 105.4 billion board feet). In order to obtain this growth goal, growing stock inventory accretions are called for. The Forest Service projections indicate that the impact of increasing requirements will reduce growing stock inventories by the year 2000 considerably below those of the present unless steps are taken to intensify management.

Although the specific forest management measures are not set forth, presumably they would include (1) reforestation of all or part of the 50-odd million acres of idle commercial forest land; (2) acceleration of growth through thinnings and other stand-improvement measures; (3) intensified insect, disease and fire protection; (4) elimination of premature clearcutting of better species of young sawtimber; (5) harvest cuttings in mature stands so as to assure maximum wood production of desirable quantity, quality, and frequency interval, and (6) shifting a portion of the present cut to poorer species and stands with resultant lowering of quality for a time. This 22.1 billion cubic foot growth goal is considered easily "realizable" and far below the biological potential. Data presented in the T.R.R. show that realizable growth would be about 27.5 billion cubic feet if these measures were applied—nearly double the present 14.2 billion cubic foot figure. Assuming that we will need only 22.1 billion cubic feet under the upper-limit figure (54 percent above the present level), we can then go on to estimate the expenditures required.

CAPITAL INVESTMENT REQUIREMENTS

In the estimates which follow, only those additional investment costs needed to obtain higher growth goals are calculated. The assumption is made that the present investment in land and timber (growing stock), and annual carrying charges, management and protection costs will continue at present levels. Therefore the estimates and forecasts which follow are only those additional inputs required for additional growth above the present 14.2 billion cubic feet. Forest development costs for roads and other improvements are considered to be either currently chargeable to harvesting or additional capital investments over and above those required for timber production. Recreational, wildlife, or watershed development and protection are not included for the same reason. It is further assumed that the level of harvesting would be kept to the allowable cut but would increase yearly as the growth potential increased. This assumption is based on the fact of excess cubic-foot growth over present cut which amounts to an annual addition of 3.2 billion cubic feet to growing stock. The increase in annual growth and cut above the present level (14.2 billion cubic feet) would be attributable to the added capital inputs. However, some shift into lower quality material may be required.

Since no T.R.R. estimates have been made of the physical input of intensified forestry measures which would be required to attain the 22.1 billion cubic-foot growth it became necessary to make approximations for this purpose. Of the six intensified forest practices suggested previously, the first three require capital investment which can be estimated, whereas the last three involve postponement of income in some cases, and modifications of harvesting practices in others with attendant additional carrying charges. Since these latter measures depend heavily upon both owner and public policy determinations, estimates of carrying costs would be of little value at this point.

This paper further assumes that full application of these minimal silvicultural measures would produce the "realizable" growth estimated in the T.R.R. at 27.5 billion cubic feet. An investment in forest practices previously set forth would involve planting of 52 million acres of nonproductive land, the application of such practices as underplanting, weedings, thinnings, and pruning to all of 114 million acres of understocked lands and approximately half of the 324 million acres of medium to well stocked stands. It would also require that adequate fire protection and insect and disease control programs be extended to all currently unprotected areas. Since no cost studies showing weighted average costs of forest planting or stand improvement measures are available, the writer obtained data from several sources and regions. While the data are admittedly estimates, they tend to be high and hence would give conservative results. Again, the reader is asked to assume these cost figures until better data become available.

While the present expenditures for protection and management are not included in this calculation, it is of interest to point out that the present level of growth (14.2 billion cubic feet) is being produced under generally extensive management at a current annual cost of $175 million according to an estimate supplied by the Forest Service. If the previous assumptions are granted, it can be estimated that the $4 billion investment in forest management made within 10 years and coupled with sufficient restraint to build up forest growing stock over the balance of this century could result in an increase of nearly 92 percent in annual wood production (27.5 billion cubic feet). Actual needs, however, were determined to be only 22.1 billion cubic feet or about 54 percent above the present level. Applying this percentage to the total estimates for the larger growth figure, an additional investment of $2.1 billion is indicated to attain our desirable growth goal. If this were to be made over the next decade it would require annual expenditure of slightly more than $200 million.

Assuming that growth is increased at a steady rate from 14 to 22 billion cubic feet per year over a 40-year period as a result of more intensive forestry, the mean annual growth rate would be about 18 billion cubic feet, or about 4 billion cubic feet above present levels. (Actually there would at first be somewhat less than 4 billion cubic feet, due to gradual curve increase in growth rather than straight-line projection). In addition, there will be accumulated in forest capital 110 billion cubic feet of growing stock. Using the stumpage value of $0.065 per cubic foot this additional 4 billion cubic feet of increased annual growth which can be attributed to intensified silvicultural treatment is valued at 1956 prices at 260 million dollars.

These growth assumptions would mean that the average annual growth would increase from the present level of 29 cubic feet per acre to 45 cubic feet per acre over the 40-year period. This increase of 16 cubic feet between the beginning and the end of the period would be at the annual rate of 0.4 of a cubic foot. Silviculturally this can be obtained by a combination of withholding growing stock, by a reduction of mortality losses, by increasing growth rates through cultural treatments, and adding more timber-producing land through planting.

ADDITIONAL PROTECTION MEASURES

These estimates do not include additional fire, insect, and disease prevention, detection, and control measures that will be required under more intensive management levels. Present expenditures by all agencies are around $80 million annually. It is estimated by the writer from data supplied by the Forest Service that an increase of $40 million, to a total of $120 million, would provide reasonably adequate protection. Such an expenditure would not be a capital investment but an increased current expense of $40 million above present levels and chargeable against current annual forest increment.

A considerable reduction in current losses in growth due to mortality (estimated by the Forest Service at 2.7 billion cubic feet or 9.6 billion board feet) may be expected as a result of this expenditure. At the assumed stumpage prices, present losses amount to $115 million per year (fire, insects, and disease account for about $65 million; the difference being due to weather, animals, and all other destructive agents.) If it may be assumed that the losses from fire, insects, and diseases can be reduced to $25 million from more intensive protection (but none of the other losses), the net value of growth saved ($40 million) would be equal to the added protection expense. This self-liquidating current expenditure could result in making additions to growing stock which should offset to a significant extent, if not completely, the need for reducing the allowable cut in order to accumulate required amounts of growing stock. If the cost of additional protection is self-liquidating (as expected), it need not be included in the investment estimates.

ECONOMIC ANALYSIS

A determination of the economic feasibility of the capital investment indicated involves analysis of several concurrent factors. If the calculated capital investment of 2.1 billion dollars is made in the year 1960 (physical and economic limitations would actually spread this over a decade) for intensive silvicultural applications which will produce an additional 4 billion cubic feet of average annual growth valued at 260 million dollars, the increase in forest growing stock (capital) during the 40-year period of 110 billion cubic feet would amount to an addition to capital value at 1956 prices of $7.2 billion in the year 2000.

The analysis then involves two factors: an increased current growth and the addition to growing stock. The first item will produce an annual gross return of $260 million. The second item will produce capital valued at $7.2 billion by the year 2000.

These gross yields do not include all additional carrying charges. If it may be assumed that protection costs are accounted for, taxes or tax equivalents are not, nor are extra administration and management costs.

To determine net returns and interest earned on the investment we may assume for purposes of this estimate that carrying charges will be met from the additional current growth at about the same proportion as on the national forests. From the

$260 million gross annual yield, 25 percent ($65 million) would be deducted for taxes, leaving $195 million. If we assume an added administration and management (but not protection) cost of $55 million above present levels, a net annual return of $140 million could be expected. The simple interest rate earned on the investment would average slightly more than 6 percent on the $2.1 billion investment over the 40-year period and actually to infinity. If the assumption that reduced mortality will offset additional protection costs is not accepted, this $140 million would be further reduced to $100 million. The earned interest rate would then be just under 5 percent. (Simple interest is used because the increased yields would be realized soon after the investment is made and during the ensuing 40-year period.)

In addition, the increase in capital of 110 billion cubic feet valued at 1956 prices would amount to $7.2 billion. These estimates appear to give a favorable ratio of costs to net yields, whereby a given capital input returns nearly 5 percent interest per year.

There are certain elements in the equation which bear further examination. In the first place, these estimates include certain policy assumptions which would withhold the thriftiest and most rapid growing stands of young timber from premature cutting (which alone would cause some rise in yields without additional capital inputs). Secondly, the cut would be thrown on to the poorest-quality trees and thus lower the average value of the current yield for several decades below those shown above. And thirdly, there would be a large volume of "in-growth" whereby young immature second-growth would cross over the line to marketability without any additional capital investment. Finally, there would be denser stocking per acre with resulting higher quality and value of yields.

In other words, some of this additional current growth and growing stock inventory would take place regardless of new capital inputs—*if* more protection and better forest management would be adopted. Then, too, the fact that present forest production is well below optimum output levels while fixed costs are relatively stable makes for a situation where a small increase in inputs will yield a large output increase.

The above additional variables with little or no basic data with which to gauge them are such that measurement [of them] would require the development of still further assumptions. Since much of the case already rests on a number of assumptions which are based on such data as are available, further calculations of this sort might well be fertile ground for those who may wish to pursue this subject.

DISCUSSION AND SUMMARY

Any estimate of the future must necessarily rest upon a series of dependable assumptions. Also, where basic data are absent or incomplete, the author obtained the best estimate possible. There is no other course; yet by their very nature, assumptions may be challenged at any of a number of points. This paper is offered mainly as a method for arriving at an approximation of the investment needed in more intensive forestry under the conditions set forth. It is to be hoped that this effort will stimulate additional constructive thought on the part of forest economists in this important area.

What proportion of our total national investment effort can properly be allocated to forestry as compared with other opportunities has been a moot question among economists. This paper has broadly estimated the amount of capital that would be needed to achieve "upper limit" growth goals of 22.1 billion cubic feet by the year 2000. Necessary measures to bring about improved cutting practices so as to accumulate additional forest capital have been assumed but policies to assure these practices have not been set forth.

The estimates presented indicate that a capital investment of $2.1 billion over the next decade plus an added $40 million in annual protection outlays for a 40-year period will produce an added 4 billion cubic feet of annual growth with a gross value of $260 million (net of 140) per year for the whole period prior to the year 2000. This study did not attempt to allocate the capital costs between private and public forests but would expect them to be somewhat proportional to ownership.

In estimating capital requirements for timber production, it is of interest to make relevant comparisons. In 1955, the United States made capital investments amounting to $56.4 billion—16 percent of our total national income of $324 billion. About $175 million was invested (i.e., 0.2 of 1 percent of our total capital investment) in forestry in the same year to protect and cultivate our national timber crop for the support of our forest-based industry which contributed an estimated $15 billion (4.7 percent) to the national income total.[7]

Thus we are currently spending 1.2 percent of the gross income generated by forestry to produce basic raw material. An additional annual investment over a 10-year period of $210 million would add another 1.4 percent to the investment cost of timber production for 10 years—which, if added to cost of forest products, would be small in relation to the indicated timber yields.

The determination of how much to invest in forestry should properly include other criteria than interest rate earned; but in any case interest rate will weigh heavily in private forestry decisions. On all forest lands, nonmonetary returns are a factor which can only be resolved through other than marketplace processes, especially on public lands. This paper does not attempt to deal with these politicosocial aspects of investment decision making. It may be hoped that these may be brought out in subsequent discussion.

REFERENCES

1. U.S. Forest Service. 1955. Timber resource review, Chapter VI, Future domestic requirements for timber. Preliminary Draft.
2. President's Materials Policy Commission. 1951. Resources for freedom, Volume 1, Foundation for Growth and Security.
3. Hartley, Wolkind, et al. 1950. America's capital requirement: 1946–1960. New York: Twentieth Century Fund.
4. U.S. Forest Service. 1946. Forest Reappraisal report.
5. Stanford Research Institute. 1954. America's demand for wood: 1929–1975. Stanford, California.
6. Nelson, Alf Z. 1956. The great guessing game. National Lumber Manufacturers Association. Also, Zivnuska, John, 1956. Timber today and tomorrow. Forest Industries Council.
7. Hartley, Wolkind, et al. Work cited.

20

HOW FOREST INDUSTRIES JUDGE INVESTMENT OPPORTUNITIES

Arthur W. Nelson, Jr.

First published in *Society of American Foresters, Division of Forest Economics and Policy, Proceedings;* pages 85–88. 1960.

BIOGRAPHY

Mr. Nelson is Vice President, Industry Affairs, Champion International Corporation. After an academic program in forestry at the University of Idaho and at Yale, he began, in 1939, a lifelong career as an industrial forester: timber cruiser for the Crossett Lumber Company; Chief Forester for the Flintkote Company; Special Resources Consultant in Champion Paper and Fiber Company; and successive positions of increasing responsibility in the Champion organization. A thoughtful interpreter of his professional field, he has published widely on such subjects as industrial forest management, land classification, watershed management, and water law. Among his activities have been the chairmanship of a Society of American Foresters section and the presidency of the American Pulpwood Association and of the Southern Forest Institute.

EDITORS' SUMMARY

Industrial interest in forestry investments is made possible by rising product prices and full material utilization. Firms typically give first priority to "essential" investments and next priority to those which promise good payout. Determining forestry payout is difficult, but is possible with the help of advancing methods, data, judgment, and computer facilities.

QUESTIONS TO CONSIDER

(1) Apropos of Mr. Nelson's opening comments, it was long expected that widespread

practice of timber management by private owners in the United States would have to await removal of the old growth, a free gift of nature as long as it lasted. Why was this expectation held? Why did it prove false?

(2) Mr. Nelson mentions the general rule that a wood-using firm must control at least a part of its raw material. Suppose that you are asked by such a firm to recommend the forest acreage that the firm should own. What factors and considerations will you take into account?

(3) Why are we unlikely to find any timber owner, private or public, managing his resources so as to maximize wood output?

(4) Review the basic model for capital budgeting as developed by Joel Dean. How is the demand for capital derived? the supply of capital? Where does one find the internal rate of return on capital? the guiding rate of return?

(5) Assume that the criterion for ranking investment projects is their prospective rate of return. What is the prospective rate on projects of priority 1 (page 155)? How is one to define "very good" with respect to priority 2?

* * *

There are many forest industries that have no investments in either timberlands or forestry. Their raw material supply is entirely furnished by others, and is presumed to be of satisfactory volume and price. There are some forest industries that have investments in timberland, but who make no investments in forestry. This form of company is declining rapidly in numbers, but at one time made up the preponderance of the industry. The third category—and the one about which we are concerned at this session—is that constantly increasing segment of the forest products industry that is finding it necessary and desirable to make investments in forestry.

I have purposely brought up this matter because, through its discussion, we arrive at the reason *why* forest industries make investments in forestry. In years gone by, when timber was plentiful and cheap, investments in both timberlands and plant were liquidated over a short period of time and the enterprise either closed out or moved to another location and reestablished. The *idea* of investing in forestry is not new, although the *practice* is. Forestry investments have been made in Europe for many years, and it was correctly predicted that such investments would be made in the western hemisphere as soon as economic conditions permitted.

A combination of factors have brought about the favorable situation which we find today with regard to investments in forestry by private corporations. Two of these are (1) a price structure for forest products which makes their growing an economic activity which can stand on its own feet without government subsidy or price support; and (2) a rapidly developing technology which allows us to use practically the whole tree in the manufacture of a wide variety of products, including chemicals.

The effect of these two factors has been remarkable. It has resulted in the stabilization of the industry, converting it from a migratory industry to a permanent industry, and it has required much larger investments in plant and equipment.

By way of illustration, green lumber can be produced with a circular sawmill for an investment of considerably less than $10,000. To use the rest of the tree, the top, the fiber in the slabs, smaller sizes of trees, species of trees not usable for lumber, and to recover some of the chemicals which they contain, investments in the order of 30 to 50 million dollars are common.

RAW MATERIAL SUPPLY

* * *

Raw material supply, once the concern only of an individual entrepreneur, is now the concern, not only of the owners and managers of a business, but of their financing institutions as well.

It is almost universally held in this country that industries having large scale investments in converting facilities for forest products must own and control at least a portion of their raw material. Although as a general rule, there are large supplies of timber sold on the open market, and forest statistics show large amounts of timber to be available, these are, *by themselves,* not a sufficient guarantee to the industry of an adequate supply at a reasonable price at all times. The following are a few of the reasons why it is not:

1. The owners of timberlands do not always sell their timber when it is needed. (a) National forest sales are made by competitive bid and usually for only a short time, although there are some long-term sales which are the exception to this statement. There is always the danger of outsiders coming in and outbidding established industries and thus upsetting the normal marketing pattern. (b) The time of marketing of an individual's timber crop may depend more on his own personal financial needs and individual income tax picture than upon the industry's need for timber. (c) Adverse weather conditions may prevent logging in many areas leaving only a few areas which are of such a geographic nature that they can be logged. Again, there is no guarantee that the owners of timber on these accessible lands will either sell it when it is needed or will even save it for sale under adverse weather conditions. (d) Some owners don't want to sell any timber under any circumstances.
2. Timberland owners are not always interested in getting the maximum productivity out of their forests, or of growing the species of trees or the kinds of forest products which a particular industry might need.

INVESTMENTS IN TIMBERLANDS AND FORESTRY

The forest industry invests in timberlands and in forestry (to make these investments as productive as possible) primarily to offset these uncertainties. Although the investment pattern of each company is different, because of size, product, and geographical location, there are a few general considerations which can be said to apply industry wide:

1. A company will limit its investment in forest land to that amount it feels is necessary adequately to protect its future raw-material supply. Because of local conditions, or individual financial limitations, many companies cannot acquire what they consider an adequate acreage.
2. Having invested in forest lands, a company will try strenuously to make the investment as productive as possible.

It is with this latter point that we shall concern ourselves for the remainder of this paper.

Investment policy is not made in a vacuum. A company has many demands for its capital and, as with any other situation where the needs exceed the supply, some system of priorities must be set up.

PRIORITIES

As I have mentioned previously, each company handles its capital budgeting in a different manner, but basically all companies follow the same broad outline of priorities. The following list of priorities may be considered typical:

Priority 1. Projects which are essential to the continued operation of the business. This includes necessary equipment replacements, projects which have to be undertaken for the safety or welfare of employees, and projects which are required in order to keep the business competitive.

Priority 2. Projects which have very good payouts.

Priority 3. Projects which are desirable but not essential (regardless of time of payout).

Priority 4. Projects of a contingent nature which might have to be undertaken as a result of possible changes in the conditions under which the business operates.

CLASSIFYING FOREST INVESTMENTS

After reading the above listing of priorities it will become clear . . . that we may have a difficult time classifying a proposed forest investment. For one thing, we do not as yet have the highly developed techniques for computing payouts that have been developed for industrial projects. Some of the reasons for this are as follows:

1. Time involved. Investments in plant projects often have payouts of a relatively few years. Forest investments, on the other hand, may have considerably longer payouts. Plant project payouts are figured on current prices. Payouts of forest investments may have to be figured at higher prices which may be expected to prevail in the future. Here we enter the realm of speculation. It has been argued that we do not make forest investments for today's wood; but rather for tomorrow's. If we expect wood to be worth more in the future than it is today, the expected future price should be used in calculating the return on the investment. Admittedly this is difficult to arrive at, but anyone seriously engaging in capital budgeting should give time and study to future price trends.

2. Lack of adequate standards of measurements of performance. We lack adequate means of measuring performance or predicting results of certain types of investments. In a manufacturing plant the increased rate of production of a new machine is definitely known and can be stated accurately. Furthermore, this same machine will produce at the same rate whether it is operated in Maine, Texas or Oregon. Returns from forest investments are not so easy to measure. They vary widely from one region of the country to another, and there has been very little factual data available to our industry in this field.

3. Lack of familiarity with the financial aspects of our industry and the results expected from financial investments. It has been only in recent years that foresters have advanced to top management and find themselves in competition for funds, so to speak, with their counterparts in the other phases of the company who are far more skilled in the areas of capital budgeting.

This situation has been largely brought about because of the fact that we have paid far more attention to forestry as a technique than we have to forestry as a business. When land and timber were cheap, foresters spent most of their time in the woods growing and protecting trees. As timber products increased in value and investments in converting facilities became larger, foresters have been moved up into top management to give better direction to the whole field of capital investments in these areas.

METHODS OF HANDLING THE FOREST PORTION OF THE BUSINESS

There are currently two methods of handling the forest portion of the business enterprise.

1. Where the industry consumes all the timber from its own lands, all of the costs in carrying the forest business are accumulated and charged on a unit basis. This charge plus the cost of logging and transportation becomes the cost of the raw material to the manufacturing plant.
2. Another way is to set up the forestry end of the business as a 'profit center'' along with the other manufacturing plants of the company which are also set up as profit centers. Here all the costs of running a forest business are accumulated and charged over the entire amount of timber grown. Timber is transferred or ''sold'' to the manufacturing divisions of the company at market, and the difference between cost of growing and current market price is profit. This profit determination allows comparison with the other operations of the company.

This type of accounting is new in the industry, but more and more companies are considering it in order to have a true look at just what their forest business is and how it is performing. It involves a determination of all items of cost in relation to production. It discloses the fact that we still do not have adequate measures for some of our cost items. For example, how do you treat $5 per acre spent on timber stand improvement? How much extra wood will it produce, and what is its value?

A large forest operation consists of many activities going on simultaneously over a wide area. Each of these activities calls for a financial decision; and since decisions in one area are related to other decisions you may well wonder if we are not attempting to take on an impossible task.

Fortunately for us, there are tools available to help us do this job.

The object here is to maximize the income from any given series of investments on the particular forest property we are concerned with. This involves the comparison of many alternatives to see which one gives the best results.

The technique of operations analysis, or operations research, as it is sometimes called, was developed during World War II to handle such problems as the Normandy invasion which had many variables. Linear programming, which is one of the tools we use in this area, allows us to solve for a maximum or minimum in problems having as many variables as 100 × 100 (in a matrix). Since electronic computers are the only means available to solve such problems, it is necessary for forest managers to become familiar with their use. Lest this sound too fantastic to you I hasten to assure you that, as in many other cases, other industries are far ahead of us in solving problems and improving their profit by the daily use of these techniques. It is high time we got started with them as well.

Almost any company of any size now has some automatic data processing equipment either on hand or on order. Those who have investigated its use in formulating management plans for a forest business have found that it is necessary to develop a completely different set of data and records, and in so doing you will uncover things about your business that you never were fully aware of before.

* * *

The forest manager faces two distinct problems in the capital-budgeting field. First, he must decide which of the many investment opportunities available to him offer the highest return and secondly, he must help his company decide which of all the investment opportunities available to the company (of which forestry is only one) offer the best opportunity for profit return.

If you will remember the four priorities of capital budgeting which I mentioned earlier, you will recall that the first priority is given to those projects which are essential to the continued operation of the business, *regardless of payout*. Items with good payouts are given a No. 2 priority. I think that this is significant for our business because it is tacit recognition of the fact that we cannot see far enough ahead to determine exact payouts on all courses of action, and that some very important ventures in research, new product development and other areas are undertaken primarily on management judgment that they are important for the future well-being of the business.

Investments in land, which is a nondepreciating item, should be treated differently than investments in plants and machinery, which do depreciate and wear out. It is obvious that the payout rates for these two types should differ. In practice, the individual judgment as outlined in Priority 1 will indicate how small a payout will be acceptable.

This is an area where the judgment and experience of the forester in management is very important. Although he cannot, in all cases, give exact rates of return for various forest investments, he is equipped by training and experience to assist management with decisions in those areas where exact data is missing and where informed judgment must be substituted.

One example of this is in the area of land acquisition. The ordinary concept of capital budgeting is that an investment which is desirable, but for which no funds are currently available, may be postponed for several years, and then undertaken. This is a valid course of action when considering the replacement of a machine or the rebuilding of a mill, but it *is not a valid consideration* when considering such items as land acquisition either through direct purchase or by corporate merger. Such opportunities usually occur relatively infrequently and no postponement of decision is possible. This type of investment choice, although heavily influenced by payouts, is also largely influenced by individual judgments as to the long-term direction that each individual enterprise should take.

CONCLUSION

. . . Our profession is coming of age, and . . . foresters are moving up to take their place beside managers of other types of enterprises. The manager of a forest enterprise may ultimately become more important than the managers of other divisions of a forest-products company, precisely for the same reason that he is somewhat at a disadvantage today. This is the inability to predict precisely the results of long-range forest investments by the same techniques as are used for short-term plant investments. Instead, more reliance must be placed on long-range economic trends in the whole forestry field which the forester, by his experience and training, should be well qualified to follow.

21

TIMBER INVESTMENTS FROM THE STANDPOINT OF INSTITUTIONAL LENDERS

Eli Ferguson

First published in *Society of American Foresters, Division of Forest Economics and Policy, Proceedings;* pages 88–90. 1960.

BIOGRAPHY

Mr. Ferguson is Senior Vice President for Corporate Planning, Equitable Life Assurance Society of the United States. He is a South Dakotan with a degree in business from the state university and several years of ranching experience within the state. Working for a time as a field man for a country bank and then with the U.S. Department of Agriculture, he joined Equitable in 1936 as Office Manager in the Farm Mortgage Department. Successive promotions led to his present appointment in 1971. A member of the Economic Club of New York, the American Farm Economics Association, and the Western Farm Economics Association, he was awarded, in 1963, an honorary Doctor of Science degree from South Dakota State University.

EDITORS' SUMMARY

Insurance companies have in recent years invested in timberland loans. They have taken great care in evaluating the collateral and the plans for its management. Typically, they have made large loans to pulp and paper companies in support of forest acquisition. Forest credit is not the answer to the small-woodland management problem.

QUESTIONS TO CONSIDER

(1) Suppose, as is widely contended, that timber growing as an independent enterprise—unaided by inflation, subsidy, or integration with other enterprises—cannot survive financially by showing a monetary profit. Consider Mr. Ferguson's ideas as expressed at the bottom of page 160 and elsewhere in his article. How can foresters be "businesslike"? What role can be played by the economics of forestry?

(2) What does Mr. Ferguson mean by "valid" (pages 161 and 162)?

(3) Why is a lender interested in the borrower's forest management plan?

(4) Why should institutional lenders such as insurance companies not wish to make very small loans?

(5) What were the values in forests which attracted institutional lenders into the forest-credit business? What do you think was the principal forest-management development in the United States which led institutional lenders to change their traditional attitude, that timber was too risky a resource to represent satisfactory loan collateral?

(6) What is the "small-woodland problem"? Why is forest credit not the answer to this "problem"?

* * *

Wood-using industries, particularly the pulp and paper companies, are no strangers to the institutional lender. They have been in again and again, and have gone out with what they wanted about every time. They came with their balance sheets and profit and loss statements and lots of other papers. But I doubt if many of them brought with them a detailed forest inventory. They didn't need to. Few of us would have known what a forest inventory really meant. If we had looked at one, we very possibly would have been misled more than led.

For many years, as all of you know, institutional lenders, as well as many of the pulp and paper companies, did not believe it was important to own any timber land. But those who did think so were acquiring timber land quietly, aggressively, and cheaply; so were some lumber companies and a few private investors. Now we know they were smart operators, but at times some must have wondered if they were not becoming land poor.

After a while, probably through the push by pulp and paper companies for more and more money to finance their ever-growing operations, institutional lenders began to take notice of the timber assets shown in financial statements at a very nominal figure.

Institutional lenders pay a lot of attention to what the balance sheet shows. They also like to get back of the figures. Assets are not always what they appear to be on paper. Timber is one which was usually grossly understated. So, it is not surprising that, as the cost of wood used in the manufacturing process increased, lenders began to inspect timber lands and form some opinion of their real worth. They still shied away from even the thought of making loans on timber and timber land as such.

In the meantime, owners of timber land were doing things. Scientific management with the objective of making a profit was developing. Your Society has heard many earnest debates on the basic conflict between growing trees for the sake of growing trees and growing trees for a profit. Foresters ceased to be principally concerned with conservation of the national forests and became businesslike in their approach.

As companies tied up more of their funds in timber lands, as stumpage increased in price, as management for a profit over the long term became commonplace, and as foresters became businessmen, institutional investors began to wonder if timber lands might not be good security for loans in the form of real estate mortgages. Forestry loans having their present characteristics have evolved rapidly in the past 15 years.

Like most lending activities, investment in timber lands requires specialized knowledge of a kind that few institutional investors had. The few courageous and enterprising ones who decided to go into this new field with the intention of staying had to develop understanding and acceptance within their own organizations, and had to work out policies and procedures for handling the business that developed. They had to build an organization which combined the talents and knowledge of businessmen-foresters, investment analysts, lawyers, and accountants.[1]

The problem then was to develop a flow of business that would support such an organization, even though some of the specialists were also occupied in other fields. The volume of strictly forestry loans is still not large, but it is sufficient to sustain the necessary organization. And it is being handled by institutional lenders with the same methodical thoroughness that has usually characterized their operations. While it still may not be known generally, the fact is that the principal institutional lenders are now meeting all credit demands believed to be sound, of forestry and of wood-using industries for expansion, development, and other valid purposes.

There are many facets to the timber business when one thinks of it from the viewpoint of an institutional lender. We cannot cover all of them, but shall try to describe briefly how forestry loans are made and the special consideration that is given the various types of properties.

FORESTRY LOAN DEFINED

Before going further, it seems desirable to say what we mean by the term "forestry loan." It is simply a loan secured primarily by a real estate mortgage on land and timber, with repayment expected to be made from the harvesting of timber. Such loans, quite obviously, have to be, and are, negotiated by trained foresters.

EXAMINATION OF PROPERTY

An institutional lender makes about the same detailed study of a forest property that an informed investor would make. . . . Local markets are investigated, and the forest management practices of the local community are observed. Salability of the whole property and of its production are considered, as well as the market for products which must be harvested in the event of fire, disease, or wind damage. Timber stands and soils are examined and mapped, using the most suitable techniques for the particular situation. Management plans are evaluated for their effect on future growth.

A lender has to be interested in the rate of return on the investment, and its relationship to annual loan charges. Some forests may be regarded as a warehouse full of stored goods, or, as it may be said, "a stay-place for trees." Others may be fully efficient production units, and others are highly inefficient. Each type may be suitable security for a loan of some amount, but obviously requires different treatment.

CONSIDERATION OF THE FINANCIAL PLAN

Forestry loans are made for many purposes, including land acquisition, plant construction, working capital, refinancing of indebtedness, improvement of forest property, or any valid interest which an applicant may have. The lender closely examines the applicant's financial plans, since it will have to be satisfied that the money requirements are adequately met and that the financial condition is sound. The lender, on the basis of careful studies of past and expected earnings, is always concerned that the total debt burden is not greater than can be safely repaid. When loans are made for plant construction and not specifically as forestry loans, the lender will consider the entire financial plan including costs of construction, expenses of manufacturing, markets, sales organizations, and predictable returns. It will assure itself that adequate supplies of timber will be available whether owned by the applicant or obtainable in the market area. Interest and other costs including taxes, management, and market risks are carefully compared with predicted returns.

RELEASE TERMS AND AMORTIZATION

Working out suitable release provisions is often the most important part of the development of the loan agreement. The lender must remain secured throughout the loan term and must have amortization in proportion to any reductions in value resulting from depletion of the property. It will reconcile expected growth with the volume and prices of timber to be released. This is why the lender is vitally concerned, when making a 25- or 30-year loan, that there is full agreement, in writing, on the plans for management of the property. Plans may call for early liquidation of inventory and reduction of the loan to balance investment and income, or liquidation may be deferred when the economics of the particular forest justify it. Efforts are always made to make the release terms and amortization fit the financial needs of the borrower and the physical condition of the property. Most institutional lenders also want some amortization, possibly 3 or 4 percent annually, to offset possible errors in appraisal and other unforeseen situations, and to give some opportunity to react to changing conditions.

Loan agreements usually provide for payment of a stated amount per thousand board feet or per cord as the timber is cut, or of a specified amount annually with provision for release of certain volumes.

INDUSTRIAL LOANS

Forestry loans of what might be called the industrial type are usually made to pulp and paper companies to finance major timber acquisitions. They may equal the market value of the land and timber when the earnings history of the company has satisfactorily covered all charges, past and projected. Several large loans of this type have been made by the principal institutional lenders. The usual pattern is for the loan to be made to a separate land-owning corporation and be secured by a mortgage on the property and assignment of a long term lease or cutting contract with the paper company which generates sufficient net income to pay loan charges.

STANDARD FORESTRY LOANS

Forestry loans which may be said to be of a rather standard type are made to all types of borrowers who own and manage timber lands. They are about equally divided between wood-using industries and individuals or firms without manufacturing facilities. Most of these borrowers employ foresters and are fully equipped to manage their land. Loan agreements often include management provisions for control of hardwood or other specific silvicultural practices where disease or other hazards may be present. They frequently include provisions for road development and other improvements. Other provisions of the owner's management plan are also included. These loans amount to 30 to 50 percent of the market value, unless increased by reason of additional credit considerations.

LOANS BACKED BY LONG-TERM LEASES

Forestry loans are available to owners of land which is leased for a long term to high-credit wood-using companies, providing the terms of the leases are not objectionable. The amount of the loan is determined largely by the minimum net rental to be received by the borrower. A loan is secured by a mortgage on the land and an assignment of the lease.

COMBINATION AGRICULTURAL AND FORESTRY LOANS

Valuable timber is often found on farms and ranches. Income can be predicted if the property is located in an area where there is a well-developed market for timber products. If wood-using industries are actively managing land in the area, farmers are apt to be better informed and are often practicing sound timber management themselves. When such conditions exist, several institutional lenders making farm and ranch loans include the timber in their appraisal and lend accordingly. Usually timber releases are negotiated when the need arises.

OTHER TYPES OF LOANS

Most major institutional lenders do not make inventory loans on trees, logs, or various forest products where the land is not owned or is mortgaged. In some cases assignment of inventory contracts may be accepted as additional collateral and the amount of loan which can be made increased.

Loans for development purposes such as new plantings, water control, road building, fire lines, and timber stand improvement are often made in conjunction with a broader loan program with an individual borrower. In a limited manner, a loan primarily for development purposes may be made to an individual borrower where sufficient income is available to carry interest and protect the forest.

Loans frequently involve various types of cutting contracts. Such contracts are examined in order to determine whether they may be injurious or burdensome to either buyer or seller, whether they are enforceable, and whether the buyer is financially able to perform. The effect on the property and the legal aspects are considered. Some contracts enhance the value of a property but they may also be very burdensome from the lender's and owner's standpoint.

FORESTRY LOANS ON SMALL ACREAGES

Institutional lenders do not generally solicit forestry loans for less than $25,000, although a few have been made down to $8,000 to $10,000 where markets are well developed and satisfactory management is predictable. Lenders have found management is not created by lending money; rather, management makes loans possible. We doubt if lack of credit is the small owner's basic problem. It is more one of fundamental economics. There is just not the income base present sufficient to attract or hold any long-term management or interest. As larger owners develop methods and markets, and bring hazards under control, small owners have benefitted by following their example. Some have accumulated lands and now have sufficient base for sustaining operations. The others have found a good market for their land if they wanted to sell.

A recent article by the Federal Reserve Bank of Atlanta[2] reports the results of interviews with pulp company representatives on the feasibility of large scale leasing of small tracts in the Southeast, with the objective of improving management and production. The consensus was that it is just not economically feasible to handle a large number of individually-owned small tracts. The cost would be too great. Furthermore, it seems that the viewpoint of owners of small tracts is too short-range to fit the requirements of pulp companies. We have found about the same thing to be true.

Many small tracts are obviously not held for their income characteristics. Where that is true, there is no interest in better management or improved income, particularly over the long term. Where they are owned for income purposes, the need for income now or soon is much greater than for income which will be available far out in the future.

While we recognize that a large acreage of timber land is in small ownerships and that, by and large, production could be vastly improved, we doubt if any real gain would come through expanded use of credit by small owners. The result very possibly would be exactly the opposite of that which is desired. Loans, if made, have to be repaid. The pressing need for money for current living expenses in addition to debt repayment would probably result in more premature cutting of timber than the opposite.

When loans have been made in other fields under these conditions, management has been unsatisfactory and frequently disappears. The loans have been costly to service and losses have been frequent. Some folks have suggested that subsidies could be used to overcome this difficulty. But such action would have undesirable side effects, such as further driving up the price of land. The taxpayer would not, in our judgment, receive any real benefit, nor would there be any significant improvement in the long-term supply of wood.

In all businesses, big or small, an owner must have some equity in order to get started. The small timberland owner, therefore, must expect to accrue from earnings sufficient capital for carrying the growing stock and land until he can afford credit. Many have done this.

Heartening progress is being made by small owners in areas where markets are well developed and pulp companies, counties, and others have provided foresters. We look for this to continue as markets and forestry services are extended to the less developed areas. The timber economy has developed rapidly over recent years. Let's not tinker with it through a subsidized small forestry loan program.

LOAN EXPERIENCE HAS BEEN GOOD

Lenders who have studied the field and who have developed qualified organizations have had successful results to date with timber investments and invite new business. A few things have become apparent as lending experience accumulates: Many timber loans are paid back rapidly; since hazards such as fire, disease, and windthrow have been given adequate consideration, no losses have occurred because of these factors; people in the timber community meet their obligations, and there have been few, if any, defaults; forestry loans have been worked out to fit many fairly complicated financial problems, and experience equips the lender to solve these problems more readily.

REFERENCES

1. Ferguson, Eli. 1956. Economic considerations in making timber loans. Jour. of Forestry (December).
2. Federal Reserve Bank of Atlanta. 1960. Gaining top production from small forest tracts. Bankers Farm Bul. (August).

22

FORESTRY INVESTMENT DECISION MAKING: AN EVALUATION

Ronald I. Beazley

First published in *Society of American Foresters, Division of Forest Economics and Policy, Proceedings;* pages 95–99. 1960.

BIOGRAPHY

Mr. Beazley's biographical note is given at Item 7, page 61.

EDITORS' SUMMARY

Three papers on forestry investment are analyzed. Investment is the process of deferring consumption in the expectation that its value will be exceeded by the present value of future net investment returns. Forestry investment decision problems include forecasting production functions and unit values, finding applicable discount rates, and weighing social and non-market costs and benefits of investment.

QUESTIONS TO CONSIDER

(1) Mr. Beazley speaks of the accumulating of timber-growing stock as an act of investment in which some consumption is postponed. Are growing-stock decisions of this sort analyzable as capital-budgeting problems? Sketch the model.

(2) Describe the financial analysis of the question whether to plant a tract of land to trees for the production of sawtimber. First, approach the tract as an isolated investment problem. Second, approach it as a small part of a going forest (timber) enterprise. Does either approach hold appeal as a guide to decision making? Can you envisage any other approaches to the planting-investment question?

(3) Can you describe a model from game theory (page 171) that would be useful in guiding the planting decision of the second question? How could the product substitutions mentioned by Mr. Beazley (page 171) be brought into the analysis?

(4) Mr. Beazley appears to draw a distinction between "risk" and "uncertainty." On what grounds may they be distinguished? On what grounds may the distinction be challenged?

(5) What is meant by "marginal costs of extending geographical wood-harvesting margins" (page 172)? Does an extension of timber-shed boundaries raise the buyer's wood costs throughout the shed or only at the boundaries? Does the price structure of the shed have a bearing on the question?

* * *

Investment, or the process of *capital formation* as it is more broadly conceived by economists, may be looked upon as a process wherein one refrains from current consumption in preference to the alternative of producing some sort of product over time, with the assumption that the present value of the expected future net returns is greater than the current value of consumption. Notice that the mere act of not consuming a stand of mature timber now, by simply reducing its annual rate of conversion into lumber, increases present investment over what it otherwise would have been. Economically, this act is *conservation,* since it defers current consumption from the present to the future by *investing,* i.e., concluding that one is better off now by not consuming now.

The same principle operates when we invest in more growing stock or plant and equipment: the current rate of consumption, this time not necessarily of lumber, is reduced from what it would have been in order to provide a greater rate of consumption in the future, or, using the previous argument, the one doing the investing does it because he feels he is better off now because he did so. That is, he expects that the present value of the net return from his alternative expenditure in the form of investment is greater than it would have been in consumption. The central core of the decision to consume or to invest is the rate at which one discounts future net returns; whether one invests in alternative A or B depends on how large the net returns are expected to be and how much each is discounted. The discount rate in turn is dependent on the risk and uncertainty in the mind of the individual or collective investor. Investors make assumptions as to future costs and prices, usually on the basis of historical data, calculate future net returns as well as possible, and then discount them to the present; if this sum is greater than what they propose to invest, then they are, roughly speaking, better off by investing. Notice that one of the very important points to decide is what the rate of discount should be. It is important to consider how investors, both individual and collective, arrive at these rates, which are often different for different investors facing the same real-world situation. But first let us consider more broadly the circumstances required for investment to take place.

Without thoroughly exploring the reasons for it, it is apparent that first of all individuals, business, and governments must save, and not hoard the amounts saved, i.e., make them available for investment. Individuals save by refraining from consumption, for various motives, one often being future economic security which the amount saved stands for. Businesses save by not declaring all their net revenues as dividends, by not paying up all their outstanding accounts immediately, by using generous depreciation allowances, by providing large funds for contingencies, and so on. Governments save and invest simultaneously, sometimes actually buying shares of firms out of current tax revenues, but more usually they save and invest by spending current tax revenues on assets such as dams, forests, and roads which will provide a stream of value returns to the public far into the future: even defense expenditures may be viewed in this light. These collective withdrawals from current income constitute saving because current consumption has been curtailed in favor of future returns and security.

Thus, a certain amount of money becomes available through each year for investment. This rate of saving may be increased or decreased by government action affecting both consumer and producer credit, and by taxation and deficit spending, just as large businesses may affect it by their pricing policies. Notice that increased taxation by government for expenditure on assets such as forests, dams, and roads is collective saving and investment.

It is apparent then, that the rate of saving and investment does not occur entirely autonomously, but to a considerable degree it may be the result of calculated decisions by government and business. Hence the *amount* of saving and the *kind* of investment becomes in considerable degree a political-economic function. However, for the moment let us assume a given amount of annual saving, including that amount of total tax revenue which is not represented simply by transfer payments. How should it be invested?

If it is assumed that the necessary expenditures for national defense and the like are made, so that the country is reasonably secure, elementary economic theory would indicate that the remaining savings should be invested by allocation into capital uses which have,successively, the highest marginal net productivities. That is, savings should be used to create that capital having the highest interest rate, a rate which just equilibrates the present discounted value of its future net returns, i.e., those above its maintenance costs, with its present total cost.

Some of the savings would be used for social investment, some for ordinary business, some for short term purposes and some for long—all in accordance with the priority indicated by the greatest expected marginal net productivity of the capital use proposed.

But in reality there are at least four major conditions to be met before such allocation can be made. One is that we know the *physical production functions* associated with various uses of capital, so it can be said that if additional physical inputs are put into any given process, whether private or public, that so many additional units of output of various kinds will accrue, and at what times they will be forthcoming. The second, and usually most difficult condition is to attach prices (or costs) to the committed future inputs and to be able to apply prices or value to the outputs. Here the investor is face to face with all of the risk and uncertainty that can occur over time due to changing purchasing power, changing tastes, changing technology and institutions, which affect both costs and returns. No matter what the cause, the greater the risk and uncertainty, the greater will be the reduction of the expected future net value of the output. Thus there will be a lowering of the present marginal net productivity rating of the capital involved.

The remaining two conditions, which often occur together, are (1) the existence of financial institutions to funnel savings into investments of highest marginal net productivity, that is, the highest uses, and (2) entrepreneurs to point out, argue for, and justify competing alternatives.

In our mixed capitalistic system, the financial institutions are many and varied; they include the major life insurance companies [see page 193] . . . as well as the various kinds of banks, trust funds, companies and corporations, and many individuals, in addition to government itself as represented by its borrowing, taxing and expenditure prerogatives.

With the introduction of the very necessary and varied financial institutions, and the numerous entrepreneurs, we can no longer envision a given amount of annual saving in total, with investment decisions being made by an unnamed and omniscient decision-maker, in accordance with the highest marginal net productivity of capital. Rather, the problem is much more complex, because the aggregate saving is in the hands of a great many financial institutions, businessmen, individuals and governments, each having different kinds of knowledge as to production functions, different attitudes toward the future and its values, and different amounts of available savings, all of which complicates the allocation problem. The result is that at a given time, we have a good deal of freedom to invest, each investor representing differing degrees and kinds of specialized technical knowledge, as well as differing value judgments as to what the various alternative marginal net productivities of capital are. These differences, incidentally, are why we have a so-called economic conservation problem: different institutions, including government, appraise different kinds of future values differently. A given business firm does not face this problem in this sense.

In hindsight, it is of course apparent that, with such a system, mistakes are made, but they are usually not large mistakes. These mistakes, or miscalculations of savings into low or perhaps negative productivity investments, are part of the cost of mixed enterprise, democratic capitalism. This is a small cost to bear in view of the great diversity of output and the degree of satisfaction which is thus generated, quite apart from the avoidance of totalitarian decision-making wherein a mistake can be cataclysmic.

In view of this brief introduction to investment, I should first like to indicate something of the kinds of forestry investment which vie for funds, then show where the papers presented today fit into the forestry investment picture, and finally, in outlining the main points of these papers, raise some issues concerning them.

Looking at forestry as a whole, it is apparent that there are five major sources of value to people stemming from it, these being wildlife, forage, water, recreation, and wood, with perhaps two others which may be looked upon separately, namely, conservation in the sense of avoidance of loss due to soil erosion and floods, and the general esthetic value of the countryside, apart from formal recreation purposes. Each of these values infers necessary investment; each may be partially competitive in terms of joint investment; and each may yield, to some degree, both private and social values, which justifies both private and social or governmental investment, and the activity of both private and public entrepreneurs. The two forms of social investment by public action which concern us are those cases where most all of the benefits yielded by the investment are difficult to appropriate and sell privately, and those in which the initial investment costs are large and the returns long deferred and uncertain.

There is still a third, "grey" area of investment where some major proportion of the joint values or benefits may be appropriated and sold privately, yet where there is still a major proportion which cannot be so handled and must be looked after through government investment. A fairly well resolved example is the private concessionaire in a publicly financed forest recreation area. There are, however, other examples in forest recreation which cannot be solved so readily and which require private-public investment cooperation or regulation.

Numerous examples could be cited, such as joint private timber values and public flood control values, private hunting and public forests and game, and public water values from large private forests. Although there are many clear-cut private and public investment opportunities in forestry, there are also many which are joint by their very nature, and which require a great deal of public-private understanding and mutual cooperation. If there ever has been a resource which required such mutual understanding and appreciation of both public and private values and of their coordinated investment needs, forestry is it!

Of the several kinds of forestry investment previously discussed, the papers which have been presented today are primarily concerned with only one: that of investment for the long-term production of value from wood. And each is concerned with only a part of the investment or capital formation process so far as wood is involved. Ferguson[1] is concerned with differing specific situations involving the appraisal of private investment opportunities for funds of private institutions. Wilson,[2] on the other hand, deals with the general question of aggregate net private investment in Canadian forest industries, while Stoltenberg, Marty and Webster[3] are ultimately concerned with determining partial physical production functions and their relevance to public investment. Each author is playing the role of an entrepreneur-once-removed by appraising and pointing up various means by which components of decisions as to certain kinds of investment in long-term forestry (and in some cases, timber conversion) may be strengthened. Each is concerned with reducing the uncertainty surrounding long-term wood production investment decisions, some of which are private and some public. They have each contributed to the reduction of the variance around the estimate of marginal net productivity of capital in given situations, which, apart from comparing alternatives, is the central responsibility in investment decision-making.

After mildly admonishing foresters for not fully recognizing forestry as an investment situation and hence for not adequately so advising forest owners, Stoltenberg, Marty and Webster go on to point out by three examples how researchers, by providing basic information, can make it possible for practitioners to assist owners in making decisions as to their investment objectives. The authors go on to say that the practitioner's task is to help the investor choose among competing investment alternatives, at least within forestry. So far, so good.

But then they say that the researcher's function is two-fold: "First, to discover and quantitatively describe the response of forest stands to forest practices; and second, to compare opportunities which are relevant to a large number of investors." One could not argue with the first function: that of providing realistic production functions, showing physically quantified responses in terms of quantity and type of product and its timing in view of additional investment inputs of all sorts, from differing soil conditions to pruning, as well as showing their marginal effect in numerous realistic *combinations*. Forestry is badly in need of knowledge of such functions: not only factor/product relationships but also factor/factor and product/product substitutional possibilities. These authors' contribution to some of these basic physical relationships is a major step forward, because without them rational investment decision-making is nearly impossible.

But it is an extremely large jump to say that the next step is to compare opportunities relevant to a large number of individual investors; except in a relative, physical sense. To expect practitioners on the basis of this physical information to describe "each investment (opportunity), including a summary of the associated costs and returns," except in an almost purely subjective value-judgment sense, is still usually impracticable. Even when we know the physical input/output relationships, which have a long time dimension, fairly well, usually we must also be able to predict future input costs and output prices facing an individual, contend with the question of the market illiquidity of the growing stock investment, handle risk and uncertainty effectively, and in general, learn to understand the behavior of various kinds of investors when faced with these additional, truly economic issues.

It is one thing to invest in an asset which may fluctuate in its market value over time, but which can be sold and the investment thus liquidated; it is quite another to invest in an asset, such as growing stock, which is largely illiquid over many years. In effect this constitutes a bet with a very long waiting period before the possible pay-off, rather than a marketable investment in the normal sense of the term. How can we develop market mechanisms which will make forestry investments more nearly approximate normal investments? That is, how can the market be adjusted to reflect future value, thus making growing stock investments currently marketable assets? Knowing production functions is a step in that direction, but still only a help and not a complete solution.

Estimation of changes in future costs of labor, production, and the myriad of possible forestry inputs over extended time periods is subject to extreme speculation, especially when great changes in technology itself are obviously possible. Changes in output prices (including the possibility of a zero price facing a given investor in a particular locality) can and do change radically relative to individual situations in the face of changes in both taste and technology, as well as industry location. What measures can be taken to reduce these uncertainties facing the individual entrepreneur-investor in forestry? It is true that various product substitutions are possible, both over time and at a given time, and such flexibility in output alternatives tends to reduce uncertainty regarding future net income, but there has been little if any serious investigation on this aspect of the problem.

And what about risk and uncertainty themselves? In a forestry investment situation with extreme illiquidity, does the time-honored opportunity or alternative rate discounting and interest system really rationally fit the circumstances in making comparisons with nonforestry investment or consumption alternatives? Even in those cases where it may be argued for, as in the instance of larger vertically-integrated forestry firms and governments, its estimation may present serious difficulties and be subject to considerable variance relative to any given long-term forestry alternative facing the firm or government. In the case of smaller, nonintegrated, potential long-term forestry enterprise investments, it seems doubtful if the concept of an alternative rate, especially regarding time preference rates for consumption as opposed to such investment, is at all acceptable. Perhaps, for example, these cases should be examined from the point of view of game theory where the participants are faced with taking long-term bets under circumstances of extreme individual uncertainty, in which the wager may include some of the owners' time and effort, or money, or both.

These are some of the basic questions of uncertainty which in most cases must be faced and to some reasonable degree resolved in the study of investor behavior and future markets, before practitioners can talk objectively of "costs and returns," or in more general terms, of the marginal net productivity of long-term forestry capital investment alternatives. I should like to stress that relative physical input/output productivity ratings are useful, and that in fact they may be completely adequate in particular circumstances where an agency or firm has a given budget and is faced with a single product maximization decision; but it must be remembered that this is really only a *partial* decision concerning investment. It constitutes a form of suboptimization, very useful in special circumstances, but not sufficient to meet the needs of decision-makers in the case of more typical *whole* investment decisions, in either private or public circumstances.

In the white pine blister rust control investment decision process previously referred to by Stoltenberg et al., the forecasts of the future expected *average* stumpage price over a large region have some validity because of the spatially aggregative treatment of the product; but for a given-sized, fixed-location owner's forest, the variance of the estimated expected future price of his stumpage would tend to be extreme, and hence of little use. Likewise, even though the selection of the $2\frac{1}{2}$ percent alternative rate (used in this instance as a social rate) may be contentious, it can be argued for at least to some extent; but for a spatially less extensive *individual* owner, whether private or public, the question often becomes almost moot. It is interesting to note that one of the major benefits from a governmental blister rust control program guided by an intensive study of this kind is the reduction of uncertainty and the increase in physical productivity which it affords many private owners, thus increasing their incentive to invest individually, i.e., raising the marginal net productivity to them of forestry capital.

Wilson's paper suggests a somewhat analogous situation in his appraisal of forest industry investment opportunities in Canada. There the main problem is not the cost of investment which firms must face in growing wood, since, on the face of it, they have a better alternative, namely that of extending supply lines further into more or less virgin areas. But in a country having partially developed resources, such as Canada, these firms are acutely pressed by the lack of infrastructure, that is, investments in social capital such as public roads, railways, and social services of various kinds, extending into the areas they would have to use. Since the major export demand for forest products from Canada is that of the United States, industry investors are faced with paying the rising marginal costs of extending their geographical wood harvesting margins, ostensibly largely by themselves, or paying the marginal costs of more intensive forestry and wood production in the United States.

From Wilson's discussion one would conclude that the marginal unit costs at these two margins are about equal in many instances, and particularly in many cases in the all-important pulp and paper industry. The two keys in deciding which way industry will tend to allocate its forestry investments would, then, seem to be the expected marginal unit costs of increasing wood supply under these two different circumstances. In the United States it will depend to a considerable extent on the marginal net productivity of forestry investment alternatives facing the industry, which may be affected considerably by long-term social investments of the kind mentioned in the previously discussed paper. In Canada the immediate problem is not so much the reduction of uncertainty and the increase of productivity in private forestry investments by social investment in programs such as blister rust control, as it is in the United States. Rather the problem is, as in any geographically expanding country, the issue of saving and investment in social capital, which makes private enterprise possible.

As Wilson points out, governments in Canada so far have provided little investment in forest resources development, with the exception of forest protection. To the extent that they do provide more investment in infrastructure available to forest industries, such as in the embryonic federal-provincial program of forest inventory, protection, and road development mentioned by Wilson, they will tend to overcome the marginal costs to industry of investing in Canada, unless of course, the marginal net productivity of long-term private forestry capital investments in the United States is still further lowered by measures which reduce uncertainty and increase physical productivity there. It would seem entirely fortuitous for there to be sufficient spatially complementary investment in other industries in Canada, such as the isolated mining example mentioned by Wilson, to materially affect forest industry investment costs in more remote areas. Short of the development of extended infrastructure, the only other alternative for further major international demand-inspired forest industry investment in Canada would appear to be the possibility of finding that long-term investments in the growth of wood nearer the market centers of the United States were cheaper to the investor than either extending supply lines within Canada or growing the wood in the United States. Of course, an alternative of this sort would have to be planned considerably in advance of the time of harvesting such wood.

Turning to Ferguson's paper, it is important to recognize that its purpose is to discuss timber loans and the problems involved from the point of view of private institutional lenders. After pointing out that some businessmen have in recent years recognized value in timber and timberland as a long-term asset, that is, recognized that it had become an economically scarce resource and hence subject to considerations of private profit, Ferguson briefly discusses the expected concomitant, institutional lending on this basis, while pointing up the specialized knowledge required of such lenders and the fact that the volume of purely forestry loans from such sources is still not large. He then defines "forestry loan," discusses the general considerations and examinations in making such loans, outlines the major types, that is, industrial loans, standard forestry loans, loans backed by leases, joint agricultural-forestry loans and other special types of loans. After this he discusses at some length forestry loans on small acreages, where lack of investment is most acute.

A forestry loan in this circumstance is one which is secured by a mortgage on land and timber, usually requiring a given amortization rate with provisions for the control of the release of mature timber in view of the outstanding debt. The usual examinations of the borrower's additional assets, liquidity and ability to pay are also conducted, in addition to a detailed examination of the forest property.

Effectively, the loan, in most cases, is made on the value of the forest as a liquidatable stock resource, that is, primarily on the value of the standing mature timber, under circumstances where it could be liquidated if necessary. The growth or forestry aspects of the forest are of importance in the sense that growth is necessary to maintain an inventory of liquidatable timber in the necessary value ratio to the net value of the loan outstanding at any point in the time of the 20- to 30-year loan period. Although, apparently loans are not made on any value attaching to the growing stock per se, it is important to the lender in terms of maintaining the inventory on which the standard forestry loan of 30 to 50 percent is made. Actually then, such loans are not made on the basis of bare ground useful for forestry purposes, nor on growing stock by itself, unless there is also currently available mature timber having a liquidatable value equal to two or three times the amount of the loan. As Ferguson points out, a loan primarily for development purposes may be made in a limited manner only where sufficient income is available to carry interest and protect the forest, which income would presumably have to come from reasonably secure nonforestry sources in the absence of mature timber to be cut.

Time does not permit further discussion of all of the other types of loans mentioned, but the industrial and small forest loan situation bear further consideration.

The industrial loan, usually made to pulp and paper companies, may equal the total market value of the land and timber, where an examination shows the company otherwise to be a good risk financially and where a long-term lease or cutting contract provides sufficient net income to pay the loan charges. Even in this case the loan could not exist solely on the basis of land and growing stock productivity since, by and large, the market value does not recognize this potential.

On the other hand, loans on the so-called smaller acreages, in the 30- to 50-percent loan category, normally face a minimum of $25,000, presumably requiring a liquidatable value of $50,000 to $75,000 for mature standing timber, which of course, eliminates essentially all of the smaller holdings. Ferguson argues that this is as it should be, even considering subsidized loan programs, and one would be hard put to disagree with him, for several good reasons. In the first place, many small acreages are held by "accidental" owners, where the forest land is a residual from some other land use, usually agriculture, hence there is little reason to believe that these owners should be interested in long-term forestry simply because they own the land. And, as Ferguson says, many people own forest land for motives other than income from wood production. In the case of these, usually lower-income owners, time preferences for consumption and leisure time alone are so high as to eliminate saving, whether in money or effort for almost any purposes, let alone for extremely uncertain forestry investment purposes. These owners almost always feel they have many better alternatives for investment than long-term wood production on their lands, when in fact they are in a position to save. Even those who do have an interest in forestry and some inclination to invest in it, find it hard to conceive what they are really getting into, quite apart from technical forestry knowledge. Obviously then, most such owners would not accept a loan, even at a zero interest rate, nor even on growing stock where the proceeds were required to be invested in wood production on the property. In the case where standing timber existed and a loan were offered and accepted, the pressing need for consumption expenditures in addition to debt repayment may very well result in still more cutting back of growing stock. Although the conditions of the loan may restrict this practice, private enforcement would be difficult. As Ferguson says, equity, at least, is required for a loan, and the owner does not have it in a depleted woodland since the market does not recognize it. Obviously loans do not constitute a solution.

In reflecting on the papers presented today, the common symptoms of long-term forestry investment problems appear to be the lack of market recognition of many forestry values, particularly those inherent in growing stock and bare forest land in the case of wood production. These symptoms are particularly acute on smaller woodlands, which constitute a large share of the total forest land.

To expect the market to recognize these values will require some further answers to questions such as these:

To what extent can reasonably reliable, locally applicable production functions be developed?

How can productivity be increased and to what extent?

Generally, what avenues of approach to the subjective uncertainty problem do we have and how may they be used?

How can complementary long-term private-public cooperation in forestry be enhanced?

What new market forms are needed?

And what will it take to sufficiently understand the behavior of small woodland owners, so that greater, relatively efficient forestry investment on their lands may become a reality?

REFERENCES

1. Ferguson, Eli. 1960. Timber investments from the standpoint of institutional lenders. Proc., Society of American Foresters, Division of Forest Economics and Policy. Pages 88–90.
2. Wilson, D. A. 1960. Investment opportunities in forest industries in Canada. Proc., Society of American Foresters, Division of Forest Economics and Policy. Pages 91–94.
3. Stoltenberg, C. H., Marty, R. J., and Webster, H. H. 1960. Appraising forestry investment opportunities—the role of the investor, the forestry practitioner, and the researcher. Proc., Society of American Foresters, Division of Forest Economics and Policy. Pages 82–85.

23

A LABOR SITUATION ANALYSIS OF THE NORTHEASTERN PRIMARY WOOD-PROCESSING INDUSTRIES

Robert S. Bond

First published in *The Northern Logger and Timber Processor* 20(8):18–19, 28–29, 31. March 1972.

BIOGRAPHY

Mr. Bond is Associate Professor of Forestry Economics in the Department of Forestry and Wildlife Management, University of Massachusetts. After graduation from the Yale School of Forestry, he worked as an industrial forester in Arkansas and then as a field forester for the state of Massachusetts. He joined the University faculty in 1956. As a National Science Foundation Fellow, he took leave during the early 1960s to take advanced studies in forestry economics at the State University of New York, Syracuse. Thereafter continuing his teaching and research in both timber and nontimber aspects of forest resource management, he became a leader in the study of forest and forest-industry labor, the topic addressed in the article reprinted here.

EDITORS' SUMMARY

Labor problems were studied in four branches of the wood-products industry in Kentucky, Maine, and Pennsylvania. Pulpmills ("chipped wood") and veneer plants ("sliced wood") generally had the advantage over harvesting operations and sawmills in respect to such indices as wage rates, regularity of employment, training, labor recruitment and retention, safety, and the working environment.

QUESTIONS TO CONSIDER

(1) Mr. Bond finds that there is a hierarchy of wood-using industries respecting labor conditions, with pulp-paper at the top, followed by veneer-plywood, sawmilling, and, at the bottom, logging. Can you think of other respects in which this same hierarchical arrangement holds? How do you explain the hierarchy?

(2) Why does the raising of wage rates increase absenteeism in sawmilling? Would you expect the same result in pulp-paper production?

(3) How would you go about identifying (page 178) the revenues generated by the sawyer in a sawmill separately from those generated by the other workers?

(4) Suppose you are planning to describe and explain labor conditions in wood-using plants within a region. Would it be preferable to examine a sample of the plants in the region, or a sample of whole communities where plants are located, or some other type of sample? Defend your choice.

(5) Apparently in the region studied by Mr. Bond, workers, employment agencies, and the public hold wood-using industry generally in low regard as an employer. Why is this? To what extent is it due to the industry's inherent characteristics? In your observation, how do other regions compare with the one Mr. Bond surveyed?

* * *

WAGE AND FRINGE BENEFITS

If the four industry groups are compared, hourly wages are most favorable in the chipped-wood industries, as are the fringe benefits. This situation is well known, and is attributable in part to the predominance of larger firms and organized labor. Other contributing factors are those cited by Slichter.[1] He states that "When high wages or earnings characterize an industry, (1) labor costs are a low percentage of total production costs or income from sales; (2) the capital investment per worker is large; (3) the value added by manufacturing per wage earner–hour is high; (4) employment in an industry has been expanding rapidly; (5) skill requirements are high."

These factors generally do not prevail in the sliced-wood, sawn-wood and timber-harvesting industries, although the sliced-wood industry exhibits some of them.

Higher wages are not necessarily the solution to gaining more productivity in any of these industries. In the sawmill operations where absenteeism was frequently a problem, it was noted that raising the hourly rate of employees often created a greater absenteeism. Kentucky sawmills and harvesting operations appeared to be particularly susceptible to such reaction on the part of workers, although it was evident in other states. There may also be a relationship between public assistance programs and the reluctance to work a normal work week at higher hourly rates.

An often heard reason for low wages in sawmilling and logging was the squeeze between the price received for the product and that paid for the raw material. In sawmills, stumpage prices rising at a faster rate than lumber prices, without compensatory increased productivity per man, ostensibly resulted in a squeeze on wage and entrepreneurial profits. Pulpwood producers use the same rationale regarding prices of wood delivered to the mill and their capacity to pay higher wages.

Sawmill labor appears to lack a sufficiently wide wage range between the lowest paid common-labor job and that of the most highly skilled employees, the sawyer or sawfiler. In the region, I estimate the average starting wage for a lumber handler approximated $1.85 per hour. I do not believe sawyers' wages would average more than $3.00 per hour. Two reasons may be advanced for justifying a wider range of wages, with the more skilled workers receiving a proportionately higher rate. One, the sawyer is entrusted with much of the firm's capital investment and two, the revenues generated by the sawyer far out-weigh those attributed to jobs having lower skill requirements.

Wages are generally responsive to competition in the labor market. However, too often men employed in sawmills or harvesting are limited in employment alternatives open to them. Employers do not always rely upon workers with few alternatives by choice, but it is often the only segment of the labor force available. Sawmills and harvesting operations afford a social good by providing employment to that portion of the work force that is not capable of obtaining employment elsewhere. Public assistance programs have increasingly made inroads on the labor supply for low-wage industries. The additional income from working is not sufficient to encourage employment.

The movement toward more fringe benefits in the sawn-wood industry is evident where competition in the labor market prevails. Many Maine and eastern Pennsylvania sawmills have life, medical, and hospitalization insurance programs, paid vacations and holidays, and some have or plan to institute retirement programs. In western Pennsylvania and Kentucky where labor of the type historically relied upon in these industries is still more readily available, fringe benefits are not prevalent.

HOURS AND SHIFTS

The standard 40-hour week is common throughout the industries studied. However, there are situations that depart from this norm.

Chipped-wood industry plants operate on a seven-day week and a 24-hour day. These operations have four shifts and rotate them so that each is always working a different set of hours and days for every rotation period. A shift wage differential is paid and Sunday or holiday work is generally at a double-time rate. Shift work is one of the few deterrents to attracting labor into this industry, but it does not appear to be critical.

A number of sawmill owners who had fairly low hourly wage rates rationalized that they offered a 44- to 50-hour work week, and thus the weekly wage earned was comparable to other employment opportunities. The fallacy in this kind of reasoning is that workers are unconcerned about the number of hours worked per week as long as their wages are at a given weekly level. Labor willing to work long hours is declining and in the future can not be depended upon to supply the needs of the wood industries. In many situations where a sawmill operator would like to operate for more than a 40-hour week, he is unable to do so because the work force does not want overtime.

Hours worked per week in the timber-harvesting industry are difficult to document because of its organization. Records of hours and weekly earnings for some Maine pulp harvesting operations indicate an ability to earn a relatively high hourly rate when being paid on piece-rate basis. Thus, a chopper or skidder operator can earn a competitive weekly wage, compared to other alternatives, working fewer than 40 hours.

TRAINING AND JOB DEFINITION

Formal training programs by wood-industry firms or associations are uncommon. Some association-instituted training has had limited success. Much of the failure seems to derive from the type of individual attracted into the programs. The screening of applicants to test their motivation and suitability for the work is often lacking, and hence the training fails to increase the labor supply. The desire to fill quotas for authorized schools for which monies have been appropriated encourages the acceptance of applicants seeking short-term benefits and not those who wish to become proficient in a skill in order to earn a living.

In-firm training has frequently met with problems. If a man is trained on the job but does not step immediately into this work he often has the opportunity to go elsewhere to employ his new skill. Therefore, employers express reluctance to venture into a training program until a particular job is available. At this juncture numerous costs may arise because of production delays, or, in the extreme, mill shut down.

The generally low skill levels for most jobs in sawmills is also a deterrent to training programs. In the woods, although skills are required, it is not uncommon to place a man on a job with no experience and a minimum of training. The implicit costs of such action too often are overlooked in the urgency of getting someone on the job.

Generally, large, administratively well-organized firms have job definitions. Job openings are filled by bids from interested employees, and seniority takes precedence in selecting men to fill these positions. The depth of skill-level classifications for any one job description permits training by those working at the upper level. In chipped- or sliced-wood plants, jobs are classified, tasks defined and a pay scale is fixed for the job.

Sawmill workers are often called upon to do other than those operations that they normally perform, in which case job definition becomes impractical. Diversity of tasks performed may be one of the reasons for a narrower spread of wage ranges in sawn-wood and timber-harvesting industries.

Where job classifications are established and mobility from a lower-paying job to a higher-paying one is possible, the labor situation is generally more favorable. An employee desiring to improve his status can look forward to better opportunities within the firm. Lack of job definition leaves the work force in an uncertain position.

Although unions are generally looked upon unfavorably by most sawn-wood and timber-harvesting firms, it appears that the definition of tasks required by unions tends to be a factor in improving labor availability. However, simply defining the tasks will not solve the labor-supply situation. Workers seeking long-term employment look for a chance to progress in the firm.

Job progression and long-term security are becoming increasingly important to the young job seeker. It is this young segment of the work force on which future production in the wood industries must rely. To continue to depend on a labor force that does not look to the future is to continue to proliferate the labor force problems that many segments of the wood industry have faced for years.

LABOR RECRUITMENT

Wood-industry firms generally rely on job applicants seeking employment in contrast to a recruitment program. Because chipped-wood industries are, for the most part, considered a desirable place to work, they can be selective in choosing employees, whereas the sawmill operator feels he must hire whoever comes along.

Very limited use of state employment services was found in the three states visited. Often if a man is referred to a sawmill for work by a state employment service, he contacts the mill with the hope of being rejected, thus remaining on unemployment compensation roles. Many stories were related of referrals who took employment and lasted a few hours to a few days.

I talked with a number of managers of state district employment offices. Their view of the timber harvesting and sawmilling industries in terms of potential employment was low. Although some managers were sympathetic, others seemed to hold the opinion that work in these industries was so unattractive they did not encourage prospective workers to seek such employment. Where new plants are being built, cooperation of public employment services may be used in recruiting the work force.

Passing the word to existing employees of sawmills or harvesting operations is the most common way recruiting is done. Newspaper advertising generally results in little response or attracts job seekers with characteristics similar to those coming from public employment offices.

Information about the background and work history of job applicants in the sawmill and logging industries is often lacking. As a result predictably itinerant workers are hired because no attempt is made to ascertain their employment history. Even though management is aware that hiring a man may be for a short period, the need to keep the operation running results in employing anyone who comes along.

RETENTION OF THE LABOR FORCE

Sawmill and timber-harvesting firms have far greater problems in labor turnover than do those producing sliced- and chipped-wood products.

Turnover in sawmills is generally high for unskilled workers for which the wages are lowest. On the average, 40–50 percent of the employees in a mill are long term. They hold jobs such as sawyer, edgerman, trimmerman, and grader. Lumber handlers have an extremely high rate of turnover. Many mill operators told of employing two or three times as many different men in a year as were used in the entire milling operation.

High turnover rates are not necessarily unique to the sawn-wood industry. Other industries with low-skill requirements have this problem as illustrated by a report[2] entitled "Labor Shortages in the Corrugated Box Industry."

Implicit in a high labor turnover rate is the cost incurred. In some cases this cost is not recognized or the firm feels that it cannot do anything about it. A few sawmill operators seemed aware of the cost incurred by high turnover rates. In an extreme example, one mill owner estimated his costs were doubled for a period of two to three months when a new man was hired to pile lumber.

Most labor turnover in sawn-wood and timber-harvesting firms is employee instituted, that is, they quit as opposed to being dismissed. The reasons for leaving employment may originate because of friction with management, but the toleration by management of poor employee working habits appears too lenient. However, there is little doubt that poor management-labor relations contribute to the turnover rate.

ABSENTEEISM

A situation closely related to turnover is that of absenteeism. In those industries in which turnover is high, so is absenteeism. One of the reasons is that management tolerates the absenteeism by not dismissing those workers who are habitually absent. Also common is failure to notify employers when absent or leaving employment.

Absenteeism may be very costly to the firm as well as to the absentee's fellow workers. As an example, one sawmill visited in Kentucky had worked only two normal 40-hour work weeks during 1969, reputedly because sufficient numbers of the work force did not report for work.

In order to insure that absenteeism is not a constraint in the operation of the mill, sometimes more men are hired than are needed. This is a wasteful practice, even to employees, as it tends to stifle incentive and detract from the wage and fringe benefits that might accrue to the conscientious workers. The cost of this practice may exceed the cost of acquiring a more stable and better motivated work force.

MANAGEMENT-LABOR RELATIONS

The lack of general concern for labor is found in those two industries where problems are greatest. Evidence of this was apparent in attempting to discuss labor with sawmill managers and the few loggers who are actually employers. These men would rather talk about problems of acquiring raw material and the function of machinery than about labor. This was true where, admittedly, labor was the greatest deterrent to operation of the firm.

Many, by no means all, sawmill operators and loggers are marginal in their general business knowledge and have little desire to discuss those facets of the business which they least comprehend. In those firms in which management is business oriented, that is, knowledgeable in the financial and marketing aspects, there is greater likelihood that labor problems are not as acute.

Unions are found in nearly all chipped-wood manufacturing firms. Some unions are found in sliced-wood plants, but even where they are not, working conditions and organization of the labor force follows that fostered by unions.

One sawmill visited has an in-shop union which seemed to be working well and could serve as a model to other mills. This union was fostered by the management against the advice of legal counsel. It, along with other aspects of this firm, created a favorable employer-employee relationship.

A past history of union organization was evident in sawmills located in the coal region of Western Pennsylvania. Five mills were organized during the 1940's by the United Mine Workers of America. Unions in these sawmills were not formed for the express purpose of benefiting the workers, but, as a necessity of marketing products to unionized mines. Because union contracts had run out, employees asked discontinuance of deduction of union dues from their pay. Lack of concern by union officials of this action seems to substantiate this unionization was purely a business arrangement and that the U.M.W. had little concern for the welfare of the sawmill employees.

Unions in the timber-harvesting industry are uncommon in the Northeast. Only in Maine, of the states visited, where pulp and paper manufacturers conduct integrated company operations, is there evidence of organization of the labor force in the woods. This unionization is of such recent occurrence that an evaluation is impossible at this time.

Management-labor relations are generally most favorable in the larger firms where there are professional personnel administrators. An effort to foster good relations with employees generally results in a favorable management-labor situation, but personalities play a major role.

SAFETY

The poor safety record of the two problem industries is well known. In terms of labor, it is the area that has received the greatest amount of attention. In particular, the timber-harvesting industry is noted for its poor safety record reflected by high lost-time and severity rates which, in turn, result in exorbitant costs for workmen's compensation insurance.

It is my conclusion that raw-material procurement methods have helped foster the safety problem. It is not clear which came first—high insurance rates or the independent logger. In any case, the small independent logger and his method of operation, in the Northeast, at least, contribute to the problem. The lack of training and supervision are obvious, even to the casual observer.

Safety in sawmills has tended to improve with changes in technology. However,a number of mills have working conditions so cluttered and disorganized that safety is at a low level. Modern mills with up-to-date labor-management practices require wearing of safety gear. Too few safety requirements are found throughout the industry. The belief that requiring safety gear will deter people from working in the mill is probably not true of the young workers entering the labor force.

Chipped- and sliced-wood industries have excellent safety records. Not only can this be attributed to safe working conditions, but to a concerted safety program. Again, the effects of professional personnel management are evident by the results achieved. The potential worker who has a choice will seek the industry with the best safety record for employment.

There is little doubt that the high costs of workmen's compensation insurance reduces returns to the worker in sawn-wood and timber-harvesting industries. Additional hourly wages could be achieved if insurance costs were reduced.

WORKING ENVIRONMENT

The working environment is a factor in getting and retaining labor. Modern-day manufacturing plants are engineered with worker comfort in mind, as well as for efficiency in production. In chipped-wood and sliced-wood plants the working environment is usually favorable, particularly in new or renovated plants. The sawmill is too infrequently designed with worker comfort and safety in mind. In newer mills there is some consciousness of these factors, but in most older mills this is not the case.

Altering the working environment is more difficult in timber harvesting operations. However, increased use of machines permits heated and air conditioned cabs, thus alleviating some exposure to weather extremes.

The working environment, particularly in sawmills where much improvement is needed and can be achieved, would be one of the greatest contributions to attracting a more stable work force into this industry. The increased production that could result from efficiency and improved attitudes of workers would, in mills with attractive surroundings, easily repay the investment in a short period of time. Management's attitude is generally the reason that poor working environments exist and this, in turn, affects the attitude of workers.

Altering existing working environments can also bring on labor problems. The attitude of the existing work force toward use of a new lunch room was unexpected. Three long-time employees quit because they were not willing to abide by the ruling that they use this area for eating the lunch they had brought. In another case only two or three employees out of fifteen or more took advantage of a new lunch room. Perhaps one would conclude that this was a poor investment. This might be true for the existing labor force, but I do not think this is true for the labor force of the future.

FLUCTUATIONS IN EMPLOYMENT

Historically wood industries have been subject to seasonal operations because of weather as well as cyclical fluctuations in employment caused by man. Timber harvesting is the industry most subject to weather, but technology has reduced most seasonal operations except for relatively short periods during the spring months.

Employment uncertainty brought on by wood-procurement methods is a major deterrent to making the timber-harvesting industry a desirable place of employment.

In the past, and even at present, farmers frequently turned to wood production as an investment and employment alternative. The supply of future woods workers with this background will be less because of industrialization of rural areas and the migration of the rural young to urban centers. Many of those who remain in rural areas tend to be less formally educated, poorly motivated, and in terms of farming a small logging enterprise are higher credit risks. Woods work is no longer experienced by even rural young people. The fluctuating employment and high risks of timber harvesting under the prevalent producer or contract system does not appear to be attractive to future generations of the labor force.

* * *

REFERENCES

1. Slichter, Sumner. February 1950. "Notes on the Structure of Wages," Review of Economics and Statistics, pages 80–91 as quoted by Sanford Cohen 1960, Labor in the United States. Columbus, Ohio: Charles E. Merrill Books, Inc. Pages 370–371.
2. Anonymous. October 1969. Labor shortages in the corrugated box industry. Labor Shortage Task Force. Personnel Administration and Industrial Commission, Fibre Box Association, Chicago, Illinois. 17 pages.

24

GAINS IN LABOR PRODUCTIVITY BY THE LUMBER INDUSTRY

H. F. Kaiser
Sam Guttenberg

First published in *Southern Lumberman* 221(2747):15, 18. October 1, 1970.

BIOGRAPHY

Mr. Kaiser is Forest Economist in the Division of Forest Economics and Marketing Research, Forest Service. A graduate of Iowa State University, he went on to a master's degree there and a Ph.D. in forestry economics at Michigan State University. His first Forest Service post was New Orleans, where his prime interest was in regional input-output analysis and in labor economics. At Lincoln, Nebraska, cooperatively with the University of Nebraska, he made a national study of forest range investment opportunities. His research in Washington, D.C., concerns the consumption of timber, range, and recreation resources in the United States.

Mr. Guttenberg is Chief Forest Economist in the Southern Forest Experiment Station of the Forest Service, where he has been employed since his discharge from the Air Force at the close of World War II. He was educated at the University of Michigan and at Harvard. Through his extensive research, his lectures, and his more than 100 publications, he has become known as a leading expert on the southern timber economy: forest management, harvesting, marketing, and utilization. He holds an award for distinguished service to Louisiana forestry and the grade of Fellow in the Society of American Foresters.

EDITORS' SUMMARY

The lumber industry, long a laggard in technology and labor productivity, exhibited remedial change during the two decades following the middle of this century, notably in the South. Many small, inefficient plants dropped by the wayside. Remaining firms improved their design and performance. Output per man-hour rose about 3 percent in the average year.

QUESTIONS TO CONSIDER

(1) Messrs. Kaiser and Guttenberg cite an average annual rise in sawmill labor productivity during the two decades following World War II. Would you expect this trend to continue? Do you know of evidence that it has or hasn't continued? Where would you look for such evidence?

(2) What accounts for changes in labor productivity in a manufacturing industry? Are they a function of workers' expenditure of energy? or of other factors? What were some of the factors involved in the case of the lumber industry?

(3) How do you suppose that per-man-hour output of professional foresters has changed over the decades? Why? In what terms do you visualize output? What about per-man-hour output of forestry professors? of forestry students?

(4) Select a state or region with which you are familiar and make a small library study, with the help of the census reports, of changes in numbers of sawmills by size class (note that size is measured in output, not in capacity). How do you explain what you observe? Can you use the principle of comparative advantage in your explanation?

(5) Do higher wage rates lead to higher labor productivity, or is the sequence the opposite?

* * *

A favorite sport of some researchers has been to single out the lumber industry as being below par in labor productivity. The prestigious National Bureau of Economic Research, for instance, has ranked lumber last among 20 major industrial sectors in efficient use of labor. While statistics for the first half of this century tend to support the claim, records for the past 20 years tell a significantly different story. Since 1950, the nation's lumber industry has changed greatly. The number of mills has been drastically reduced and labor productivity in the surviving mills greatly improved. Many new mills have been erected, others modernized, and manufacturing firms generally have raised their annual output. Breakthroughs in milling technology and improved forest resources in some regions have also helped. The result has been industry-wide gains in labor productivity, with the pace set in the South. Today the southern industry, in particular, is probably in a better position to sustain growth and development than at any time in the past half century.

PRODUCTIVITY TRENDS

Productivity of an industry is customarily measured by changes in output per man-hour. Expressing output solely in terms of labor does not imply, of course, that labor is the only factor contributing to changes in productivity. Technological innovation, rate of utilization of capacity, and scale of output are also important.

In the estimates given here, output was expressed in physical units to eliminate problems associated with price changes. Complications associated with changes in the characteristics of lumber products and logs sawn over the years were ignored in the analysis. Also ignored was the increased merchandizing of residues—pulp chips, green sawdust, planer mill dust and shavings, and bark. Some of these changes reduce productivity and some raise it; on balance, the productivity estimates given are likely to be conservative. Basic data were obtained from the periodic Census of Manufacturers and the Annual Survey of Manufacturers.

1967 PRODUCTIVITY EXCEEDED 1954

The calculated output per man-hour for sawmills throughout the United States in 1967 exceeded the 1954 figure by 49 per cent. The average annual rate of increase of 3.2 per cent was a marked improvement over the 1.2 per cent in the first half of the century. Regional differences were also found. The South led the nation, with an annual increase of 3.4 per cent, followed by the West with 2.9 per cent and the North with 2.3 per cent.

Gains in labor productivity are usually associated with industry-wide increases in output. Rising output creates a favorable climate for innovation. It often spurs new thinking about technology and encourages purchase of the newest and most efficient equipment. This traditional pattern does not fit the lumber industry. Lumbermen improved productivity by steadily reducing the work force while annual output remained about the same. Total man-hours by the industry's workers dropped 46 per cent from 1954 to 1967.

A key factor leading to the reduction in the labor force was elimination of small portable mills. In the past, the number of mills in operation fluctuated in response to cyclical economic conditions. As the market became favorable, numerous small mills would become active, then quickly shut down when the market soured. In that era, entry and exit from the lumber industry was relatively easy. Capital requirements were not large, particularly for small mills, and sawmilling required relatively few highly skilled workers. Except for a few key positions, much of the work could be done by semiskilled and unskilled labor.

The frequent changes in number of mills in operation created unfavorable conditions for improving productivity. With large numbers of mills entering and going out of business as economic conditions changed, banks and other sources of funds were reluctant to provide long-term loans needed by the industry to make plant improvements. In addition, permanent sawmills did not benefit as much as they might have during periods of rising lumber demand. An upward shift in demand was usually met in part by the in-and-outers that used old equipment and were inefficient in terms of labor productivity. These operations were profitable as long as the lumber market remained favorable, but they deprived established firms of a chance to accumulate capital for future long-term investments.

MAJOR REORGANIZATION

Relative stability of the lumber market since 1954, however, brought about a major industry reorganization. The fluctuations that occurred in the market were mild compared to former periods and provided incentives for efficiency by driving out inefficient producers. The Bureau of the Census estimates that the number of mills decreased from 46,000 in 1954 to 25,000 in 1967. While the nation's mills declined by 46 per cent, the decline was a more drastic 80 per cent in the South. Most of the mills terminating operations were small. They lacked the willingness or ability to compete when the emphasis was placed on efficiency rather than flexibility.

Another key factor leading to the reduction in the labor force was the pressure created by rising wages. Contrary to most predictions that increased mechanization and automation would create substantial unemployment, manpower has become scarcer. Growth of employment in the service sector is the major cause. In education, for example, emphasis is placed on reducing the ratio of students to teachers. Over all, the number of people providing services increased 70 per cent from 28 million in 1947 to 47 million in 1967. Substantial unemployment has been avoided because consumer preferences for more and more services have expanded jobs in numbers equal to those saved in manufacturing plus the new additions to the labor force.

Over all, hourly labor costs rose in the lumber industry at an annual rate of 3.9 per cent between 1954 and 1967. Legislation contributed to the rising cost of labor. Minimum wages increased from $0.75 in 1954 to $1.60 in 1968. The South was affected more heavily than other regions because legislation pushed up pay for common labor much faster than it would probably have risen otherwise. In addition, workmen's compensation, social security, and other mandatory "fringe benefits" have added to labor costs.

It is clear, therefore, that improvement of labor productivity in the lumber industry was brought about by the creation of an environment that favored continuous substitution of machinery for manpower. To meet the rising costs of labor, mill managers intensified their effort to improve manpower efficiency. Mills were redesigned with emphasis on mechanization. For instance, handling lumber on conveyor systems and forklifts became commonplace. Many mills also reduced the need for extensive handling by simplifying their product line. An extreme example is the small, highly efficient mill that produces only studs and pulp chips.

As a result of improved productivity, the lumber industry has avoided substantial increases in unit labor cost. Over the period from 1954 to 1967, unit labor cost for lumber manufacturing increased at an average annual rate of 0.7 per cent, which compares favorably with the average for all manufacturing of 1.5 per cent.

Part Three

FOREST MANAGEMENT

25

THE MULTIPLE PROBLEMS OF MULTIPLE USE

John A. Zivnuska

First published in *Journal of Forestry* 59(8):555–560. August 1961.

BIOGRAPHY

Mr. Zivnuska's biographical sketch is given at Item 2, page 16.

EDITORS' SUMMARY

Forest management for multiple products is increasingly demanded by the public. Such management is difficult because of our uncertainties, especially about physical relationships among products and about product values. This uncertainty fosters controversy. The successful manager works with both; he is sensitive to political as well as economic influences.

QUESTIONS TO CONSIDER

(1) A timber production function may be thought of as a set of equations or tables, empirically derived, in which growth or yield is related to such variables as tree species, site quality, and stand age, condition, and stocking. How are we to visualize a production function for timber and forage which takes their interactions into account? How about a production function for all the principal types of forest outputs? Consider problems such as holding the numbers of variables to manageable proportions and devising suitable units of measure.

(2) Mr. Zivnuska mentions on page 194 the difficulty of arriving at an interest rate (guiding rate of return) for expressing the cost of capital. Take the case of a comparatively simple one-product industry: Christmas-tree growing. List the chief considerations that you, as a consultant, would take into account in recommending a rate to a firm in this industry. How would your problem differ if you were advising a multiproduct forestry firm?

(3) If the decisions of multiple-use forestry are, in the last analysis, political, then it is helpful to understand the political analogue of the market. What is this analogue? Explain your view. What is the relationship of political science to other social sciences?

* * *

What do we mean by "multiple use"? In a restricted economic sense, the term simply means that forests and wildlands have more than one use—that the typical forestry enterprise produces more than one product. In itself, this is certainly no novelty in our economy. Indeed, it is the multiple-product enterprise which is representative of our entire economy, and the single-product concern which is the rarity.

Thus multiple use might be interpreted as a slogan term for multiple products. These multiple products may be the joint products of traditional production economics—two or more products resulting from a single process, such as the production of peeler logs and sawlogs from the harvesting of a tree. They may also be the products of production processes which are separate except for the sharing of some managerial input—an area in which traditional economics is somewhat hazy.

But multiple use is something more as well as something less than an economic term. It also describes a particular philosophy and method of land management—a method closely identified with the approach of the U.S. Forest Service. The basic concept here is that efficiency in the management of the varied resources of forest and wildland areas for various purposes is best obtained under a single coordinated administration. The administering agency or enterprise must be empowered to respond to *all* of the values and costs involved. The resulting land use pattern is characteristically an intricate mosaic of uses, both complementary and competing.

In terms of this concept of management, current arguments that the National Park Service or other preservation agencies are engaged in multiple use simply because the lands are watersheds as well as recreational and aesthetic areas can only be termed specious. Such arguments, of course, simply illustrate the dangers of excessive reliance on slogan terms.

There are other views of multiple use as well. One enjoyable definition, at least salted with truth, was given by David Brower[1] of the Sierra Club:

> The multiple-use concept . . . was a political scientist's dream. . . . It could establish a protective cordon of interest groups that could be played against each other on the periphery—and at dead center all could be calm. Are the miners asking too much? Just point this out to the grazers, loggers, water users, and recreationists.

The central problem of economics has been said to be the allocation of scarce means among competing ends when the objective of the allocation is the maximization of the attainment of the ends. Certainly the key problem in multiple-use management, as in any multiple-product enterprise, is the optimum allocation of the land and the other economic resources among the various possible uses. And it is important to realize that with the passing of the wildland frontier we are increasingly confronted with competing uses rather than complementary uses. As Charles Connaughton,[2] president of the Society of American Foresters and regional forester for California, has said:

> The application of the multiple-use concept in forest land management is most easily obtained where there are ample accessible public forests to satisfy public demand. The

real problems arise when demands of people exceed the available resources of the land as is becoming increasingly apparent in the West.

In short, the question of resource allocation becomes important only when resources become scarce and take on value.

ECONOMIC MODELS

The economic model for resource allocation in a multiple-product situation is clear enough. Maximization of the attainment of ends will be achieved when resources are allocated among the multiple uses or products in such fashion that the marginal returns per unit of input from all uses are equal. If the last input applied to forage production is returning more than the last input applied to timber production, then total returns can be increased by shifting resources from timber to forage. In such a marginal calculus, full recognition must be given to value and cost interrelationships among the uses. These complementary and competing relationships result in substantial operational complexities in the application of the model. If only uses internal to the management unit are considered, this model will define the allocation of a fixed budget which maximizes the internal rate of return. If alternatives outside the management unit are recognized through including the constraint of the guiding rate of interest . . . , then the model will define both the optimum budget and its optimum allocation.

More elaborate models have been presented. For example, in the first issue of *Forest Science,* G. R. Gregory[3] developed the multiple-use concept in terms of iso-cost and iso-revenue curves. Others have proposed the techniques of programming as the analytical approach to the problem. Such model building is an enjoyable intellectual exercise and may shed further light as to the precise relationships involved. It is not the route, however, to solving the dilemma of multiple use.

A basic problem to be faced is the very low level of knowledge of the physical relationships and values and costs involved in the various uses which can be made of forests and wildlands. All economic models which have been proposed require some knowledge of production functions, cost functions, and value functions. By a production function we mean simply the various possible physical outputs which can be achieved from various possible combinations of inputs. Here our knowledge is rudimentary at best. The cost function appears to be a less difficult problem, requiring only knowledge of factor costs for the input combinations. The value function, however, is again a major difficulty as a result of the extra-market nature of certain of the outputs resulting from wildland management.

We recognize intuitively that great importance must be attached to these extra-market consequences of land management. Thus the land manager is brought to grips with a general problem in resource allocation of increasing national concern. The market economy is a significant and powerful social institution because it provides a useful basis for comparing values which would not be commensurable in the absence of market institutions. Increasingly, however, resources are being allocated to areas such as education, public health, and public safety which lie wholly or in part outside the market economy. The problem of making valid comparisons between values having market expression and extra-market values is as yet unresolved. Economic research must be directed to the quantification of such extra-market values, but there may well be a basic lack of commensurability between market values and extra-market values that is not fully susceptible to economic analysis. Thus reliance must also be placed on the development of appropriate social institutions for controlling such resource allocations.

A brief discussion of timber, recreation, and watershed influences will serve to illustrate these problems.

TIMBER MANAGEMENT

Even their severest critics (who seem to be increasing in number) attribute to foresters a considerable knowledge of timber production, but actually such knowledge is surprisingly fragmentary. Our silvicultural prescriptions tend to be on an "all or nothing" basis. A particular set of practices is recommended to develop a fully stocked plantation. But what are the probabilities that these practices will lead to stocking of various degrees ranging from zero to full stocking? And if these practices are modified in some degree, how are these probabilities changed? By and large, research in forest management has not yet even been directed to such questions, but until we have some idea as to the answers we can be in no position to make refined analyses of economic alternatives.

Among factor costs the greatest difficulties lie in the question of the cost of capital, or, in simpler terms, the rate of interest to be used in financial calculations. This long-discussed problem almost certainly has no single or enduring answer. All factor costs, of course, are subject to major uncertainties as a result of the great length of the production period.

Timber values are interpreted by the market place, and good evidence of current values can be obtained even though the timber market is subject to major imperfections. Again, a very high degree of uncertainty is attached to any estimates of future values as a guide to current production decisions.

MANAGEMENT FOR RECREATION

Knowledge of the management of wildlands for outdoor recreation is even less developed. One major aspect of this problem is so obvious that it often seems to be overlooked. Recreation is not a single, consistent use of land, but a whole group of widely varying uses, often sharply conflicting with one another. Their number is legion—the noisy outboard enthusiast *vs.* the solitude-loving canoeist; the ski resort development *vs.* the beauty of the unmodified natural scene; cabin sites *vs.* campsites; the naturalist with his love of the forest primeval *vs.* the deerhunter appreciative of the more bountiful herds on cutovers and recent burns; and on and on. Thus within the classification of recreational land management we again face all the problems of multiple products and the need to achieve an intricate mosaic of uses.

* * *

The question of the value of the output in recreational management has been much more widely discussed than the production process itself. What is the economic value of recreation? What is it worth?

In looking at these questions, the first thing to recognize is that the extra-market nature of outdoor recreation is institutional in origin. It results from the public policy decision to provide such recreation either free or at nominal charge. There is nothing mysterious about recreation per se which makes it an extra-market value or an intangible. Commercial recreation is widespread throughout our economy. We have perfectly good market expressions of the value of the chance to attend a football game or an opera, to week-end at a resort, to ride a roller coaster, or to own a great work of art. Furthermore, in at least some areas of the United States there is an active market for small forest properties for recreational use. The values of land adjacent to Lake Tahoe are certainly not determined by their potential as timber sites! And some of our ranches in the Coast Ranges are now obtaining more income from leasing hunting rights than from livestock operations.

Thus outdoor recreation as an extra-market service is primarily an issue on public lands. But this makes it an issue of great concern in the West. According to the recent progress report of the Outdoor Recreation Resources Review Commission,[4] 22 percent of the total area of the West is recognized as nonurban, publicly designated recreation areas. These 170 million acres include 71 percent of the total of such recreational areas in the United States (excluding Alaska and Hawaii).

Many different attempts have been made to express the value of such recreation. Perhaps the crudest and the most common are the various estimates of total expenditures by recreationists. For example, a survey conducted for the U.S. Fish and Wildlife Service[5] estimated that in 1955 some 25 million American anglers and hunters spent nearly $3 billion for 500 million days of sport. But what does this mean about the value produced by allocating resources to provide hunting and fishing opportunities? Absolutely nothing. These expenditures were for food, lodging, travel, clothes, guns, rods, and similar items. These values account for the three billion dollars, and there is nothing left over as a return to the recreational use of the land.

Certainly the particular groups serving the recreationists are benefiting from these expenditures, and data on expenditures may well be highly useful in gaining the support of such groups in lobbying for larger budgets for recreation. Just as the motel owner may invest in a swimming pool to increase his volume of patronage and net earnings, so the recreational service groups in an area may support investment in recreational facilities. But from a broader public standpoint, all that this is achieving is a transfer of expenditures from one group to another; there is little if any net gain to society from this level of effects. The social case for public support of recreation must rest on the values to the users, not the increased profits of certain recreational service industries.

Recognizing the lack of meaning in the simple expenditures approach, various economists have attempted to use these data in more sophisticated ways. For example, Trice and Wood,[6] following a suggestion by Hotelling, have used expenditure data in an effort to estimate consumer surplus. Crudely stated, the idea here was that since the recreation area has values which attract somewhat distant users at fairly high expense, then all closer users who get to the area at lower expense are enjoying some sort of surplus. There are a number of practical obstacles to such an approach, but the entire concept founders on the fact that the notion of consumer surplus rests on an invalid interpretation of the Marshallian demand curve. This literally involves a footnote to economic theory and does not deserve exploration here. The essential point is that any estimate of consumer surplus cannot be weighed against market values as a valid guide to resource allocation.

A more recent effort is an ingenious approach to deriving a demand curve proposed by Marion Clawson[7] of Resources for the Future on the basis of travel zones and expenditures data. Unfortunately, this approach involves combining nonhomogeneous items in developing the quantity variable and ignoring the highly important differences in alternative recreational opportunities characteristic of the user zones in developing the price variable. Thus again the model does not appear valid for its intended use.

All such efforts to derive the economic value of publicly supported recreation seem sure to end in frustration because of a failure to identify the exact nature of the value to be estimated. The services provided by a recreational area or opportunity are consumers' goods. Unlike producers' goods, such consumers' goods are not purchased to produce income; instead they are purchased for motivations which may not be known, but whose force is presumably reflected in market price.

It is not the function of the economist to attempt a direct appraisal of the "worth" to people of consumers' goods. When the motive on the demand side is anything other than the production of cash income, the economist's or appraiser's function is only to appraise the force of the motives, not the motives themselves. The economist may be able to estimate the costs of such goods and in some cases he may be able to predict the price which people will pay for them, but it is not his role to estimate what they "should" pay.

In short, the economic value to society of public recreation areas is determined by what society is willing to pay to have such areas. The economist would do better to look at such evidence of society's decisions than to attempt the inappropriate task of appraising the benefits obtained from a consumer's good.

Such an approach places sharp limits on the role of the economist and does not promise any general solution. Analysis of actual transactions involving public investment in recreation could, however, provide useful guides to administrative decisions. To date very little has been done in exploring this approach.

WATERSHED MANAGEMENT

Similar uncertainties and lack of knowledge are found in the area of watershed management. While research has established certain highly important principles of watershed relationships, quantitative data of the production-function type are almost wholly lacking. The problem is made the more difficult by the complex nature of water yields. Water must be considered in its three dimensions of quantity, timing, and quality. In general, there appears to be a tendency for improvements in timing and quality to be achieved only at some loss in quantity.

From the economic standpoint, it is helpful to identify two separate functions of watershed management. One is the influencing of the total quantity of water and the other is the reduction of damage from erosion, siltation, and floods.

At the risk of oversimplifying the question of water yields, it can be noted that in California the average annual water yield is 71 million acre-feet, while the ultimate requirements are estimated at 46 million acre-feet. The water problem of the state is one of storage and distribution, not of quantity. Oregon and Washington are even more abundantly endowed with water. With the exception of certain special areas such as the Central Basin and Southern California, or of unusual drought, the marginal value of water on the watershed appears to be zero. On most of our forest lands the popular conception of squeezing more water from our watersheds would probably not be economically useful.

The reduction of damage from erosion, siltation, and floods, however, is commonly an important function of land management. The measures required are quite different than those for increasing water yields, and fortunately are usually compatible with other uses managed on a permanent, productive basis. To the extent that special measures are required, the value of the reduction in damage can serve as a guide to resource allocation.

In this problem of valuation, engineering data can provide some upper limits. The value of vegetation management to control such damage cannot be higher than the cost of engineering practices (debris basins, impoundments, etc.) to achieve equivalent effects. There is always the possibility, of course, that the economic value is considerably less than this maximum and that no special programs are justified. Efforts at the direct estimation of such values have generally been blocked by the gross inadequacies and uncertainties of knowledge of the physical relationships involved.

THE PROBLEMS OF PUBLIC DECISION

From all of this, the conclusion emerges that in the issues of multiple use of forests and wildlands, the economist is better at asking questions than of providing answers. The asking of valid questions is in itself a useful function and can help in coming to sound decisions. But the administrator must come up with answers now, however fragmentary and inadequate the basis for reaching his decisions. In doing this, his problems are not eased by the anomalies of our public policies affecting land use.

In fiscal 1960 the Forest Service received $140 million from timber sales, $4.5 million from grazing fees, nearly $2 million from mineral leases, a little over $1 million from recreation, and about $0.5 million from miscellaneous sources. This, however, cannot be interpreted as an indication of the values produced by these forests. Both on the national forests and on other public lands, the various products move toward the consumers under very different policies. Timber is appraised for sale at its estimated full market value and is sold for as much more than this as competitive bidding can produce. Forage is sold through the medium of grazing permits, which have generally been held to involve a fee structure below market value. Recreation is provided either free or at nominal charges. By and large, water simply flows from the land, and prevention of damage produces no cash.

Essentially these differences reflect historical accident and administrative problems rather than carefully considered economic policies. Through prolonged usage and acceptance, they have set in motion a variety of economic forces and the development of vested interests in their perpetuation. For example, the benefits of the less-than-market value fee structure for grazing have long since been capitalized into the values of the participating ranches, and thus have been transformed into costs to the present owners.

The widely acclaimed Multiple-Use Act of 1960 (Public Law 86-517) does not appear to have shed much light on the administrator's problems. This act states that "it is the policy of the Congress that the national forests are established and shall be administered for outdoor recreation, range, timber, watershed, and wildlife and fish purposes." This does not establish any priorities among uses, with the listing being simply alphabetical in order—although some skeptics have noted that, except for the modifier "outdoor," recreation would appear second rather than first in order.

The act then proceeds with some pious admonitions for "the most judicious use of the land" and for "harmonious and coordinated management" and firmly states that multiple use is "not necessarily the combination of uses that will give the greatest dollar return or the greatest unit output." The major degree of administrative discretion implicit in such a Congressional mandate may well be wise policy under the circumstances, but it leaves the primary issues of multiple use wholly unresolved.

In the public sector of the forest economy, multiple use is clearly an instance of incremental policy making. We move by a series of small steps, with action guided by reaction. Public control is exerted primarily through the appropriation bill, which itself is largely conditioned by the previous appropriation measure. Economic analysis is inadequate to provide clear-cut answers to the problems of resource allocation. The public forest administrator appropriately is responsive to political stimuli as well as to economic stimuli.

A successful labor arbitrator once stated his test of a successful arbitration. If labor was happy and management angry, the arbitration was poor; similarly, if labor was angry and management happy, the arbitration was poor; but if both sides were equally angry success had been achieved. Under this criterion, the Forest Service seems to have been remarkably successful in several recent controversies.

In all seriousness, controversy is the medium for multiple-use decisions in the public sector. In the market place of the private sector, costs are usually borne by the individuals receiving the benefits. With the extra-market products of the public sector, quite the opposite is characteristic. Costs are borne primarily by some groups, while benefits are enjoyed by others. Conflict between local interest and national interest is almost constant. Controversy is inevitable in such circumstances and serves to bring out forcefully the varied interests affected by the decisions to be made.

David Brower was sound in his allegation that in multiple use the Forest Service is constantly balancing one interest group against others; he was wrong in inferring that the position of dead center is one of calmness and inaction.

THE PRIVATE SECTOR

Multiple-product decisions must also be made in the private sector of the forest economy. Here the criteria of the market place can be given primary, but not sole, recognition. Commodity production tends to be dominant since it is the primary source of cash income.

Costs and benefits accruing to others are not of immediate concern to the private owner. Thus watershed-management aspects rarely are primary considerations on private lands. Under the complexities of water law, exact title to water is not always clear, but it certainly does not belong to the owner of the land on which it happens to fall. Nonetheless, the private owner cannot wholly ignore watershed effects, since the political consequences of watershed influences may have pronounced economic effects on his operations.

In recreation the situation is highly variable. The more intensive and commercialized forms of forest recreation are well established on private lands—resorts, cabin sites, organization camps, and so on. Impact on land values is making itself felt increasingly. In the less intensive forms of recreation such as camping, hunting, or picnicking, the private owner has been discouraged from much participation by the competition from public recreation priced at less than cost. The efforts that are made at such management are commonly charged to public relations. Study of current trends in recreational use suggests that both the free use of public lands and the free use of private lands deserve serious restudy, and that the issues involved are rapidly increasing in urgency. Certainly the desires of our highly urbanized, recreationally minded society are going to have a tremendous impact on private forest lands in the years ahead.

And while he may rely on the criteria of the market place, the successful private forest land manager must be highly perceptive of the changing needs and desires of society, both in material goods and in services. His production efforts will bear full fruit only in the somewhat distant future, and, as society changes, the signals called by the price structure will also change.

THE ROLE OF THE FORESTER

From all of this it follows that the successful forester must be far more than a skilled technician. Society's growing concern with the multiple uses of the forest resource will test the forester's claim to professional status.

In reaching a pragmatic and workable solution to the multiple problems of multiple use, the forester occupies a central position and he can validly claim a key role. Through education and experience he is qualified to judge the complex interactions of the production functions for the several uses of forests and wildlands, and he is aware of the various values resulting from management. He sits at the focus of all the forces of supply and demand socially relevant to the final management decisions. Thus, if he approaches his problems with full comprehension of his responsibilities and of the society which he serves, he has a more valid base for contributing to management decisions than any individual or group not situated at such a focus.

Decisions affecting the multiple uses of forest land cannot be made by standard formulas or rules learned by rote. Neither will economists nor other research specialists develop neat analyses providing all the answers. Instead, the forester must work in uncertainty and controversy, the heat of which will reflect the growing importance of the resource for which he is responsible. In this very real sense, multiple use is more the symbol of the problems we face than a simple method for their solution.

REFERENCES

1. Brower, David. 1959. Crisis in the Northern Cascades: the missing millions. Sierra Club Bul. 44(2): 10–15.
2. Connaughton, Charles A. 1959. Impact of population on forest management. Proc., Society of American Foresters. Pages 2–6.
3. Gregory, G. R. 1955. An economic approach to multiple use. Forest Science 1(1):6–13.
4. Outdoor Recreation Resources Review Commission. 1961. Progress Report. Washington, D.C.: U.S. Government Printing Office. 85 pages. Illus.
5. U.S. Fish and Wildlife Service. 1955. National survey of fishing and hunting. Dept. of the Interior, Circular 44. Washington, D.C.: U.S. Government Printing Office. 50 pages.
6. Trice, Albert M., and S. E. Wood. 1958. Measurement of recreation benefits. Land Economics 34(3):195–207.
7. Clawson, Marion. 1959. Methods of measuring the demand for and value of outdoor recreation. Resources for the Future, Reprint No. 10. 36 pages.

26

MULTIPLE-USE DECISION MAKING—WHERE DO WE GO FROM HERE?

R. S. Whaley

First published in *Natural Resources Journal* 10(3):557–565. July 1970.

BIOGRAPHY

Mr. Whaley is Head of the Department of Landscape Architecture and Regional Planning in the University of Massachusetts. His concern with planning and with broad applications of social science to resource problems has developed in the course of a well-focused career: a tour of duty with the Forest Service's Division of Forest Economics and Marketing Research; a period of service at Utah State University, where he held the Headship in Forest Science and teaching appointments in economics and agricultural economics; and, most recently, the position of Associate Dean in the College of Forestry and Natural Resources, Colorado State University.

EDITORS' SUMMARY

A problem of resource allocation on public lands is the lack of value data. Decisions can be made without explicit values, but economists have the obligation to work up such data, and methods are available for the purpose. Multiple use is best approached by first defining the goal of management and then deriving values consistent with the goal.

QUESTIONS TO CONSIDER

(1) What are the characteristics of those forest outputs which are not marketed and thus fail to have a market value? That is to say, why do these outputs not pass through markets?

(2) What are the characteristics of outputs which, though they pass through markets, are not fully evaluated thereby?

(3) Mr. Whaley writes (page 202) of the "intertemporal value" of resources. What does this mean?

(4) Why is forestry in the United States so largely socialized, with government exercising controls over the resource and owning a great deal of it outright?

(5) What differences do you see between the multiple-use management problems of the public firm as described by Mr. Whaley and the multiple-use management problems of the private firm??

(6) What is a "noneconomic goal" for resource allocation (page 206)?

(7) It has been contended that all methods for estimating value are, at bottom, methods for estimating opportunity cost. Review the methods cited by Mr. Whaley and by Mr. Zivnuska. Do you agree with the contention?

* * *

Considerable attention is being focused on the problems of making multiple-use resource-allocation decisions about public lands. Private businesses from the corner grocery to General Motors seem to allocate their capital resources without encountering any particular conceptual problems. The criteria for making investment decisions—such as pay-out period, rate of return on investment, or present net worth—are appropriately used. However, these tools of private business are deemed inadequate for making resource allocation decisions about federal lands.

UNIQUENESS OF RESOURCE ALLOCATION ON PUBLIC LANDS

Problems of allocating resources on public forest lands are of special interest because of two characteristics peculiar to multiple-use decisions: (1) their "publicness" and (2) the lack of data for traditional investment analyses.

First let us examine the problems associated with "publicness." A central concept of economics in the capitalistic world is that the price system, operating through the market place, balances supply and demand, efficiently allocating scarce resources among competing uses. The market system, however, does not result in an optimum allocation of resources in all instances. We, as a body politic, have decided that many natural resources are exceptions and have removed some decisions regarding these resources from the traditional market system, putting them under public control.

Such decisions need not be arbitrary. There are criteria that may be used to judge the legitimate "publicness" of any resource allocation decision. These criteria arise from the failure of the market system to efficiently allocate resources.

The first criterion occurs in conjunction with substantial third party benefits. That is, some goods and services such as education, defense, mental hospitals, etc., offer benefits beyond those that accrue to the direct recipient of the service. The market system, which expresses only the demand of the direct beneficiaries, tends to underestimate the values received from such services.

The second criterion arises from the fact that some commodities are associated with indirect social costs. Choice examples are water and air pollution. If these social costs are not or cannot be absorbed by the producer of the commodity or its consumer, they may have no influence on the price of the commodity, and they then would not be weighed by the market system.

A third criterion for public intervention occurs with technical monopolies. Postal service, telephone service, and transportation are classified by some economists as technological monopolies, where government may have to subsidize production of the commodity to insure a supply at a price that meets the demands of society.

A fourth criterion arises from substantial differences between the time preference of society as a whole and an individual's time preferences. It rests largely on the greater ability of government to absorb uncertainty in investments in such things as basic research.

Elements of several of the above considerations are relevant to the supply of multiple products from the nation's forest lands. For example, the conservation issue involves the consumption of limited resources over time. It is argued that the time preference of individuals is too limited to weigh adequately the intertemporal value of resources, the use of which straddles several generations. It can also be argued that substantial third-party benefits result from the production of water and recreation on forest lands.

This article, however, argues neither for nor against government ownership of vast acreages of forest land. Rather, its purpose is to point out the resource allocation problems that result from public ownership. Regardless of which criteria best justify a public supply of forest-oriented goods and services, the conclusion for investment purposes is the same. Many of the commodities do not command a well established market demand. No market-established prices can represent the values of recreation and water in investment analyses. Despite the established markets for timber and forage in many parts of the country, there has been little study of how closely administered prices of forage approximate market values, nor of how federal timber sale appraisal procedures affect the market price of timber from either federal lands or competing private lands. Thus, the major problem for multiple-use decision makers relative to public lands is the lack of data that are needed if the benefits from the production of certain commodities or combinations of commodities are to be evaluated. For some products we have no measure of market value. Whenever demand for goods and services from public resources is totally, or even partially, free to reflect consumers' desires, any mispricing is quite apparent. If overpricing exists, disuse of the resource develops; with underpricing, "overuse" may emerge. This process, however, offers a decidedly imperfect substitute for market values. Nevertheless, some possible approaches to natural resource allocation or investment decisions can be proposed.

DECISION-MAKING TECHNIQUES

The most common approach to the dilemma of making decisions without value information is to *avoid the problem*—that is, use a method that does not require value data. Two forms of this approach are:

1. Establish physical production goals at least cost.
2. Maximize physical output for a predetermined level of expenditure.

Hundreds of illustrations of these two procedures have occurred in "public forestry" during the past several decades. The forester, for example, may have had a specific budget item for tree planting, and within the limits of his budget he tried to plant as many acres as possible. As a result, the least promising acres were often planted first, because areas such as the poorly producing mid-western sand flats offer inexpensive planting opportunities. Thus, criteria that do not require value data may result in improper investment priorities from any kind of benefit-cost standpoint.

There have been some interesting recent applications of investment criteria that make only limited use of product values. U.S. Forest Service researchers attempted to develop planning models for multiple-use management and devised the imaginative Resource Allocation Models. These models have been successfully used with linear programming techniques to determine least-cost solutions for prescribed multiple-use goals. The computerized linear-program solutions have allowed consideration of extremely complex problems involving many different kinds of costs and physical outputs.

Though these Resource Allocation Models presently offer the best solution to multiproduct output decisions on Forest Service lands they do not incorporate the important policy issues involved in setting appropriate production goals. Solutions to these models depend upon first setting the physical production goals. That is, water production, timber cut, or animal-unit goals must be determined as inputs into the model. The land manager must determine the optimum output from his lands. Thus, even the newest refinements in using cost-minimization criteria for solving multiple-use decisions do little to guarantee an optimum solution based on measures of public welfare.

Still another way to avoid the value problem—which unfortunately has been used too often—is to claim that no economically rational solution exists. Some would advocate that an uninformed decision in the political arena somehow is superior to other decision-making techniques. In truth, however, the best political solution can only be achieved with information regarding benefits and costs.

A second possibility for solving public investment or allocation decisions is what can be called the macroeconomic approach. If a major role of public resource utilization is economic development in its broadest sense, then the techniques of simply minimizing costs or maximizing differences between benefits and costs may not be appropriate to investment decisions on public lands. As Kenneth Boulding states, "the great hiatus in economics . . . is a real link between price theory of any kind and a theory of economic development." If the goal of public resource use is economic development, why not deal with the problem more directly and look at the impacts of certain allocation decisions on such variables as regional or national income, regional or national employment and economic stability?

The stated goals of public land management have not explicitly included economic development as a central issue. However, it is implicitly included in justifying certain programs—for example, stabilizing the livestock industry or protecting a certain locality's lumber industry. Thus, employment and local or regional income considerations do seem to influence public land allocation decisions.

Little or no research has been done concerning the regional or national economic impacts of alternative forest land uses. Only recently has some study been directed toward measuring and predicting the impacts of dams and other water developments on surrounding communities. One thing apparent from these few studies is the extreme difficulty of accurately measuring the regional impacts from even multimillion dollar projects.

The size of the region over which impacts are expected has considerable influence on our ability to identify changes due to specific investments. If the relevant region is large, for example a state, one might logically conclude that most measures of macroeconomic variables lack sufficient sensitivity to assess changes resulting from the relatively small investments that characterize forest lands or from shifts in land-use patterns. If, however, we are interested in measuring impacts on smaller units, such as communities, the techniques available, which include input-output analysis or economic base analysis, do not seem particularly appropriate for measuring the economic interrelationships that exist within small rural areas. The strong economic dependence of the region under study on distant urban centers tends to cloud the intraregional economic picture.

The macroeconomic (or regional-analysis) models for making multiple-use decisions seem to have three major shortcomings. First, they are not sufficiently sensitive to measure changes associated with small investments. Second, they do not come to grips with the major policy issue of how much should be invested in the various kinds of development that are possible on public forest lands. At best, we can set criteria that require maximizing the level of employment for a given budget or obtaining a given level of employment at a minimum cost. These suboptimization techniques do not solve the problem of how much money should be invested in various development or use combinations on public lands.

Third, in dealing with aggregate figures for income or employment, the problems of income distribution are often ignored. An apparent increase in regional income may equate with decreased income and employment in other regions. Changes in land-use patterns may generate interregional flows of income or they may change the relative contributions of the public and private sectors in supplying resources.

Considering the above shortcomings, most economists would agree that the best approach toward ranking alternative land uses would require some attempt to *evaluate the difference or ratios between benefits and costs*. If evaluation of benefits in relationship to costs is the appropriate criterion, then many analytical models come to the foreground. Benefit-cost ratios, internal rates of return, and joint production models equating marginal rates of substitution between goods, are all methods that can be used to compare benefits and costs of various investment schemes.

Though the mechanics of performing these kinds of analysis are relatively simple, they have been little used in analyzing public investments. The attempts by the Corps of Engineers and Bureau of Reclamation to apply these methods could at best be called incomplete, unsatisfactory efforts to compare benefits and costs. The dissatisfaction associated with this use stems from the difficulty of assigning a quantitative measure to benefits derived from nonmarket-supplied goods and services. In many instances two of the most important products of water development (and likewise from forest development) are recreation and water for domestic use. Yet neither of these products has an established market value which can be plugged into investment analyses. Methods that approach multiple-use decisions from a profit maximizing standpoint have therefore been little used because of the lack of value figures for many of the benefits. This lack has promoted considerable recent research on the problem of resource values. Most of this has dealt with problems of recreation valuation.

STATUS OF RESOURCE VALUATION

Before examining the status of resource valuation, it is important to clarify precisely what kind of value we are seeking. Many of the critics of current research in resource valuation are not fully aware of the problems of setting *a value* on a particular resource use. Those critics seem to assume that every good or service has an inherent value peculiar to it and that the researcher must find this single value for each resource use. This concept of a single inherent value for each commodity is false, since every good and service has several values. Each has a value in exchange, that being the number of goods that can be obtained by means of giving up or exchanging one unit of the commodity in question. Each good or service also has a unique value for each individual consumer. This is the amount that the individual's psychic welfare is improved through owning or consuming the particular commodity. A good has a third value that equates with its cost of production.

The fallacy is therefore obvious in an assumption that a particular resource has only one unique value and that the researcher has but to gaze into a crystal ball to find this heretofore hidden number. Rather, determining a value for a particular type of recreation or for the domestic consumption of water is a problem solved by arriving at an index number (expressed in dollars) that approximates one of the above measures of value. Therefore, the many values of a particular resource may each have a possible application in some resource allocation model. The only valid grounds for criticizing a particular proxy value determined through research are: (1) that it is an index of a value not applicable to a particular allocation model or (2) that through a flaw in concept or methodology, the index is not an accurate approximation of the value being estimated. Many researchers can and should be criticized, however, for not explicitly stating just what kind of value they are trying to approximate. Without this definition, it is impossible to evaluate the prospective usefulness or accuracy of their estimates.

Most of the research currently directed at valuing nonmarket-supplied resources has been devoted to putting a dollar value on recreation. To date several general kinds of approaches have been applied to the problem. These have included:

Expenditure method—measures the value of recreation in terms of the total expenditures on recreation.

Gross national product method—attempts to measure the contribution of recreation to GNP.

Consumers'-surplus method—attempts to determine the willingness of individuals to pay for various quantities of recreation. Instrumental in this method is developing a hypothetical demand curve for recreation.

Cost method—uses the cost of supplying recreational facilities as a measure of the benefits derived therefrom.

Monopoly-revenue method—uses the estimated revenue that would be obtained by a monopolist owning the recreational site as a measure of benefits.

Market-value method—uses fees charged at private resorts as a proxy value for the value of public-supplied facilities.

Apparently, there is no dearth of ways to evaluate recreation. These methods or modifications of them can be used in valuing other resource uses. Yet there has only been limited success when the calculated values are inserted in resource-allocation models. Although we have made inroads at developing individual resource values, we have yet to develop *value systems* which allow analysis of complex combinations of resources and resource uses. Even though each of the above valuation schemes has its appropriate use in isolated circumstances, their application in resource-allocation models must be evaluated on the basis of (1) their appropriateness for measuring benefits in terms of the optimization criteria of the allocation model, (2) the comparability of all measures of value in the allocation model (it is impossible to approach an optimum solution if cattle, timber, recreation, and water are all measured by different indices of value), and, (3) whether the value scheme is empirically quantifiable.

CONCLUSION

An orderly approach to multiple-use decision making requires a reorientation of research toward a broader approach to the development of resource-allocation and investment models. The several kinds of values we have noted imply a need for equally as many kinds of allocation models, each using a different value criterion and aimed toward a different kind of policy goal. The current problem in multiple-use analysis is that we use only one or two allocation or investment models and restrict ourselves to one valuation system. We start with a valuation system based on exchange or market values of a few goods and services and then attempt to force all goods and services into a like mold. When some uses have not fit this mold, many resource managers have thrown their hands into the air and claimed that it is obviously not an economic problem.

At this juncture it is important to know just what economics and the economist have to contribute to multiple-use decision making and conversely, what are the limits of economics. It is presumptuous to assume that economics and economists should or can make allocation decisions on public forest lands. These decisions most often include noneconomic goals and require inputs in addition to economic data. To assume otherwise would be to ignore the complex ecological and hydrological interrelationships that influence land management. However, achieving even noneconomic goals generally costs society money directly or indirectly in the form of alternative opportunities forgone. Economists therefore have a contribution to make in supplying data inputs to improve the knowledge base on which these decisions are made.

More directly, the economist can perform at least three services to the decision maker. The tools of economics (1) can tell how best to use a resource to maximize an economic goal; (2) can identify the costs of sacrificing an economic goal to achieve a noneconomic one; and (3) are useful in organizing a procedure that will minimize the economic costs of achieving a complex goal. The appropriate role of economics in multiple-use decisions therefore depends upon the particular goals of public resource management.

Our current orientation to solving multiple-use problems seems to have the proverbial "cart before the horse." We are concentrating on the quantification of values without a clear-cut definition of how derived values will be used. A more logical approach involves three steps, the order of which is critical. Step one must be a realistic and explicit statement of goals for the development and use of the public resource in question. Are these resources to be managed on the basis of some efficiency criterion, regional growth and stability criterion, national growth criterion, physical output criterion (i.e., a conservation goal expressed in terms of physical output per unit of time) or societal welfare criterion to be evaluated in the voting booth? With an explicit statement of resource management goals, the second step is to develop a valuation system which produces a set of indices related to the measurement of benefits. This value system should recognize the three criteria previously mentioned as means of evaluating value systems. The third step is, of course, the application of the allocation model and its associated value system to multiple-use decision making.

27

THE CHANGING SHAPE OF FOREST RESOURCE MANAGEMENT

William A. Duerr

First published in *Journal of Forestry* 65(8):526–529. August 1967.

BIOGRAPHY

Mr. Duerr's biographical sketch is given at Item 4, page 30.

EDITORS' SUMMARY

Forest management over the next few decades will undergo technological change that will force adaptation in labor and wood use. Computerized systems will tend to centralize decision making. Rising uncertainty will spotlight faith as a management guide. Emphasis on quality of living and on consumers will bring multiple-use forestry led by teams of specialists.

QUESTIONS TO CONSIDER

(1) What is the process for making qualitative predictions such as are made in this article? The predictions are all answers to the question, What will happen in this or that specific case? Each therefore has involved the use of a model, into which information has been placed in order to derive an answer. Select one of the predictions and identify the model and information that may have been used.

(2) The predictions are old enough so that one can form ideas about their validity. Which predictions have come true or promise to do so?

(3) We frequently run across the sort of notion expressed on page 208, that change is accelerating. Has this been true of forestry? In what respects? Design a quantitative empirical study of changes in the rate of change (d^2x/dt) in some aspect of forestry in recent decades.

(4) How does one measure economic growth? social growth in general?

(5) What connection do you see between prediction number four and the industrial hierarchy perceived by Mr. Bond (Item 23) in respect to labor problems? Explain the connection.

* * *

Since our ultimate professional interest as foresters is in the management of our resources, our ultimate interest is in the shape of the future. For management is decision making, and decisions cannot be made about the past—or about the present, either: Only the future is subject to decision. The context of management lies in the future.

With tomorrow standing here in the wings, I am encouraged to listen to its prompting and to undertake some predictions. I propose to consider the circumstances of forest resource management in the years ahead. My purpose is to make explicit my interest, as a forester, in the future.

I will be talking about a future that lies variously from 10 to 60 or 70 years hence. For the most part, I will not specify when I expect a prediction to materialize. In other words, I will be speaking about directions of change and will be treating of rates only by implication. My thoughts will be expressed mainly from the viewpoint of industrial forestry or of public forestry.

Just to be orderly about the project, I will number my predictions as I go along. There will be 13 of them. The number is happenstance, though when I stop to think of it, no other number would suit the present rash venture so well.

1. CONTINUING CHANGE AND GROWTH

My first forecast is really more of a premise. It is that change will continue, that the pace of it will be faster as time goes on, and that the aspect of change which will most directly concern us as foresters will be social development and notably the economic growth of the United States. Economic growth will take the form of increases in the national output. I am speaking of the national output in real terms, expressed in dollars of constant value. In recent years, the total has been rising at the rate of about three percent annually. The national product per capita has been rising, too, at a slightly lower rate. Such rates of growth will be at least maintained in the future, so that the absolute amount of growth will get larger. This is assuming that we will escape the destruction of wars and other catastrophes.

The notable feature of economic growth will be an increase in the value of human labor. All kinds of labor will benefit: not only common labor, but skilled workers as well, and even the professional and managerial classes. The value of labor will rise because national output per man-year will have risen, and higher output will mean higher income, which is the measure of labor's value. People will be earning more and more wages or salaries or other payments for a year's work. Employers will be incurring more and more cost per man-year for the use of this increasingly precious resource, and still their demand for the services of the gifted and trained person will be higher than we have ever known.

2. EXPANSION OF RESERVOIR OF TECHNOLOGY

My second forecast concerns the origins of our future economic growth. Some of this growth will of course stem from the upsurge in population. Again, some will arise from enlargement of our national stock of capital goods such as machines, buildings, and forest resources. However, by far the greatest share of economic growth, possibly half of the total, will come from expansion of knowledge and its application to the arts of production and consumption. That is to say, economic growth will stem primarily from improvements in technology. Technology in the future will be, as it is today, our prime resource.

One can expect technology to grow by means of our national effort in research and development, and to be transmitted through bigger and more effective educational programs. Our national standard of life will be a function of our research and education. Here I mean to include both formal education and the various kinds of continuing education, such as self-improvement and on-the-job training.

3. LABOR PROBLEMS

Third, I predict that the rise in labor productivity which will characterize national economic growth will tend to proceed at a slower than average rate in silviculture, logging, and the traditional forms of wood-using industry. That is to say, while labor is becoming scarcer and most industries are finding means to economize in its use, equally efficient means will be hard to devise within much of forestry. Managers can respond initially in one of two ways: They can reduce the labor intensity of their programs, cut down the number of man-days of work put into their land or their other capital, and thus run the risk of reducing the physical abundance of their products. On the other hand, they can maintain labor intensity and pay the wage bill by raising prices, thus reducing the economic abundance of their products.

What a choice! In an effort to escape it, society will strive in various directions. One direction will be to increase the efficiency of labor use in forestry and its related industries by introducing new applications of genetics and engineering in the process of production. To the extent that such moves succeed, we will see retirement of submarginal lands from silviculture analogous to the retirement of submarginal agricultural lands. Another direction in which society may strive is toward greater government subsidy for forestry operations. Still another is toward replacing forest goods and services by substitutes that can be produced with a less lavish outlay of resources.

4. ADAPTATION IN PROCESSING OF RAW MATERIAL

My fourth prediction concerns the manner in which the wood-using industry will adapt itself to national economic growth and to rising labor values. It will adapt by turning more and more away from those forms of wood raw material that require human attention and judgment and care and, above all, handling. At the same time, industry will make correspondingly greater use of raw materials that are uniform and that lend themselves to being moved by belt or pipeline. This means a flight from the log and the board, where the size and quality of the individual piece dominate the production process. It means an ever wider adoption of the wood chip and particle as raw material for reconstituted products, notably silvi-chemical products. These developments will be abetted by concurrent changes in consumption technology, in which consumers will seek more efficient and easy means for satisfying their material wants.

The implications for timber management leap at once into our minds: final tree harvest at a relatively young age, an emphasis upon mechanization in silviculture, a de-emphasis of intermediate silviculture treatments. Here will be the means for simultaneous physical and economic abundance—that is, the means for keeping somewhat abreast of the national trend in labor productivity.

5. AUTOMATION

Fifth, I foresee that our efforts toward technological development and toward the saving of labor will lead to considerable automation in resource management. The most startling developments will come where the dimensions of management can be measured in terms of numbers.

I think that within the coming decade, research efforts now under way will have succeeded in describing portions of the forest resources and the process of resource management as systems of variables measured in numbers. Research workers will have written the mathematical equations and the tabulations that estimate how the variables are related, one to another: how, for example, the allowable cut of timber would be affected if certain lands were set aside for recreational use—and how the income of the region would change if recreational use were substituted for timber use on certain lands. We can picture equations and tabulations and other necessary data stored in a computer and the computer programmed so as to predict the consequences of the alternative courses of action described to it. We can picture the resource manager reaching for the telephone to obtain from the computer stored information—for example, about the extent and success of his forest plantations. We can picture him asking for a forecast of the forest conditions, allowable cuts, employment, revenues, and so on that would follow if he should plant this recently cutover tract or allow it to restock naturally.

What I am describing is not management by a computer. It is the performance of drudgery—of intricate and lengthy calculations that we would scarcely undertake by ordinary clerical means, but which the computer can accomplish in the twinkling of an eye, once it is programmed for the purpose. The manager is freed for the performance of his highest function, decision making. Furthermore, he is supplied with the best available bases for enlightened decisions.

6. CENTRALIZATION OF DECISION MAKING

Sixth, I foresee that as a consequence of automation, more decisions will be made centrally. It will be not only possible, but also efficient and wise for some decisions now made on the ground to be made, instead, in the office. And by the same logic, some of the decisions now taken in the local office will in the future be taken in the central office. The trend toward centralization will be an inevitable consequence of our expanding knowledge, our social development, and our search for economy in the use of human labor.

I think that an increase in centrally made decisions will not be accompanied by any reduction in the decision load of on-the-ground and local managers. What will happen is that the total flow of questions to be deliberated and acted upon will be measurably enlarged; the volume processed centrally will march in step with advancing decision technology; and the share carried locally will change, not in size, but only in character: The manager on the spot will have the chance, as never before, to focus upon those decisions that can best employ his personal knowledge and judgment.

7. INCREASING UNCERTAINTY

Seventh, I want to say something still further about the consequences of rapidly expanding knowledge. Some of the consequences, to be sure, are obvious. We acquire the means to develop computerized simulation models such as I mentioned. In general, it becomes possible to describe more things in terms of numbers, to make forecasts that are more reliable, and to achieve in some respects more rational decisions, more reasoned management. But at the same time some oddly contrasting results show up.

Not long ago I revisited for the first time a research station where I myself had worked some 15 years earlier. I was impressed by the knowledge of the resident research team concerning the conditions and behavior of the forest. They knew a great deal more than we had known when I was there. Evidently research had been paying off at compound interest. However, what impressed me even more was the great doubt that existed in the minds of the team about many facets of almost every question regarding their forest. When I was at that station, we thought that we were getting nicely on top of the job. Now, 15 years and reams of knowledge later, the team was beginning to appreciate the extent of its ignorance.

Evidently society's store of knowledge is a mere blob in the universe of the knowable. The perimeter of this blob is manned, like a stockade, by defenders who are, at the same time, pushing their lines forward and outward. The greater their success, the more extended their lines of defense and the more gigantic the opposing arrays of the unknown.

Here, then, is the forecast that I come to on this point: The more swiftly our knowledge grows, the faster our ignorance, too, will increase. Paradoxically, the better we learn to foresee the future, the more overwhelming will become our uncertainty.

8. RELIANCE UPON FAITH

My eighth prediction, then, follows inevitably. If there is to be an uncertainty explosion, some powerfully effective armament against uncertainty is going to have to be brought up. And clearly the armament in question is something additional to knowledge—to science. I predict that for defense against uncertainty in resource management, we will lean more and more heavily upon faith.

9. REVIEW OF BASIC TENETS

My ninth prediction concerns the steps that will be taken in forestry to make faith a more effective management tool. I foresee that we will earnestly review our tenets of faith: We will study the doctrine of multiple use, the doctrine of the necessity of wood, the doctrine of sustained yield, and all the many others on which we lean for support in our decision making. Our effort will be to enlarge the scope of our faith to match the growing scope of our questions about tomorrow. And our effort will be to keep our faith constantly under revision so that it will continue to reflect not only the best lessons from the past, but also the best dreams for the future.

Here, of course, lies another paradox: that as change becomes more swift, we will be forced to seek refuge in institutions which are rooted in the past and thus are hostile to change. It is a paradox which we feel keenly in resource management. The forestry profession is self-conscious about its traditionalism in a world which so largely places its premiums upon what is new rather than upon what is old. I think the paradox will be, not only inescapable, but actually welcomed and cultivated in resource management, where the timespan of our interest is so very great.

Will it seem odd for a public forestry agency or a private forest-land-owning concern to write into its management manual a major section headed, "This Is What We Believe"? I think the section will be one of the most useful in the manual.

10. IMPORTANCE OF QUALITY

Prediction number ten: As our social and economic development progresses—and assuming, all the while, that we escape destructive catastrophes—the national energies will be turned to an increasing extent from the arts of production to the arts of consumption. There will be a cultural revolution. In the course of it, the national ethic which esteems work will be at least supplemented by another, which will exalt discriminating consumption and creative leisure. People's interests will turn to the quality, rather than the quantity, of living.

Incidentally, I expect the members of the professional and managerial classes to do a great deal of thinking and talking about leisure, but to enjoy less than the average share of it. I foresee that their briefcases will have bulged to the point that two per person, one under each arm, will be the norm.

I foresee in the natural-resource field a good deal of effort being devoted to consumer education and related research. In formal education, the schools and colleges will develop courses in the intelligent and efficient use of goods and services. We can think of such courses as analogous to our studies of music and art appreciation. Extension programs will carry the same information and points of view to persons who are out of school. Research will be focused upon product design and methods of utilization and upon the design of living. The technology of consumption will come in for a share of the attention now lavished upon production technology.

11. SHIFT IN FOREST PRODUCT MIX

Prediction number eleven: In the course of cultural revolution, the changes will continue, which are already under way, in the product mix from forest lands. The mix will shift in favor of watershed, wildlife, aesthetic, and recreational values. The change will take place at some expense of timber values, both on multiple-use lands and on limited-use lands. The latter will include the retired lands which have proved submarginal for the new silviculture.

12. FOCUS ON THE CONSUMER

Twelfth, I forecast that forest land managers will adapt to cultural revolution by recognizing that the ultimate consumer is their customer, their problem, and the purpose of their work and by accepting the thought that their field is social science and not simply or primarily biological and physical science. The forest manager will, in the course of his education, be exposed to the social sciences, such as social psychology, anthropology, and economics. In the course of his career, he will be exposed as never before to the ultimate consumers of his products.

13. MULTIPLE-USE MANAGEMENT, NOT MANAGERS

Thirteenth and last, I predict that in the future, we shall have multiple-use land management, but not multiple-use land managers. I expect land managers to be single-use people by training and by at least much of their experience. Multiple-use management will be performed largely as it is today: by the team of specialists captained by a former specialist. Surely the team members will have acquired, partly in school and partly on the job, an appreciation and a tolerance for the specialties other than their own. They will have learned about the biological and social systems of which the forest is a segment. Thus they will be able to see the forest and its management in entirety and in context. They will have an attachment for multiple use, and multiple use will be maintained as it is today: primarily an ideal and an abstraction.

I recognize that the "land manager," the man of many parts, is an appealing answer to the problems of excessive functionalism such as beset the management group: fragmentation, poor communication, and failure to see the whole picture and the whole purpose. However, the land and its users are too complex a system, and the complexity is growing at too rapid a pace, for the system as a technical subject to be grasped by an individual. Indeed, the day for this did once exist, but now has gone by. The man trained from the outset as a generalist in resource management is apt to become simply a superficialist. The future holds more and more specialization, since it holds greater and greater knowledge. The land-management team of the future, and the forest-management team, will be notably larger than today.

Therefore I predict that we shall achieve the land manager of many parts in the same way that we shall reach our other technological achievements in tomorrow's complex world—that is, collectively.

28

SOME CONSIDERATIONS FOR OPTIMIZING PUBLIC FOREST RECREATIONAL DEVELOPMENT AND VALUE

Ronald I. Beazley

First published in *Journal of Forestry* 59(9):644–650. September 1961.

BIOGRAPHY

Mr. Beazley's biographical sketch appears at Item 7, page 61.

EDITORS' SUMMARY

Psychosocial forces are causing outdoor recreation activity to burgeon. What are the values in outdoor recreation? How do they arise? Who receives them? How estimate them? How should outdoor-recreation development be financed? How can the supplies of outdoor-recreation opportunity be suited to the demands, especially in the East?

QUESTIONS TO CONSIDER

(1) Consider the large changes in outdoor recreation activity described by Mr. Beazley on the first pages of his article. What changes have these, in turn, caused in forest-land use? in forest resource management? in forest managers? in forestry education?

(2) What does Mr. Beazley mean by the "value" of outdoor recreation? Does he use the term in the same sense throughout his article? For example, compare his use of the term in the list on page 217 with that connected with Figure 28–1, page 219. Propose a definition which, inserted near the beginning, would make the article easier to understand.

(3) How do you explain the large professional interest in identifying the "value" of outdoor recreation? What is it that people want to evaluate, and for what purpose: recreation? recreation opportunity? resources used in recreation?

(4) What is wrong with charging consumers for the use of recreation opportunity, just as for most other things? Will there not always be consumers for each of whom the price of something is "more than he will pay" (page 222)?

(5) Is Mr. Beazley's time-allocation scheme (pages 224–225) applicable to increasing the supply of resources other than recreation opportuntiy? What are the characteristics of resources which lend themselves to the use of the scheme?

* * *

The average annual percentage increase in visits to the various park systems, refuges, and national forests has been in the range of 10 to 20 percent . . . on a sustained basis in the post-World War II period. Municipal parks have had a somewhat lower average rate of increase, about 4 percent, while some reservoir properties have experienced an average rate of almost 30 percent.[1]

The major parks and forests approximate a 12–15 percent compound growth rate in number of visits, which is an extremely high rate of increase when compared to other indicators of growth, such as population, with a 1.75 percent rate and real national product at about 3 percent. Why should outdoor recreation be so much higher, and on an apparently sustained basis?

Some of the reasons typically suggested include population growth itself, changes in work environment, changes in hours of work and in vacations, changes in the distribution of income, increasing urbanization, and the general increase in the ease of travel. The consistent rate of family formation, recently about equal to the rate of population increase, also has a positive effect, since outdoor recreation is compatible with small groups and especially children. It is also likely that our increasingly liberal attitudes toward leisure, and the apparently increasing status value of an outdoor vacation are conducive to increased participation.

Both a continuing increase in disposable per capita income and a changing distribution of income in favor of skilled and semiskilled workers are permitting more and more people to participate if they wish. Longer weekends and paid vacations also have acted permissively, as do less puritanical attitudes toward leisure. Similarly, cheaper, more convenient highway, bus, and air travel in effect is bringing forest recreation areas closer to the population, thus permitting more participation. But the *desire* to participate and thus to make use of these permissive circumstances for outdoor recreation, instead of something else, appears to be caused by another factor. It seems to be a psychosocial urge *to participate generally,* and particularly in an outdoor environment. Americans have always had the urge to see what is over the next hill, but increasingly since World War II they have also wanted to become more of a part of what is over that hill; and in large degree this consists of the natural environment of forests and water. No doubt increasingly concentrated, and sometimes monotonous work environment, associated with increasing urbanization, heightens the need. However, it is not the purpose of this discussion to attempt to quantify or even identify the complex of variables at work which form this psychosocial factor sustaining the rate of change in taste toward the forest and outdoors. But it is reasonable to assume that tangible and intangible changes in environment, particularly those stemming from an increasingly complex society, associated with higher levels of education and perception, form the basis of what is presumably the cultural change taking place. Apparently there is more to life than fleeting glimpses of the spectacular from a tail-finned automobile, and a house in the suburbs stocked with a T. V. and home appliances—P.T.A. meetings, clubs, and cronyism notwithstanding.

Although the 12 to 15 percent rate of increase in participation is likely to taper off eventually, there is no good reason, in view of the causes, to expect it to decline in the near future, barring unexpected emergencies. For some time to come it would appear that both the permissive circumstances and the variables contributing to an increasing desire for forest and outdoor recreation will behave so that they will continue to support a substantial rate of increase. Even the prospect of short-term recessions would not be expected to have a depressing effect because less expensive forms of outdoor recreation can be substituted readily for more expensive ones, or they may be substituted for other more expensive leisure time pursuits generally.

The implications for foresters and land managers of such an expanded rate of increase in participation in outdoor recreation are, first, that larger numbers of people will want to participate more so in the future, and secondly, that there will be a greatly increasing use-pressure on desirable land, facilities, and services. Marion Clawson has estimated, perhaps conservatively, that there will be 10 to 40 times as many annual visits to the national forests alone by the year 2000.[2] The use of land, facilities, and services for this order of magnitude of visits on most of our outdoor recreational lands will cost billions of dollars per year.

These realistic estimates raise some important questions, not only because we are dealing in figures which would account for a substantial portion of our expected national income in future years, but because a great deal of our total energies and culture as a nation will be involved.

Some of the questions are these: What kinds of values, as individuals and social groups, are we receiving from recreation? How should they be paid for and by whom? What will tend to control the optimization of these recreational values accruing to all concerned?

To consider possible answers to such questions intelligently, it is necessary to examine these values, show how they arise, and to see who receives them, which is the main purpose of this discussion. However, before considering forest recreational values specifically, and how they arise, I shall discuss briefly the general notions of value, production, distribution, and consumption.

Consider how economic value arises, in the sense of some good or service, which will yield satisfaction to people. First, there must be a technology—a method and organization for producing and distributing the product. Secondly, there must be a method of paying for it, or financing it, since sacrifices are involved in its production. Thirdly, it must be capable of being used or consumed over time to the greatest degree possible, if we are to get the most from our sacrifices. Whether we get the most out of something for what we put in, and whether we put the right amount in, depends on our understanding of how it can be produced, distributed, paid for, and consumed. It is most important that we know what is being "consumed," in the sense of the kinds and amounts of benefit the consumption stands for.

In the case of many consumer goods—pencils, clothespins, shirts—individual enterprises produce and distribute them on the usual profit-motive basis, and these goods are paid for by consumers individually because they are consumed individually, with the individual receiving essentially all of the value.

On the other hand, the defense system, which provides a basic service, is feasible to organize and operate, that is, "produce and distribute," only socially; in this case nationally, if it is to have any real value. Furthermore, it can be paid for only socially, that is, jointly, and not just because of its method of production, but also because its value can be utilized only jointly.

An example of an in-between case is the postal system. Conceivably it could be organized on an individual enterprise basis, but national organization and operation are much more efficient, particularly in terms of the certainty of the service. However, this service is paid for, very largely, by consumers as individuals, just as they pay for the services of private enterprise, because a great deal of the value is easily identified on an individual-use basis. We produce and distribute this service socially, and pay for it largely individually, though partly socially, because there is some joint or social value over time. We use this mixed system because we think it yields the greatest value or satisfaction for all concerned.

Our question is, into what sort of "production, distribution, payment, and consumption" system does public outdoor recreation fit?

To partially answer this question we shall first examine the values of forest and outdoor recreation.

VALUES FROM FOREST AND OUTDOOR RECREATION

Outdoor recreation is important because of its initial, direct effects on individuals. It is important also because of its subsequent, indirect effects on our society as a whole, in both short- and longer-term senses.

The direct effects of recreation may be considered as:

1. Greater physical health.
2. Better mental hygiene, i.e., inspiration, hope, optimism.
3. Educational and cultural effects, i.e., greater understanding of one's environment and a better ability to enjoy it.
4. The immediate satisfaction of consumption, i.e., the immediate fun of participation.

Each of these commonly agreed-upon effects may be considered as future streams of benefit or satisfaction potentially available over time from the use of the resource. As such, each can be thought of as the present worth of the stream of benefits discounted back from the future to the present, just as we would consider the present worth of a future stream of dollar income. Thus each represents a present *utility value* of expected future benefit, just as the discounted present value of future net dollar income represents present capital value. As we shall see, these direct, individual values represent only a part of the total value which accrues, because the social value has not yet been accounted for.

Now when one talks of value, there are a dozen concepts of it which may be used. But apart from the notion of the present utility value of expected future consumption, just mentioned, we need to be concerned with only two other concepts of it. These are individual versus social values, and market or common money values.

First, it will be convenient to consider the individual and social values involved. Besides the fun of participation in recreation, the individual receives directly the benefits of better health, better mental outlook, a deeper education, and is better able to interpret and enjoy his total environment. Notice that he may not be conscious of the health, mental hygiene, educational, and cultural values, or if so, often to a relatively small degree. It is the utility of immediate participation, simple fun, which tends to involve him with the other values, just as does the floating toy in a little boy's bath—he tends to enjoy himself and come out as a much more acceptable little boy, to both himself and others!

When an individual decides to engage in outdoor recreation, it is difficult to measure how much of each of the four personal benefits or values mentioned above is affecting his decision to participate—probably the satisfaction to him of immediate participation or fun is predominant, though. He may also derive a good deal of reflective or vicarious enjoyment in the future, yielding him still more satisfaction, which may bolster his decision to participate again.

On the other hand, society, in the sense of the national community, or many of the smaller groups, also tends to benefit. It benefits as a whole over time, in at least two ways.

First, we as a society, are better off in terms of our general frame of values because of the *diffusion* of individuals' direct benefits of improved mental attitudes, better physical health and deeper education. We are happier, more pleasant, more stimulating and more resilient over time, and we suffer fewer cases of poor mental and social health, including crime. That is, society's immediate income of satisfaction is increased, in addition to the direct benefits received by individuals.

Even if that were the end of the social value, we would be better off for it. But a second social value accrues by means of an increase in income and output from the economy generally; the only requirement for this conclusion being that there is a range of social complementarity between leisure time and gross national product, the latter being the result of what we commonly term "work." An increase in income results from people as a whole being better stimulated and more satisfied, who work better, produce more, innovate more and who are more creative over time, even though many of them may not have engaged in outdoor recreation. The question then becomes, *how much* work and *how much* leisure are needed to arrive at the most desirable combination.

But, may it not be argued that these benefits are in fact the same ones which accrued to the individual, even though he may not be particularly conscious of their cause? And isn't the social value simply the sum of individuals' returns of value? Of course, benefits in these same senses do accrue to individuals, but the social value is not the sum of individuals' immediate receipts, as it were. The social value, which it can be argued, is large and important, comes about because the effects of outdoor recreation on individuals, conscious or not, persist, and through time constructively affect a thousand and one interactions of the relations among individuals and groups in the process of normal social intercourse. Consequently, the total value is far greater than the sum of the immediate parts which accrue directly to individuals, the difference being the social value. Apparently then, it is necessary to first recognize these kinds of values to decide how the corresponding benefits should be paid for, whether by the individual or social groups, if individual and social benefits are to be equitably optimized. They must be measured, appraised or judged as well as possible to decide simultaneously how much outdoor recreation there should be (which depends on leisure time and taste for such recreation) and who should pay how much.

MEASUREMENT OF OUTDOOR RECREATIONAL VALUES

Presently we cannot measure consistently *any* values or benefits in an ultimate, comparable sense. Any attempts at measurement are, of course, relative, and in many respects imperfectly relative. This does not mean that partial, imperfect measures should not be made: quite the contrary, since decisions must be made now and in the future. Even imperfect measurement and incomplete value frames of reference can be extremely helpful in making better decisions.

What of market values? It is usually possible to measure reasonably well what the market value of outdoor recreation would be in a given situation. This *market value of participation* found from market price may be measured by *demand analysis* in terms of dollars.

Consider the typical classroom market demand curve, which represents a ranking of less and less willing buyers, *DD* (Figure 28-1). At any given level of price, *PP*, there are some *least willing buyers* who will just purchase some of the commodity. This quantity, in addition to the amount purchased by all the more willing buyers, who would have bought at higher prices, just equals the quantity *Q* which is sold at the demand price *PP*. Notice that the dollar value, the product of *P* times *Q*, values *all* of the output in terms of its dollar value to the least willing buyers. Consequently, it does not measure all of the dollar value accruing to the more willing buyers. Furthermore, since individual consumers do not, and should not, consider social values when acting

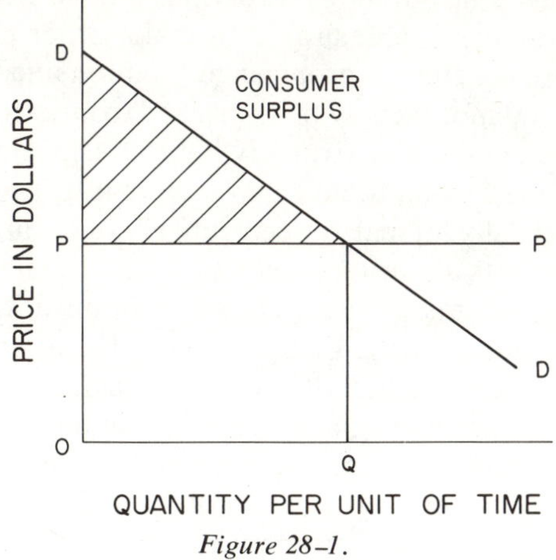

Figure 28–1.

simply as consumers, these values are ignored in the market demand value. Those individual benefits of which buyers may not be fully aware also tend to be ignored in market value measurements.

It is apparent then, that *market or demand* value of anything does not measure even its relative value in terms of dollars for many individuals or to society, let alone its value in some terms of ultimate satisfaction or utility. Market value measurements probably ignore all of the social value, and a good deal of the individual benefits of mental hygiene, education, and culture, as well as some of the health benefits from forest recreation.

Actually, the function of market demand values is to guide the allocation of more or less resources into the production of a given product. In this function the least willing buyers are the buyers who count, because they largely are the ones who would buy more if more of it were produced and offered at lower prices. Market demand values tend to be good indicators for resource allocation *where the great majority of the benefit received is individual in character and is fully recognized by the consumer, with respect to both immediate and future effects,* as in the case of ordinary commodities. Obviously these criteria are not met in the case of outdoor recreation.

Because market price ignores to a great extent many important values involved in outdoor recreation, only some small part of the additional individual and social value of more outdoor recreation is indicated by the price least willing buyers would pay. Consequently, dollar or market demand value cannot be used as *the* major criterion governing the allocation of resources into outdoor recreation. However, despite the fact that it is less than an ideal measurement, estimated market prices is still a useful one. As we shall see, it can be helpful in making decisions as between individual and public burdens of payment for outdoor recreation, as an indicator of fairness or equity, and as an aid in optimizing resource use.

Great art, education, hospitals, roads, and many other so-called "community services" fall more or less into the same category as outdoor recreation, wherein many social values, and unrecognized values to the individual are involved, often having long-term implications for the future of society. Actually, practically all products and services are similarly involved with the same market demand value problems to some degree; it just happens that outdoor recreation does not lie near either end of a social-individual value spectrum, but somewhere in between.

Where, in the instance of outdoor recreation, a good deal of value is not recognized by the market system, it is desirable to make other partial analyses of value through psychosocial approaches to preference, opinion and attitudes.[3] Even unconscious notions of value and motive may be appraised to an extent, and such work in the outdoor recreational field could be extremely helpful, especially since relatively little has been done. But let me hasten to add that *any* analyses presently cannot hope to provide a complete, internally logical or even entirely rational system of value appraisal. Even with our best efforts at various kinds of value measurement using disciplined approach, the overall problems of value appraisal in outdoor recreation will leave a considerable amount of uncertainty. Consequently, the attendant issues of resource allocation, financing, and use, which depend on these appraisals and forecasts, will continue to require large amounts of informed common sense and sensitive value judgment on the part of those making the necessary decisions.

Benefit-cost analysis, wherein the ratios of identifiable dollar benefits and costs are compared among alternative government projects, is an aid used in making these value judgments. But it is only an aid since many values, particularly indirect values occurring over time, social and otherwise, cannot be measured properly.

There are also a few other, sometimes misleading, attempts to measure and compare outdoor recreational values, of which we should be aware.[4]

One of these, *consumers' surplus,* represents the amount of money value indicated by the area above the price line and below the demand curve in Figure 28–1. It is the amount of money each buyer would have been willing to pay had he been required to do so, in order to purchase the amount he did buy. In other words, consumers' surplus here represents the monetary "bargain" one gets when the price one would have paid is above the market or least-willing buyer price. The main use of its possible estimation is to attempt to show the net benefit to the public in dollar terms, above total costs, of a given forest recreation area.[5] In this sense it may be useful as a partial measure of the net public benefit of the area, but it still would not indicate the unrecognized benefit to individuals nor the social benefit. And we could not compare this use to alternative uses of the area unless *their* consumer surpluses were also measured. Furthermore, there are some inherent conceptual difficulties in measuring it, although these may not be insurmountable.

Second is measurement of *gross dollar volume of business or income* connected with outdoor recreation. In this case the dollar values to least-willing buyers of many products purchased in association with outdoor recreation are added together. Of course, essentially all of this money would be spent in some other way by the people concerned, in the absence of the recreation area under consideration; a good deal of it for the simple requirements of day-to-day living. Thus gross expenditure does not measure the marginal demand value as dictated by prices least-willing buyers would pay for the use of the resource itself. In itself, it does not give any clue to the dollar value of the recreational output in question. In other words, we are dealing with the wrong thing if we are looking for a measure of either the total dollar importance of outdoor recreation, or a criterion as to the justification of further expenditures on outdoor recreation.

One other often-used method is *value added* by recreation-related expenditures, which amounts to gross recreation-related expenditure less the cost of materials. It *does* show volume of net new market business or income value in the community, which may be compared with alternative uses of the land. But, again, value added does not indicate a market demand value of the resource itself. And, as in the case of gross expenditures, it may well show up in about the same magnitude elsewhere, in the absence of the local land-use in question. In all of these cases, i.e., of consumers' surplus, gross expenditures, and value added, social values are, of course, ignored.

AVENUES FOR FINANCING PUBLIC RECREATION

Having considered something of the size of the recreation issue, the kinds of values involved and their importance and measurement difficulties, let us briefly consider the logical avenues for financing it, where by "financing" I mean nothing more than rational methods of payment.

Obviously some of the cost, particularly for certain kinds of outdoor recreation, where the participation values to the individual are obvious, should be paid directly by individuals. The exchange could be efficient, and equitable or fair, to those involved. But what of those benefits which have a definite individual component, wherein the individual is largely unaware of the individual benefits he does receive, e.g., better individual mental and physical health? One's first reaction would be to say, "Raise the price of participation," but this is an irrational and thus unacceptable suggestion because it is tantamount to saying "Charge him more than he will pay!"

One long-term solution would be to educate people concerning these values of which they are presently unaware, so that they would be willing to pay for them, i.e., in effect to shift their demand curves for participation (Figure 28–1) such that they would pay more for the *same amount* of participation. This is what people attempt to do by various means and for various motives in commercial advertising—now we can see the fine difference between education and advertising! Another more immediate approach would be to tax more widely and more heavily those items used in participation, such as skis, camping equipment, binoculars, etc., on which people *would* pay a higher price to cover the tax, and then transfer the revenue to the agencies providing recreation resources. In other words, broaden the present Federal Aid in Fish and Wildlife Restoration programs which are financed from excise taxes on certain fishing equipment, firearms, and ammunition. Such a plan would be practicable, and equitable as well, since only those who *did* participate would bear the tax burden.

Apart from either direct or indirect payment by *individuals* who participate in outdoor recreation, the other method of payment is from the general revenues of the various local and state governments, as well as the federal government. Payment through such agencies for some major part of many kinds of outdoor recreation is justified, since, as we have argued, a great deal of the value involved is social in nature. Quite apart from the usual argument of the impracticability of both the production of outdoor recreation and the collection of payments by private owners, payment of a major part of the cost by government is necessary to more nearly optimize the social value of outdoor recreation.

Regarding private owners of large land areas having considerable social recreational value, it would appear justifiable that they receive payments from government, to the extent that they make their lands generally available for public recreation. There would seem to be little reason for the subsidization of social values by the private sector of the economy, although this happens to some degree in the absence of transfer payments from government. Private payment by corporations and companies for social benefits simply tends to increase the prices of the products of these firms, thus putting some of the burden of paying for social values inequitably on the shoulders of those buying the particular products in question. It may also unjustifiably restrict the sale of these products to a degree in view of their increased prices. Since the money which companies use for this purpose must come from somewhere, the other outcome, in the absence of product price increases is, of course, a tendency to force down the prices of the goods and services used in production by the firms involved. This has similarly undesirable consequences. Furthermore, in the absence of transfer payments from governments to the companies concerned, it is to be expected that companies would under-allocate resources for the development of social values from their lands, unless, of course, a great deal of public-relations value to these firms was involved. This again is an inequitable method in view of the purpose involved. Various methods of effecting transfers are possible; perhaps the simplest would be special rebates or tax concessions in view of socially desirable outdoor recreation expenditures by the firms concerned.

RESOURCE ALLOCATION

I have no intention of attempting to get into that fascinating field concerned with the actual physical combination of forest recreational resources on individual areas, nor to be concerned even with the general issues of design, master-planning, organization and management, for obvious reasons.

Rather, I shall consider a few of the major aggregate allocation and use problems involved.

One of the most striking is the extreme *maldistribution* of our forest and outdoor recreational resources relative to the present (and likely future) population distribution and density. We now have about 50 percent of our population in the Northeast, where there exists presently only 14 percent of our recreational land, whereas about 4 percent of the population is in the western mountain region, which encompasses about 28 percent of the recreational land.[6] Effectively, if not literally, we need to "create" outdoor recreational space relative to the heavily populated areas.

One way to do this is to make serious attempts to deliberately plan our total highway, railroad, airline, and water transportation systems with ease of reaching outdoor recreational resources as one of the main objectives—not simply to try to maximize the ease of inter-city travel, or the simplicity with which one can go from coast to coast. If such a thought sounds grandiose, recall the greater and greater numbers of billions of dollars of annual expenditures expected to be involved in the direct costs of outdoor recreation.

Another way of "creating" recreational resources is through direct invention and innovation. We have for many years approached the design and handling of our recreational resources in simple, sometimes obvious time-tested ways. How can resources of this kind, particularly near populated areas, be physically combined in novel ways so as to, say, double the capacity to provide satisfaction which they presently have? To do so will require some careful study of how, on the actual site, satisfactions are in fact generated and how resources can be better planned physically to create these satisfactions. Obviously such an approach will require that more money and capital go into research and into the areas themselves in terms of landscape, design, facilities, and control. Such use of capital to create simple capacity, and effectively increase the quantity of the recreational resource would well pay for itself, at least up to a point.

Another problem in recreational resource allocation is long-run *flexibility,* both spatially and over time. For example, the composition of our population has changed. Presently we have both more young people and more older people than we used to have. Each of these groups wants different *kinds* of forest and outdoor recreation *now,* and within reason, *in the vicinity* of where they live. Usually these changes are not sudden; most can be predicted considerably in advance and planned for. Flexibility in use over time and space is needed in order to be able to change the qualities of the resource and insure its most beneficial use.

Analogous to the maldistribution problem is what may be termed the *institutional allocation problem.* Numerous public and private institutions or agencies literally do the allocating of this resource, yet many have diverse capabilities, prerogatives and formal obligations, which no longer hold much relation to the obligations in resource-use planning and allocation which they should be able to accept. A good deal of inter- and intra-agency redesign and reorganization for the efficient allocation and management of recreational resources is implied, if we are to meet the emerging national outdoor recreation problem, particularly in view of the somewhat abstruse, yet important nature of the values which such allocation largely determines.

Finally, I would like to mention one other point which is also concerned with resource creation and allocation: that is what may be termed the *time dimension of the allocation*. By this, I mean simply the amount of time per year during which the resource can be exposed to people, that is, the amount of time they can be in contact with it. Quite obviously one cannot have rationally allocated a resource unless the customer can use it! Consequently, the more use he gets from it the better it has been allocated. If a given recreation area is used twice as long per year, effectively it is about the same thing as having allocated twice as much resource.

One way to increase the allocation in this sense is to keep the facilities open longer: more hours per day (perhaps with an assist from artificial light), more days per week, and more weeks per year. But a more productive method of resource creation would be to change some of our currently accepted customs in the use of time which may act to severely diminish the amount of satisfaction people can derive from the recreation resource. We could adopt customs in the use of time which would give people the opportunity to visit these recreation sites when the areas are *not* overcrowded. By simply substituting Mondays for Saturdays as the "day off" on a weekend, for about half of us, to go along with Sunday, the resource could be effectively enlarged by up to one-third. The same argument would also hold for a huge number of other resources, such as roads and transportation facilities generally.

If the two-day "weekend" were staggered throughout the week, the use possibilities could be multiplied by hundreds of percent. Such a scheme is certainly attractive from the point of view of increasing the enjoyable use of recreation resources themselves, as well as the associated resources. On the other hand, it would tend to disrupt the time sequence of religious custom by having "one day in seven" occur every day. While this may put an additional burden on the clergy (but perhaps not an undesirable one), it would lighten the burden of cost for places of worship. The fully staggered weekend would also cause employers some initial problems in labor and production scheduling. But it is quite conceivable, once the transition were accomplished, that aggregate efficiency of production in the use of both capital and labor would be increased, particularly when one considers the salutary effect of a relaxed, uncrowded weekend!

A more immediately acceptable way would be the scheduling of vacations jointly by employers and employees, from the point of view of optimum use of recreational resources. Such additional use of the resource could be achieved to some considerable degree, particularly if regional and national outdoor recreational facilities, as well as climate, were considered in scheduling vacations. Education of employees by joint employee-employer groups as to the value of such a system, in addition to a current annual scheduling program to show which periods are overcrowded, would tend to make the system feasible. Planned regional coordination among major employers also would be highly desirable.

Regional and national vacation-impact planning has another obvious implication, apart from greatly improved recreation resource use. That is the possibility that production may be somewhat reduced in some parts of the economy because of such scheduling. The question is, how much scheduling can be done before the loss from the nonrecreation parts of the economy becomes greater than the gain in recreation resource use?

In short, effective recreational resource use could be at least doubled simply by arranging for some people to be there when others are not, and perhaps by extending the current hours of use.

OPPORTUNITIES FOR INCREASING DEVELOPMENT

It becomes apparent that optimum acceptable recreational resource development depends in large part upon the kinds of values involved, their identification and the estimation of their implied demands, and the development and articulation of the avenues for paying for these values. It depends as well upon the degree of imagination, planning and coordination which recreation resource allocators and managers are capable of exhibiting in the actual allocation-management process over time. And more so in the future, it will depend on our social flexibility in the conventional use of time. Optimum development, then, means maximum satisfaction in use of present and potential outdoor recreational resources, subject, of course, to higher demands on our time and money for still greater satisfaction and more important additional values per dollar of cost, which may be available from other parts of the economy.

The optimum development problem is complex. It has many simultaneous relationships; it is dynamic and hence requires continuous appraisal. There are many subtle, extra-market values involved which are nonetheless important and require our most careful attention and evaluation.

First we must identify important, sometimes elusive individual and social values. Then the problem is to infer how much of each kind of value would be forthcoming, so they may be paid for in equitable proportions by individuals and by the various levels of government. It is important to recognize that if we continue to expect individually recognized values, which individuals would in fact pay for, to be paid for socially through government simply because government must manage the resource, we will not optimize development and the values involved. Market demand measurement is obviously implied. If, conversely, we were to charge such high prices to individuals that many people would be deterred from participation, we may seriously impair important social values. Clearly, extra-market value measurements and judgments are implied.

Notice that the question here does not primarily concern income distribution, nor does it have much to do with public versus business expenditure. Rather it is one of arriving at the optimum proportion of the total cost which should be paid for directly by participants and by governments. These proportions may differ with differing situations, but only when the "correct" proportions are adopted in each given circumstance can we expect total satisfaction and thus resource development to be optimized.

More efficient, well articulated, and flexible methods of paying the government share of the cost are, of course, inherent in the optimizing of forest and outdoor recreational resource development. Without adequate methods of financing, this bottleneck could largely nullify other efforts. Partly it is a matter of public education. But largely it rests on recognition of the problem of value identification and consequent financial organization of governments so that they may, having perceived the values involved, appraise them and act cooperatively in their financing; if necessary, through regional recreational authorities and compacts, in addition to our present levels of government.

All of this is going to take a good deal of continuing research in the social sciences as well as in the various physical arts and sciences related to the out-of-doors. But even more so, it will require planning and coordination among all of the agencies and firms involved to make use of such research, some of which is presently available. Then it will take more planning and more organization—more intelligent anticipation of the future and more preparation for it. Provision for continuing research, coordination and planning appears to be a necessity, no matter how well *ad hoc* commissions and boards may do their jobs. Granted, we could reach a degree of planning intensity where a higher level of it would no longer pay, but in view of the complexities to be faced, we are a long way from that point now.

No longer is the goal just to do things big—to take sporadic giant steps and then ignore the problem until it again represents a crisis. If we are to envision the United States as becoming a large, beautifully conceived, flexibly designed outdoor environment capable of responding to changing tastes and demands for all the uses of land, continuing forethought and planning are required.

REFERENCES

1. Clawson, Marion. 1958. Statistics on outdoor recreation. Resources for the Future, Inc. 165 pages.
2. ———. 1959. The crisis in outdoor recreation. American Forests (March and April).
3. Dana, S. T. 1957. Research in forest recreation. U.S. Forest Service. Washington, D.C. 36 pages.
4. Clawson, Marion. 1959. Methods of measuring the demand for and value of outdoor recreation. Resources for the Future, Inc. 36 pages.
5. Hotelling, H. *In* Prewitt, R. A., et al. 1959. The economics of public recreation. Land and Recreational Planning Division, National Park Service, Washington, D.C.
6. Clawson, Marion: The Crisis in Outdoor Recreation. Work cited.

29

ECONOMICS IN UPSTREAM WATERSHED MANAGEMENT

M. M. Kelso

First published in *Society of American Foresters, Division of Watershed Management, Proceedings;* pages 120–123. 1962.

BIOGRAPHY

Mr. Kelso has been long and widely known as an agricultural economist, notably a land economist: teacher, researcher, administrator, and practitioner. A native of the northern Great Plains, he commenced his professional career in Washington, D.C., in the Departments of the Interior and of Agriculture, in the latter serving for four years as Chief of the Division of Land Economics. During World War II, he managed a cattle ranch in eastern Montana. Thereafter, he joined the faculty at Montana State College, where he became Dean of Agriculture and Director of the Experiment Station. In 1958, he accepted a position at The University of Arizona. He is now Professor Emeritus of Agricultural Economics in that University.

EDITORS' SUMMARY

Watershed management is a sociocultural art supported by the abstractions of physical, biological, and social sciences. The art addresses itself to human behavior toward the watershed, including economic behavior. Watershed economics concerns choices in production made in consideration of the benefits and sacrifices for the group represented in the decision.

QUESTIONS TO CONSIDER

(1) How do you distinguish between *science* and *art*?

(2) Mr. Kelso characterizes watershed-management science as comprising abstractions from physical, biological, and social sciences. What does he mean by "abstractions"? Is not

his statement true for each branch of forest resource management? Cite examples, paralleling the discussion on pages 228–229.

(3) How does Mr. Kelso view *art* as being related to science in resource management? Can you think of illustrations from areas other than watershed management?

(4) Mr. Kelso states (page 231) that the economics of production concerns (a) what, (b) how much, (c) by what means, (d) for whom, and (e) when. Do you have other items to add to the list? Does the same list apply to the economics of consumption?

(5) The need for choosing is ascribed (page 231) to scarcity. Is this the same as saying that choices are mutually exclusive? Are there no choices to be made among alternatives which are not mutually exclusive?

(6) Because of the perceptiveness and central importance of Mr. Kelso's view of resource management, it will be very useful to you to summarize his ideas in your own words, making sure that you understand each idea which you summarize.

* * *

I begin by calling attention emphatically to a duality that necessarily exists within any subject-matter discipline that aspires to any name implying management, such as watershed management. Any such discipline is, at one and the same time, both a science and an art.

As a *science,* it aspires to the description of functional relations among those variables found to compose the watershed, and to the prediction of future states of the watershed derived by the application of these functions to data representing its present state. As an *art,* it is an application of science, intuitions, experience, and value judgments to the purposeful manipulation of the physical-biological-social complex that composes the watershed toward the attainment of some more or less clearly or vaguely formulated goal. As a *science,* its goal is to understand in order to predict. As such, it is objective and unemotional. Only the value judgments peculiar to science—the commitment to truth and objectivity, faith in the regularities and repeatabilities of natural phenomena, the ability of man to understand and direct the universe, and the desirability that he do so—only these value judgments enter into watershed management as a science.

As with any science, the demands placed upon the science of watershed management necessitate abstractions from reality. The finite abilities of man do not allow him yet to play at being God in his understanding and prediction of phenomena and at the same time stay within the rules of science. These necessary abstractions to which the science of watershed management subjects reality are most conveniently classed as physical-, biological-, and social-science abstractions.

The *physical-science* abstractions in the science of watershed management are concerned with the descriptions and functional relations of nonliving processes—geology, soils, hydrology, soil and water development, productivity of resources, etc. These abstractions can but need not be divorced completely from the biological and social attributes of the watershed complex. The *biological-science* abstractions introduce the descriptive and functional attributes of life into the phenomena—birth, growth, and decay of organisms. These biological-science abstractions can but need not be divorced completely from the social, but they cannot be divorced to the same degree from the physical attributes of the watershed for biological phenomena cannot exist except within a physical environment.

The *social-science* abstractions in the science of watershed management introduce man, not as a biological phenomenon, but as that peculiar psychological animal that uses tools and possesses memory, a highly developed communication mechanism, a penchant for group behavior and, most important of all, goals or ends-in-view relative to which he makes choices among alternative courses of action through the use of reflective intelligence. The social-science abstractions add memory, selective choice, goal fulfillment, institutionalized behavior to the biological-physical complex. Though I have spoken of social-science abstractions in the study of watershed phenomena, it really is not possible to abstract the social from the biological-physical, even for the purposes of science. A society is composed of biological men living in a bio-physical environment. But, as a social-science abstraction, watershed management must, nevertheless, abstract from the bio-physical attributes of the watershed by the device of assuming some values for these parameters and holding these values constant for each analysis of the social variables.

Recalling my earlier reference to science as being characterized by the derivation of functional relations among variables and the prediction of the future states of systems containing those variables, the social science content of watershed management science can now be characterized. It is abstract and analytical; its objectives are to discover and describe what the social variables are in the watershed complex, to plot their functional relations with one another and with the matrix of bio-physical variables, and to predict the social and, hopefully, the bio-physical states of the watershed complex. The social science of watershed management asks such questions as these: Why do men use a watershed like they do? How would they like to use it? What values and goals do they hold toward the watershed and its use over time? How do they go about trying to realize these goals of watershed use? How do their goals and their instruments for reaching these goals change with time? Why? What will they do with, on, and about this watershed 10 years, 50 years, 100 years hence?

Let us note here, parenthetically, that this scientific study of the values, goals, and processes of men toward the use and management of watersheds includes study of watershed management scientists, of watershed managers, and of other public groups in addition to such study of the direct users of the watershed.

So much for watershed management as a science. Recall my argument that, as a science, it must be an abstraction, hence "unreal." Further recall my assertion made earlier that watershed management, as is any management discipline, is an art as well as a science and that the two are not synonymous though closely related they may be. The *art* of the management of a real-world watershed is not an abstraction, cannot be unreal in the sense that science is unreal. As an art it encompasses science but is so much more than science; it is, besides, intuition, experience, and value judgments on the part of its practitioners, private and public, individual and group, applied to the complex that is the watershed without abstraction, in complete reality. As an art it must be concerned not with how the variables interact in some state of abstract simplification, but with how they really are and how they all—physical, biological, social—interact together without benefit of the idealizations, the simplifications, that science must use. As art the discipline asks these questions: What goals do we wish to or ought we to attain in the use of this watershed? How should we go about attaining them? What are the strategic variables in the complex which, if properly manipulated, will attain the desired goals? And what is the right, the best, way to manipulate them to attain the desired result?

This, which is the *art* of watershed management, involves, without distinction, without favor, the physical variables, the biological variables, and the social variables in all their complex web of functional relations. The abstractions of physico-biological science in the hands of watershed scientists may predict that on a given watershed less grazing by domestic and game animals will increase long-run production of silt-free water. But so what? This bit of stimulating information may be of interest only to another watershed scientist. In the *art* of watershed management one can't "abstract" the social sciences away. How do you get people to graze fewer domestic animals or to take steps to reduce the grazing of wild ones? And, worse, how do you know that the benefits from more or purer water over a longer time at the cost of less domestic and/or wild animal products over the shorter run are what the dependent people want or ought to want? And, still worse, how much more or purer water at the expense of how much less domestic and/or wild animal product in the eyes of whom is a good trade? As an *art* watershed management cannot abstract from the social.

The *science* of watershed management is partial and incomplete just so long as the socio-cultural variables and their functional interrelations are not as well known and understood as the physico-biological. The *art* of watershed management is naive and immature so long as its practitioners are not as familiar and conversant with the socio-cultural as with the physico-biological phenomena of the watershed complex. In fact, I'll be more blunt—watershed management isn't a physico-biological science at all! It's a socio-cultural art related to man in his environment and the purposeful manipulation of that environment to attain ends that are man's—and at the risk of argument—that are not nature's, not God's, not mine, not yours, but ours in some collective sense. Watershed management in its usual sense as a science is really the study of physico-biological processes in and on watersheds—as such it's not management at all!

THE SOCIO-CULTURAL CONTENT OF WATERSHED SCIENCE AND MANAGEMENT

Let us now dissect out of that complex called a watershed those attributes that are socio-cultural and see something of what they are and a little of their functional relations. First we will note that all those attributes called socio-cultural have one common characteristic: they are all in one way or another concerned with human behavior toward the watershed. Next, such behavior will be seen to be sometimes that of individuals acting independently and sometimes that of people acting as groups or under group direction—what the social scientist calls institutionalized behavior. Both forms of behavior toward the watershed can be classified into these four classes: (1) the sociological: nonpolitical, noneconomic behaviors; (2) the political: the behavior of men in and toward government; (3) the institutional: group controls over and facilitating individual behavior; and (4) the economic: the manipulation of productive attributes of the watershed to maximize some net value product.

We will bypass the political, the sociological, and the institutional and consider further only the economic aspects of watershed science and management. (Note, I am, as a scientist, already abstracting from the totality of reality.) Hence, let us now examine the identity and the functional characteristics of economic variables in the watershed complex.

Concerning any watershed, if man has any concern for it at all, he must make certain basic decisions concerning it: (1) what to produce from it; (2) how much of each to produce; (3) by what technical means to produce them; (4) for whom they should be produced; and (5) when they should be produced. These are the basic economic questions. Whenever any question pertaining to watershed management is couched in terms implying any one or more of these five general questions, that question is an economic one. Now, I ask: how far can you get in watershed management without the science of economics and its participation in that art!

As an empirical science, economics, in its application here, is a study of men in the process of making these five decisions on and about watersheds. As an empirical science its desire is to explain and predict man's behavior in this regard.

As a normative science, economics asks questions that have to do with what the manager of watershed resources ought to decide relative to their use. What goals of use are most desirable and how can they best be attained? (Note the normative words "desirable" and "best" in that sentence.)

Such empirical and normative economic questions can be asked about individual, private users of watershed resources as well as about institutional managers of the total watershed. Each is making choices about the output mix that he considers it best to obtain from the watershed resources under his control and how best to obtain them. Such choices by the institutional manager are no less economic than are those of individual, private managers.

The central characteristic of these economic questions is that they demand that a choice be made between alternatives of outputs from the watershed, of inputs into the watershed, and of technical means of exploiting the watershed. This necessity for choosing implicitly demands some criterion for deciding upon "right" choice. Economic science, then, is concerned with the problems of choice from among alternatives of watershed management in the light of some criterion of right choice.

Why this need for choosing? Just one thing—scarcity. There's not enough of any one or all things to satisfy all of man's wants so he has to choose; hence, the central place of scarcity in economic science. Such scarcity demands (1) choosing between alternative possible products that might be obtained from our resources today but, also, (2) choosing between products today and the same or alternative products in the future. So we have choice between products at some moment in time and between different points in time. But choose we must.

It is pretty obvious to any of us that if we go about reducing deterioration of a watershed, it may demand that less meat, wool, and/or game animals for human use be produced and that hence some choice is required as between "animals" and "conservation." But it is not quite so obvious that if extensive capital structures are required, another choice is also demanded, viz., that between putting current purchasing power into these structures for the conservation of resources for tomorrow and putting that current purchasing power into current consumption. What has actually occurred in a case like this is that resources have been diverted from want satisfactions of today for other satisfactions of tomorrow. We are inextricably tangled up in a choice problem no matter how much we try to wriggle out of it or try to ignore it.

Since we must choose, how do we know how to recognize the right choice? In other words, what is the criterion for choosing? Economics provides one criterion (or one family of criteria) from among the many that control total choice. (Remember that as an economist, I am consciously abstracting from reality.) The economic criterion is the maximization of present satisfactions over dissatisfactions when such satisfactions are measurable, quantifiable, in some reasonable, defensible way.

Satisfactions are benefits; dissatisfactions are sacrifices. Hence, one criterion is to choose so that benefits exceed sacrifices in some maximum amount. Satisfactions in the area of economic choosing are revenues (incomes); dissatisfactions are costs (outgoes); the criterion in the world of private choosing is to maximize incomes over outgoes. In the world of social or collective choosing exactly the same demand exists but is not so clearly seen. The criterion is to maximize benefits over costs.

In each case the first term, income in one case, benefit in the other, denotes the magnitude of the satisfactions that can be obtained from each considered alternative. The second term, outgoes or costs, denotes the magnitude of the satisfactions that are sacrificed or given up by virtue of choosing each of the considered alternatives. The criterion is to maximize the difference between satisfactions secured and satisfactions sacrificed in each alternative. This is what the benefit-cost comparison boils down to: satisfactions secured, less satisfaction sacrificed. Economics in watershed science and management provides an analytical framework, a *modus operandi*, and a fund of expertise for determining this magnitude as a guide in watershed management decision-making.

In deciding between alternative watershed uses, it makes some considerable difference who the decider is. There isn't just one answer—not even one right answer. We, of course, recognize that each individual exhibits his own peculiar brand of cussedness and of ignorance that will bring about individual differences in choosing among watershed programs. But I wouldn't argue these are economic factors although I remind you that they are part of the socio-cultural content of watershed management. There are, however, several aspects of this individual difference in choosing which are economic and on which economic science can cast light in watershed management. We will consider only one.

I have pointed out that the economic content of choosing in watershed management relates to maximization of benefits over costs. Of course, it is implicit in this formulation that only those benefits and those costs incident to a watershed plan that actually impinge on the decision-maker are relevant to the act of choosing between alternatives. Those benefits or those costs that impinge on persons other than the decision-maker simply don't count in his act of choosing. The "broader" the base of the deciding unit, the more inclusive of the whole of the benefits and of the costs of a watershed management plan will it be; the narrower its base, the less inclusive. But, and here's the significant thing, the degree of exclusion (or of inclusion) of benefits will change at a different rate than will the degree of exclusion (or of inclusion) of costs as we move from broader to narrower-based deciding units.

An individual firm in choosing between two alternative plans of watershed use more than likely will find that, although all the costs which it incurs under either plan will rest on it, a larger part of the benefits resulting from one of the plans will devolve on beneficiaries outside the firm. It will, then, choose the alternative in which a greater share of the benefits created redound to it.

On the other hand if the decision-maker is, at the other extreme, an agent for the entire nation, all costs and all benefits incurred from any plan will impinge on it and hence such a decision-maker will choose that alternative wherein all benefits are a maximum over all costs.

These two plans of watershed development can be vastly different. They differ because in one the developments will be such that the benefits created will be heavily concentrated on a narrowly based recipient—such as a single firm, or a local community—whereas the other plan of development will produce benefits whose impact will be widely diffused over a wide area—the whole nation, for example.

Consequently, decision-makers selecting among alternative plans will come to quite different decisions as to what is "right" and "best" on a given watershed simply because the results, to be beneficial to one of them, must be highly concentrated in their incidence whereas to the other they can be widely and thinly diffused over a wide area. Such decision differences are honest and objective—not selfish or visionary nor the result of cussedness and ignorance. This is a purely economic difference that grows from the fact that positive results from watershed development aren't benefits to those who can't capture them and that consequently they aren't, for all people, there to offset costs incurred for their creation.

This difference in the economic consequences stemming from the degree of diffusion of the benefits ranges from the most concentrated, when the decision-maker is an individual person or firm, through increasing degrees of diffusion from local community to state or nation—and to continent and to world for that matter. Herein lies the reason why local areas and whole states and even whole regions seem always to be pressuring the national government for investment in development programs within their boundaries—the costs impinge on the whole nation (diffuse) but the benefits will be relatively highly concentrated within the local area, state, or region. As a result the benefits may greatly exceed the costs to the local, more narrowly based unit although the benefits may fall far short of the costs to the more broadly based unit, say, the nation.

CONCLUSION

What has been said can be digested in the following eight points. 1. Watershed management is both a science and an art. 2. As a science it is and must be abstract and unreal. Socio-economic abstractions relative to watershed science are no less abstract and unreal, though vastly more complex than are physical or biological abstractions. As a science it is partial and incomplete so long as the socio-economic variables and their functional relations are not as well known and understood as the physico-biological. 3. As an art it cannot be unreal as science is unreal. As an art it must encompass the totality of that booming, buzzing confusion which is reality and which embraces the physical and the biological and the social, and which requires not only the idealizations of science but also intuitions, experience, and value judgments in its decision-making. 4. The economic content of watershed science and art is concerned with questions of what, how much, by what technical means, for whom, and when to produce from the watershed's resources in order to maximize some net benefit criterion. 5. The net benefit criterion provided by economics is maximization of measurable satisfactions secured over measurable satisfactions sacrificed. This is the meaning of the benefit-cost comparison. 6. The comparison of satisfactions secured to satisfactions sacrificed may be between satisfactions of different character at one point in time or between similar satisfactions at different points in time or both. The latter comparison involves the complication of time discounting of values. 7. The economic benefits relative to costs for any given watershed program will differ as between decision makers depending on how broadly based the decision unit is. 8. Economics in watershed management as science or as art provides an analytical framework, a *modus operandi,* and an expertise for determining the magnitude of measurable costs for alternative programs of use and development of a watershed as a basis for scientific prediction or for the art of management.

30

THE ROLE OF ECONOMICS IN FORESTRY PROGRESS

William A. Duerr

First published in *Forestry Chronicle;* pages 116–121. June 1961.

BIOGRAPHY

Mr. Duerr's biographical sketch is given at Item 4, page 30.

EDITORS' SUMMARY

A major role of economics in forestry progress is to guide forest management decisions. These may be viewed as comprising five steps: (1) specifying the decision maker's goal, (2) identifying the alternative means to his goal, (3) weighing these alternatives in light of the goal, (4) making the choice, and (5) reviewing results as a check on data and procedure.

QUESTIONS TO CONSIDER

(1) Forestry economics and management obviously have wide areas of overlap. Economics is sometimes referred to as the science of management. Thinking of them as subjects rather than activities, what are the primary distinctions between the two?

(2) Goals are referred to (page 237) as "not absolute." What is an absolute? Are there any absolutes in the decision process?

(3) How does one discover the goals of a forest owner with respect to his property? What are the sources of error in the process?

(4) Why should any forest owner want wide plantation spacing or a lightly stocked timber stand (page 239)?

(5) Under what circumstances may there be no alternatives and thus no decision to be made?

(6) Why should an alternative be weighed "from the viewpoint of a firm's *entire* business" (page 239)?

(7) If an action leads to a result different from the goal, how can one tell whether the fault lies with the action or the goal? Think of specific examples.

* * *

In talking with you about the role of economics, I should like to stress just one part of the subject. I should like to dwell upon the relation of economics to the forest-management decisions of a firm. By a firm I mean an industrial concern or a farmer or some other individual or company—or, for that matter, government as a land manager. For such a firm, how may economics contribute to better decision-making in woods management?

Consider the process of making a decision. To view it analytically—and to clothe it in an orderliness which, I grant, it lacks in many cases—there are five steps in the process. The first, surely, is to identify the goal—what the firm aims to accomplish with its decision. Reaching some goal is the whole point of the matter. Second is the step of identifying the possible means to the goal: the likely alternative courses of action. It is the existence of alternatives that makes a decision necessary. The third step is to conjecture the pros and cons of each alternative and thus its net value to the firm. How well does each alternative promise to fulfill the terms of the goal? Then comes the fourth step, the decision itself, which is to take the data thus far derived and add some art and a bit of the self of the decision-maker. Here the sequence may appear to end, but there is a fifth step, which is to follow up after the decision, to see how well it turned out, what went wrong and why, and how to do better next time.

Now that I have one of life's complex problems all neatly simplified into five points, I propose to devote the rest of my talk to each of the five in turn. The five steps in making a decision or in making and carrying out a plan: One can look on them either way.

THE FIRM'S GOAL

Let me begin with goals. My concern for goals comes from the very nature of economics. Its subject is the use of resources to meet human goals. Economics takes engineering and biological questions and rephrases them in terms of human purposes and decisions. For example, it takes the biological question, *Can* we grow trees here?—a question that concerns site and species and methods of culture—and asks, *Shall* we grow trees here?—a question that additionally concerns human aims and the values they generate. To grow trees here may be feasible but still not desirable in view of the values (the costs and benefits) to man. And lest I be misunderstood, I will repeat that these values which govern resource use arise out of the firm's goals. Values in general exist only because goals exist. And each particular goal generates a particular set of values and thus a particular program of resource use. For a simple instance, the firm whose goal is to produce veneer sets a higher premium on large timber than does the pulp and paper firm, and tends to plan for longer rotations.

Goals, of course, are not absolute: They are not independent of the means for reaching them. Where forest site quality is low and thus big timber hard to produce, landowners' product goals tend to be shifted toward small sawlogs or pulpwood.

Now one of the interesting things about goals in forestry, as in life at large, is that they differ so greatly from person to person. Solon Barraclough, who made a pioneering study in New England to discover the purposes for holding forest land, found that there were nearly as many purposes as there were owners. Some owners, but not a majority of them, had various goals connected with timber-growing, or at least the cutting of timber. Others were mainly interested in the recreational services of their land. Many had mixed goals; some apparently had none at all. A few had unusual goals—or at least unusual ways of stating them—like the man who had bought a wooded residential property big enough so that he could stand on his porch and cuss at the top of his lungs without fear of offending any neighbors.

Thus goals differ in respect to the products or services that owners want from their land. Many owners are guided by other than a money-profit motive. And a few owners are fortunate in being able to manage land, not for any material result, but simply for the pleasure of working with growing things.

From the fact that goals vary, it follows that whereas silviculture is a matter of biological potentialities, *good* silviculture is a matter also of the manager's aims: What is good practice for one owner is not necessarily so for another, even though their forests are identical. Only where nature is very grudging and inflexible does she alone dictate forest practice. A friend of mine whose father owns a small woodland told me the other day that the public service foresters in that area had drawn up a plan for the woodland. He went on to remark that the service had been rendered with a minimum of fuss and bother: The foresters made the whole plan while the family was away vacationing. Father had not needed to trouble his head about it. No; and the chances are that he will not in future, either, give much heed to the plan. He will store it in the kitchen cupboard and pursue his interest as best he may. What sort of plan is this, anyway, that was made without consulting the woods owner? Clearly, one feature of it is that the foresters and their agency have substituted their own goals for those of the owner, perhaps without realizing that this was what their procedure amounted to. Even where forest practice is publicly regulated, the owner has some latitude to express his own purposes in his management. The case I cite was an example of thoughtless naivete on the part of dedicated and sincere foresters—and a good example of time wasted.

I think that economics may contribute to the work of the forestry profession by calling attention to the goals of forest owners and encouraging foresters to understand them and take account of them. I predict that foresters will in future study more and more closely the goals of forest owners and managers—farmers and other small woodland owners, forest industry, and government too. A number of helpful outcomes will flow from such study. One will be specific knowledge of goals in specific cases where there are management plans to be made. Here a by-product of the forester's inquiry will be to push owners and managers into careful review of their goals—their policies. Another helpful outcome will be specific knowledge of the goals of whole classes of forest owners, which should be indispensable in any educational or regulatory program to bend their forest practices toward the public interest. Still another helpful outcome of studying goals will be a general appreciation in our profession for the human aspect of forestry, the human variables, which may outweigh physical variables in forest management. We are all familiar today with terms such as "silviculture of the black spruce type," or "silviculture of nothern hardwoods." Some day we shall be using such terms as "silviculture of a pulp and paper firm" and "silviculture of a prosperous dairy farmer."

I understand that A. R. C. Jones, in planning for the demonstration areas in Macdonald College's Morgan Woods, near Montreal, is thinking of three groups of plots, to be managed from the viewpoint of three different classes of timber owner. At one extreme would be a demonstration for the institutional owner who is in position to follow an intensive program of selection management for large, high-grade northern hardwoods. At the other extreme, Mr. Jones would manage even-aged pioneer hardwoods on a short rotation as a hint of what the typical small owner might constructively undertake. Thinking such as this is going on here and there, and ever more widely, in our profession. It breaks with the old tradition of prescribing the same sort of silvicultural pill, compounded from the pages of our European pharmacopoeia, for everyone in sight regardless of what ails him, if anything. It bows to the fact that people differ as greatly as forests do. And incidentally I believe that it offers a clue to a saner and more fruitful professional life for the farm forester, whose pills are so often coughed up by the patients.

WHAT ARE THE ALTERNATIVES?

Let me turn now from goals to the alternatives for getting there. These alternatives are biological and engineering procedures. They include alternative forest uses and combinations. They include alternative programs for each use—for instance, in the case of timber, alternative species, regeneration methods, degrees of stocking, and intensities of protection and culture.

I believe there are three contributions that economic thinking may make to our knowledge of forestry alternatives and especially to research on the subject.

One contribution is the idea of the input-output schedule: the numerical statement of what it will take to accomplish specific jobs, or the statement of results expected from successive levels of effort in some line of production. A familiar example of such a schedule is a yield table, which relates output of timber growth to input of forest site and growing stock. Another is a statement of pruning or weeding or other cultural output related to labor input. Another is a cost-loss schedule for fire control.

The input-output idea can usefully be extended to other problems in forest management. I am thinking, for instance, of the problems of multiple-use forestry. How can the output of recreational services best be measured, and what is its relation, on specific tracts, to the inputs and outputs in timber production? That is to say, what is the silviculture of forest recreation? Even less is known about this question than about the silviculture of water production, which is another matter that may lend itself to input-output analysis.

Schedules that estimate the output associated with successive levels of input are congenial to economic thinking because they are so well suited to the analysis of resource use. Suppose that one can place, on the inputs and the outputs, values that are appropriate to the firm's goals. Then if one knows how much the output would be raised by an additional input, he can tell whether the input would be justified: whether the result would justify the effort. And in general he can estimate how much input—what intensity of management—is justifiable. This is the essence of the firm's economic analysis. Input-output schedules are a means for shifting forestry practice a little bit further out of the realm of art toward the realm of science.

A second contribution of economic thinking to the study of alternatives may come through emphasizing the effect that costliness of operation has upon the future of resource management—emphasizing the growing problems faced by the forest economy in its competition for resources, especially labor. The solution of these problems lies in the direction of technological progress: reducing the inputs required for a given output. Thus the issue regarding input-output schedules is not only one of learning the facts of production, but also one of discovering ways to improve them: to brighten the alternatives. My mind turns to A. M. Koroleff and his work on the use of labor and other aspects of efficiency in logging. As the practice of silviculture is extended and its costs become more at issue, I am sure we shall see foresters working with like zeal to strengthen silvicultural technology and also to improve the status of silvicultural labor.

A third contribution of economic thinking to the study of alternatives is linked to recognition of human variables in forest practice. It may be good practice for some owners to carry a timber stand of normal stocking; but others may find it advisable to carry less. Again, for some owners a four-foot plantation spacing may be best, but for others, a wider spacing may be preferable for the same species and site. The fact of the matter is that the human variable prompts us to take an interest in quite a wide range of management intensities and to discover the input-output relationships at the lower levels of intensity as well as the higher: Economic thinking prompts us to delve into such questions as "subnormal" yield and wide-spaced-plantation development.

WEIGHING THE ALTERNATIVES

Now I have heard it contended that weighing alternatives, the third of my five steps in planning and deciding, is comparatively simple if one can just have a clear, reliable statement of the alternatives themselves, with all the inputs and all the outputs considered. But this contention belittles two jobs that have to be done: the job of working out methods for weighing alternatives and the closely related job of deriving values to assign the inputs and outputs. I should like to talk with you about these jobs, because I think they represent a contribution that economics can make—and will make—to forestry progress.

When I was in forestry school, I recall that we studied quite a number of long, complicated formulas for regulating the timber cut, determining the rotation length, and that sort of thing. Some of the formulas were constructed around the notion of an identical series of events that would go on and on within the forest, cycle after cycle, forever. We memorized these formulas for the final examination. However, I do not recall having made much serious use of them in the nearly thirty years since that occasion.

I like to think that economics takes well to simple procedures. For example, it tends to play down formula approaches. It leans toward approaches that permit listing the pros and cons of an alternative from the viewpoint of a firm's entire business and applying judgment to their interpretation. Again, it tends to shorten up the planning period so as to concentrate upon the near future. The advantage of putting first emphasis on a short period is that it raises the resiliency and lowers the risk of the plan. An example of short-term planning is the use of the financial maturity idea as a guide to timber marking. The idea is to look ahead one cutting cycle and to mark those trees that do not promise to grow in value at an acceptable rate, or guiding rate. There may well be opportunities for devising short-term planning methods to apply to our varied management problems. The task will be to keep enough weight at the same time on the inescapable long-term aspects of forestry.

The essence, I believe, of a good planning device, aside from simplicity, is that it should be able to take account of the firm's goals, whatever these may be, offering guides to decision-making that are in keeping with these goals. Here economics has a bad reputation to live down. Classicial economic theory was built on the idea of economic man, an abstract person who was not only fully rational, but was positively obsessed with the desire to maximize his revenue or minimize his cost in terms of market values. Beyond the sphere of the market, all was void. Thus classical economics considered only a part of life, a part of man.

Abstraction is a useful scheme for solving problems. But if they are real problems, the abstraction must be slackened off in the course of the analysis and removed as fully as the human mind will allow before a solution can be claimed. Forestry deals with real problems, with ordinary problems. It concerns whole people and forests, and not parts. And so the goals we must attend to in forestry are whole goals and not parts. It will not do to assume that forest owners all seek the highest money profit, any more than it will do to assume they can all practice intensive silviculture.

Procedures of economic planning have only started to bring real goals into consideration. One approach is to take a given goal and then plan for maximum net forest revenue within the constraint of the goal. Thus a retired man may wish to manage his woods for an even flow of income. A timber program is then drawn up which will maximize the flow without reducing the current amount of it. A trouble with this method is that it treats the goal as an absolute, not admitting of any judgment whether the attendant sacrifice of future returns—or of present returns—is too great a penalty to pay for the sake of regularity. Nevertheless, the method is useful.

Another useful method for realistic economic planning is one conceived by Henry Vaux at the University of California. The method ensures that the future decision shall be consistent in certain respects with past decisions and thus shall further the firm's aims, no matter what they are, so long as past decisions did so. The method has been tried out on the problem of developing a guide to the reservation of timberlands as public parks. Timber values forgone when past reservations were made were computed per unit of recreational use of the park. The resultant implied value of recreation was then used to judge whether a certain proposed forest park should in fact be reserved, considering its timber values and its prospective recreational use. That is to say, one could know from past experience how much recreational use would have to be made of the area to justify reserving its timber values. This planning scheme, admittedly imperfect, is yet an imaginative, forward move toward accounting for complex goals. It is a move, also, toward the difficult and important question of non-market values in forestry.

The guiding rate of interest on forest investments, which I mentioned in speaking about timber-marking guides and financial maturity, is another device for fitting decisions to actual goals and for reflecting values appropriate to these goals. Interest is far and away the principal cost of growing and maintaining a forest. Consequently the rate of interest is the main economic force, and in many cases the only such force, that sets one silvicultural alternative above another in the order of choice. Now interest has a market rate, just as the wage of labor has, or the price of wood. But one need not use this market rate to guide the firm's decisions. Instead, one can adjust the rate to allow for the special circumstances and aims of the firm. And one can adjust it so as to make it consistent with past decisions that the firm considers successful and wise.

REVIEW

Now I should like to pass quickly over the fourth step which I mentioned at the outset: that of actually taking the decison. Economics has, as I see it, no special contribution to make to this step, beyond what it may do in the supporting steps. And I should like to turn to the last step: reviewing and evaluating past decisions. Here I believe that economics has a good deal to offer. Tracing the effects of past action—upon market values, income, employment, earning capacity, and the like—is a central and familiar function of economics. Furthermore, economics has concern for the records that support decisions and help evaluate them: not simply the accuracy, but also the form, the usability and simplicity, of these records. But a more basic role of economics in this connection is to help suggest the criterion for judging a successful decision.

I gather that the western Europeans, who have a longer history of silvicultural decisions behind them than we do, take great interest in this matter of a success criterion. I gather that their main interest has been in calculating the net income from the forest. This calculation is no easy thing. In fact, I doubt if we could altogether agree on how to make it.

Perhaps a more widely useful criterion of a good decision, or of good management generally, is how near the result comes to the goal.

To be sure, I may have a circularity here. Under the heading of consistency, I let the goal be inferred from past decisions. And now I may be using the inferred goal in an effort to test those same decisions. There is no test in such a case. However, the test comes when fresh evidence—fresh values externally determined—are brought in to open up the circle.

If no fresh values are forthcoming, then the firm's goals take on the aspect of social goals: subjective, always evolving, never fully enunciated, always being adapted to current changes in climate, always moving out ahead of decisions, so that success is a comparative thing. To evaluate an evolving, closed system of goals, values, and decisions is to evaluate the direction in which the whole system is moving. It is to evaluate, for example, successive changes in forest use and the allowable cut of timber on a large landholding. The changes influence the forest and the community in ways that must be judged in the very last analysis on ethical grounds. But economics, through its familiar function of tracing effects, is able to contribute much to this judgment.

Before I finish, I want to ask myself what sort of impression I should leave with you, concerning the scope of forestry economics. I notice that I have not used the word "dollars" very much if at all in my talk, and on the other hand I have brought in a few ideas that may seem improper to the subject of economics. In my own view of the subject, I take guidance from the great British economist, Alfred Marshall. He wrote,

> The less . . . we trouble ourselves with scholastic inquiries as to whether a certain consideration comes within the scope of economics, the better. If the matter is important let us take account of it as far as we can. If it is one as to which there exist divergent opinions such as cannot be brought to the test of exact and well-ascertained knowledge, [let us remember] always that some sort of account of it must be taken by our ethical instincts and our common sense, when they as ultimate arbiters come to apply to practical issues the knowledge obtained and arranged by economics and other sciences.

Forestry is compounded of economics and other sciences and of arts. Our concern is for forestry—and for economics only so far as it may serve our profession. To my mind, the role of economics in forestry progress is to grow important and unobtrusive, to integrate itself and lose its identity in our ordinary and common work.

31

FOREST GROWTH GOALS IN A PRIVATE ENTERPRISE ECONOMY

G. Robinson Gregory

First published in *Journal of Forestry* 53(11):816–821. November 1955.

BIOGRAPHY

Mr. Gregory is George Willis Pack Professor of Natural Resource Economics at The University of Michigan. In the first phase of his professional career, he was a Forest Service researcher who took time out for graduate study in economics and then joined the faculty at Ann Arbor. His major research during these years, reflected in his Ph.D. dissertation and in the article reprinted here, concerned goals for the timber economy. In the second phase, he cultivated an interest in international and development economics as applied to forestry. Here he has acted as consultant with FAO in Rome and with a number of the developing nations. He is the author of a textbook on forestry economics and has accepted related assignments in the Society of American Foresters, the International Union of Forest Research Organizations, and the Public Land Law Review Commission.

EDITORS' SUMMARY

National growth goals (production goals) for timber are valuable guides to resource-management programs. It may be possible to set such goals by a time-jointness method, in which timber supply and demand are estimated for each of a succession of future periods, recognizing the interrelationships between successive supply—and perhaps also demand—functions.

QUESTIONS TO CONSIDER

(1) Are forest growth goals such as discussed by Mr. Gregory part of a decision process like the one analyzed by Mr. Duerr in Item 30 and implied by Mr. Worrell in Item 11? That is to

say, are there alternative sets, or chains, of ends and means (i.e., sets of goals) from among which a choice is made on the basis of the applicable values?

(2) What are the alternative goals in Mr. Gregory's scheme? What are the adjacent links in the chain?

(3) What is the relation to the end-means chain of Mr. Gregory's time-jointness approach, which also involves a chain?

(4) By what value system does Mr. Gregory weigh his alternatives? Is this system appropriate to the social question at hand? In what ways would you attempt to modify the system?

(5) Consider the reverse of the question raised in the paragraph at the foot of this page: How can one decide the growth goal unless one knows approximately what our action programs should be? Is this not as sensible a question as the other? If it is, how does one resolve the apparent difficulty?

* * *

A preceding article was devoted to a critical analysis of the framework within which the Forest Service has established growth goals, or production goals, for American forestry.[1] Two major shortcomings of that framework were disclosed: Costs of production (the supply factors) had been ignored, and the time interrelationships of stumpage production had not been systematically considered (the approach was static). As a result of the analysis, it was suggested that forest growth goals be redefined—that they represent planned patterns of stumpage output over time instead of output targets for some arbitrarily chosen date.

Growth goals of the proposed type would serve as logical guides to major forest policy decisions. For this reason, if for no other, it seems highly desirable to build a valid framework for their formulation. This is the first objective of the present article—to sketch the essential features of a suggested methodology for growth goal determination. A second objective is to examine by discussion the basic features of the method.

WHY GROWTH GOALS?

Prior to presentation of the method itself it seems desirable to explore briefly a question basic to production goals of any type. This question is simply, "Why develop goals? What are they good for?"

The past provides some help. Previous growth goals have been used as a foundation for proposed action programs. For example, the last analysis called for an annual production of 18 to 20 billion cubic feet, including 65 to 72 billion board feet of saw timber.[2] The action program proposed to facilitate attainment of this goal was embodied in three major categories of recommendations: (1) public aids to private forest land owners, (2) "public control of cutting and other forest practices on private lands sufficient to stop forest destruction and keep the land reasonably productive," and (3) expansion, and intensified management, of national, state, and community forests.[3]

Details of the action program are, of themselves, unimportant to a discussion of growth goals. What is important is the fact that the goal was the "springboard"—the "raison d'être"—for the entire action program. The question—How can growth goals be used?—might be countered by another question: How can one decide on action programs in forestry *unless* one knows approximately how much wood the nation's forests should be growing?

Yet this is at best a partial answer, for with a specific growth goal of x billion cubic feet one might still ask, Why a goal of x billion? Why not half—or twice—this much?

The complete answer cannot be found in growth goals per se. Production goals of any kind are useful because they establish a physical output that must be attained if other objectives are to be met. Production goals are secondary goals; they are derived from some more basic objective. For example, with profit maximization as an ultimate objective, the entrepreneur can derive an *output schedule*—a production goal—which will permit (though perhaps not assure) maximum profits to be achieved. Production goals are working objectives rather than ultimate aims. They are performance guides—intermediate goals that must be achieved in order that basic objectives might be attained.

Goals can be set without reference to either production technology or the political and economic realities of the world. Such goals, however, can seldom be rational guides to production. Moreover, they can be downright dangerous if made the basis for policy decisions: their attainment may require major and undesirable changes in our social or economic way of life. Obtaining general approval of such goals may be a prelude to gaining acceptance of previously disputed policy, for once the goal is accepted logic may force acceptance of the methods proposed for its achievement. It is, in fact, entirely possible for goals to be formed as a result of established policy instead of being used as guides to policy establishment.

This article is not concerned with goals springing *from* policy decisions. It is concerned solely with the derivation of production goals which may serve as logical guides to policy formulation. A prerequisite for such goals is a rational, logical method for their derivation.

THE TIME-JOINTNESS APPROACH

The approach to be developed is based on the economic theory of joint production and has been called the "time-jointness" methodology. Its essential features are relatively simple; actual derivation of specific quantitative goals may be more difficult.

The principal advantage of the time-jointness method stems from the fact that its use permits specific recognition of time interdependencies. A definite *period* of analysis is selected and divided into a number of *intervals*. A *period* of 50 years, for example, might be divided into five ten-year *intervals*. The output of each interval would then be viewed as a separate product, and the central problem becomes one of discovering the particular combination of products that would permit attainment of the ultimate objective, whatever this latter might be.

The steps through which this product combination can be determined may be summarized as follows:

1. A set of interrelated supply functions (equations or curves) is determined. If five intervals had been decided upon, there would be a set of five supply functions—one for each interval. In forestry, these would necessarily be related because of the time interdependencies involved in stumpage production.
2. A set of functions (equations or curves) describing the demand for the product is determined. Again there would be a demand function for each interval, or (using

the same example of 5 intervals) five functions in all. These may also be related through time.

3. The final step is simply the simultaneous solution of the two sets of functions. This gives the set of equilibrium outputs—a pattern of output over the entire period which is determined, essentially, by the intersection of the supply and demand curves.

The specific result achieved by application of this series of steps will, of course, depend upon the working assumptions with which one begins, and the objective toward which the growth goal is oriented. The presentation which follows is predicated upon three basic assumptions which should be held in mind throughout the development; discussion of their applicability follows the sketch of the model. The key assumption is that forest production will be carried on in this country with the primary objective of making a profit. The second assumption is that the stumpage-producing portion of the forest economy can be characterized as operating under essentially purely competitive conditions. The third assumption is that the area of land to be devoted to the production of forest products will remain approximately constant.

With these three assumptions as a starting point, the problem of growth goal formulation can be restated as one of determining the particular pattern of stumpage output through time that will maximize the present net worth of the nation's forest land. We thus rule out, for the time being at least, problems raised by the many nontimber values associated with forest land.

The growth goal problem, and the basic ideas of the time-jointness approach, are developed in Figures 31–1 and 31–2. Figure 31–1 presents the usual static approach to supply and demand integration for any particular instant of time.

Supply is represented by a curve showing the quantities of the product that producers will place on the market at various prices. Increased quantities can be furnished only at increased prices: increased stumpage production requires an increase in the "inputs" of management—man hours of thinning, pruning, fire protection, weeding, insect control, under-planting, etc.

Demand is similarly represented, though here the curve is based on consumer psychology and plain common sense. The demand curve shows that, other things

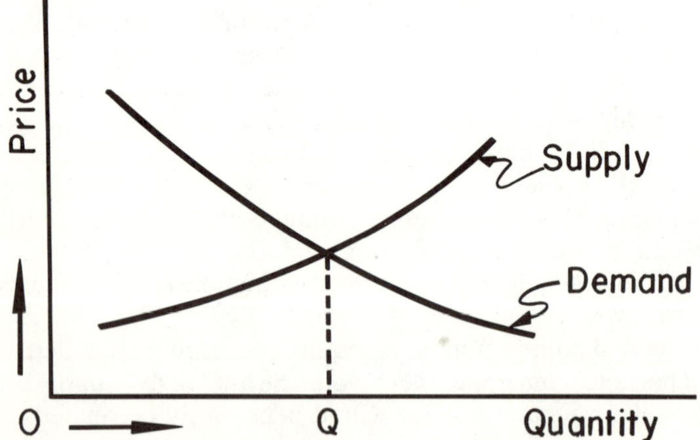

Figure 31–1. The usual static approach to supply and demand integration. Time interdependencies of supply cannot be demonstrated in this model.

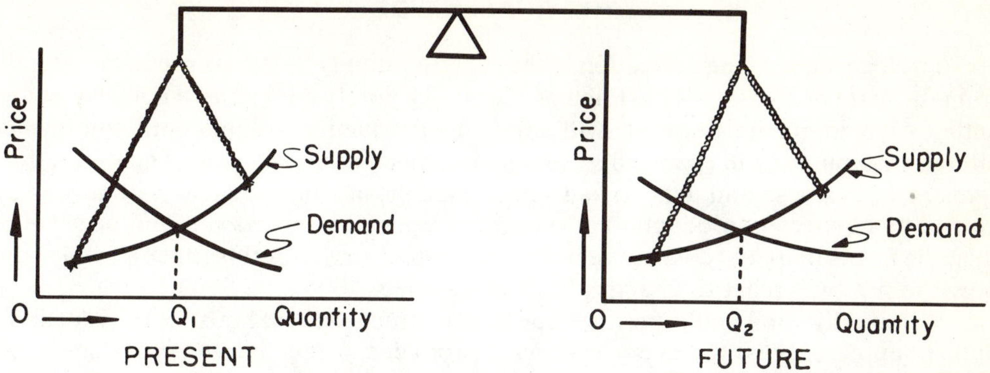

Figure 31–2. An overly simple graphic model for growth goal determination. The supply curve of the present is "tied" to the supply curve of the future.

being equal, people will buy more of the product at lower prices than they will at higher prices. This relationship holds true whether the goods happen to be one that is consumed directly, such as bread, or one that is used to produce other goods, such as steel or wood.

The particular quantity, Q, which will satisfy both of these curves is determined in the market, for producers will produce no more than they can sell at a profit, while consumers will take only a certain quantity at any one price.

But the physical supply of stumpage can be changed only over fairly long time intervals. It takes time to realize the benefits of increased intensity of management. Furthermore, *future* stumpage production is dependent on *present* growing stock, and therefore on *present* harvest. The supply curves of different time periods are not independent. They are interrelated due to the basic *time interdependencies* of the production process. The simple, static model of Figure 31–1 is not applicable to forest growth goal determination.

Figure 31–2 represents an attempt to illustrate the kind of model we need. The quantities Q_1 and Q_2—the growth goal, or pattern of stumpage output over time—are again determined by the intersections of demand and supply curves.

But the two supply curves are not independent: a change in one will have an effect upon the position of the second. The connection illustrated is direct and simple; a decrease in the quantity offered now means more to be offered in the future. And in general this would be correct: Less harvest in the present means an increase in growing stock which, in turn, makes possible greater production in future intervals. But any forester could cite conditions under which this might not be true. There should be a series of springs and pulleys in the connecting mechanism so that while a shift in one would usually be reflected in the other, offsetting reactions would also be possible. Furthermore, some mechanism making it possible to change the *shape* of the curves should also be provided.

It appears that a simple, diagrammatic model cannot serve for logical growth goal formulation. An alternative is to derive mathematical functions to replace the geometric curves, thereby permitting an approximate portrayal of the relationships among several variables. The growth goal model requires a series of interrelated supply functions, and perhaps a series of interrelated demand functions as well, for demand may also be connected over time. Solved simultaneously, these two sets of functions will give us the required growth goal.

THE SUPPLY SIDE

The logic underlying derivation of the required supply functions proceeds directly from the first two assumptions. Each producer in a purely competitive industry, when acting so as to maximize profit, will adjust his production volume until any further alteration in output will change his costs just as much as his revenue. Therefore, if he remains in business and acts so as to maximize profit, the producer will always be guided by changes in production cost—by his *marginal cost curve.* And because of this, the firm's marginal cost curve is also its supply curve, for this function will show how the firm will react to changes in product price.

Admittedly, one could hardly hope to construct marginal cost curves (or individual supply curves) for every stumpage producer in this nation. Fortunately for national growth goal determination we do not need each *individual* supply curve. Instead, we must derive a set of *industry supply curves.* These latter may be regarded as marginal cost curves for the nation's stumpage factory—the curves or functions that show how total production costs will change as we change the output of the various time intervals. To derive these functions requires that we go back to the physical basis of forestry and ask how, from the given area of forest land, the output for the different time intervals might be changed.

One way, of course, is to change the level of forest management—to change the inputs in one or all of the intervals being considered. The production of any product, whether it be corn, automobiles, washing machines—or stumpage—requires that certain materials or inputs be combined in particular ways. And this is true whether man plays a part in the production process or not. Man had little to do with the combination of variables that made possible the vast pine forests of the Lake States, though he played an important role in the establishment of the extensive aspen stands that now cover much of the same area. Yet the very existence of these stands is proof that certain biotic variables did react in a particular way.

The growth goal model, then, requires a knowledge of the relationship between inputs and outputs of the different time intervals. But it requires more than this, for (as we have previously seen) the planned output of any one interval may also be changed by changing the harvest of other intervals. The physical data required for growth goal formulation, therefore, include information on "output-output" relationships as well as on "input-output" relationships.

While this means that the task of obtaining and organizing the physical data may be somewhat complex, it scarcely implies that it is impossible. As scientific foresters, we are beginning to identify the important variables that influence forest production. Furthermore, we are beginning to learn how some of these can be altered in either quantity or timing so as to affect the final forest product, standing timber. This knowledge is the physical basis of all forestry; silviculture is concerned entirely with the identification of these variables and with their manipulation so as to achieve certain results.

These physical relationships are fundamentally important, but in an economy such as ours they are not the sole determinants of production methodology. While it is physically possible to grow bananas in Detroit, one nevertheless cannot buy Detroit-grown bananas in the corner grocery. In a profit-oriented economy physical data must always be combined with economic information to determine practical methods.

The economic data needed to complete the supply side of the picture are those concerning costs of the various physical inputs. And, since timing as well as quantity and price of inputs is important, some method is required to make the values of inputs at different intervals directly comparable. By using the familiar technique of discounting to a single point in time (the present will usually be chosen) this aspect of the time problem is readily managed. Then, knowing the physical relationships between input and output, and having the discounted costs of the inputs, we can construct a function showing the total cost of producing stumpage over the entire period *in terms of the outputs of the different intervals*.

From this total cost equation can be derived the marginal cost functions for the production of stumpage during each interval. These latter expressions make up the needed set of supply curves—a set of equations specifying the change in cost as a function of the ouputs of the different intervals.

To summarize the supply side of the growth goal model: both physical and economic information are necessary. On the physical side we must estimate (1) the inputs required to produce stumpage on the nation's forest land; (2) how variation in these inputs will cause associated changes in the output or the yield of stumpage; and (3) how changes in output or harvest in any particular interval will cause associated changes in the output or harvest of different intervals. On the economic side we must estimate the expected cost rates (or prices) of these inputs. These data are combined into a set of interrelated marginal cost functions of forest production. This is our set of interrelated industry supply functions.

THE DEMAND SIDE

During the past decade a great deal of new knowledge has become available in the general area of demand analysis. Economists working with other industries have accomplished much along lines of quantifying demand relations for their particular products. There is little to suggest that quantification would be any more difficult in forestry than in other fields.

The particular problem at hand is the determination of a set of demand functions for forest stumpage—a set of functions quantifying the relationship between the price of stumpage and the quantity purchased. Many factors besides price influence the quantity of stumpage purchased, and the identification of these variables would form a most interesting part of any demand analysis. The number and exact nature of the variables that would appear in these functions can scarcely be known until work has progressed considerably beyond the current stage. We know for certainty, however, that two key factors would be quantity and price for each interval being studied. Whether either quantity or price during one interval is a significant determinant of price and quantity in other intervals is again a matter for conjecture—conceivably they might be. The end result of the demand analysis would be a set of stumpage demand functions in terms of quantity and price. Other significant variables such as national income and population would be determined outside the forest economy.

THE INTEGRATION OF SUPPLY AND DEMAND

The final step in the time-jointness framework is purely mechanical. Given a set of supply equations in terms of quantity, and a set of demand equations in terms of quantity and price, the problem is simply that of finding the set of outputs which simultaneously satisfies the two equation systems. By themselves neither system is determinate but we know that in a purely competitive system profits will be maximized when marginal cost equals price. By substituting the series of marginal cost (or supply) equations for the price variable in the demand functions, thus making marginal cost equal to price, we will be left with a system of equations in which the only unknowns make up the set of quantities to be produced in each period. There will be as many equations as unknowns, and (unless there are some peculiar equation types) the system can be solved simultaneously for a series of specific quantities of stumpage output.

These quantities are very particular quantities, for if all our estimates of values, costs, etc., are correct we will plan to produce in each interval exactly that amount of stumpage that can be sold at a price just high enough to pay for the last one thousand board feet of logs producers are willing to grow. This is the forest growth goal for a free-enterprise economy.

As a summary of the technique the essential steps might be repeated: (1) integration of physical and economic data through a set of interrelated supply curves, (2) determination of a set of (perhaps interrelated) demand curves, and (3) determining the set of equilibrium outputs. Under our assumption of pure competition in the sale of the product, this is also the set of outputs which will maximize the present net worth of the nation's forest lands.

METHODOLOGY: OBJECTIVES

One of the major questions about the proposed method concerns that of objectives: Is profit maximization a suitable objective toward which to orient national growth goal analysis?

The forest economy of the United States has been in the past, and certainly will be in the future, subjected to many influences, not all of which are conducive to private enterprise. Yet it seems clear that if our forest growth goals—regardless of methodology—are to be set realistically, the fact that about 75 percent of our commercial forest land is in private ownership must be recognized. This means that a growth goal for American forestry must at least be compatible with the basic objectives of private ownership.

It is believed that the basic objective of private forestry in the United States has been, and will continue to be, the maximization of profit. This does not mean that every private owner is out to squeeze the last nickel from his forest acres—only that decisions made concerning private forest land are based, by and large, upon the profitability or nonprofitability of the action. Facing the alternatives of a profitable or an unprofitable action, the private entrepreneur will choose the first: given the choice between two actions, both of which are profitable, the private entrepreneur will choose the more profitable of the two (other things being equal). And, in so far as we assume continuity of existence, maximization of profits over time is a logical objective of the forest entrepreneur. This can be achieved by maximization of the present net worth of the property.

PRIVATE VERSUS PUBLIC OBJECTIVES

Is the *private* objective of profit maximization compatible with policy objectives of a *public* agency concerned with the welfare of the nation as a whole? To answer this, let's start by granting the original assumption—that stumpage production is purely competitive. Then each forest owner, when operating so as to maximize his own profit over time, will plan to produce so that the marginal cost of stumpage production will just equal its selling price. Since price will be at the level of the intersection of the demand and the supply curves, all those who want stumpage will be able to get it if they pay the cost of producing the last required unit: that is, consumers can get as much stumpage as they are willing to pay for, whether this happens to be fifty billion, seventy-two billion, or one hundred fifty billion board feet per year. But if stumpage production is an industry of increasing cost (and there is little reason to suspect otherwise) the more stumpage consumers get, the higher its price is going to be. This relationship between production, consumption and price is what gives consumers in a competitive economy a "reasonable choice of available materials." With each of their purchasing dollars they cast a vote for the production of the product purchased. The system then totals up these votes and allocates the various resources of the nation—land, labor, capital and management—in such a way that the "voted" amount of each product is produced.

Viewed nationally—which is the way a public agency should look at the picture—stumpage production actually does not depart greatly from purely competitive conditions. There are over four million owners of forest lands in this nation, only one of whom conceivably controls enough forest land at the present time to materially influence stumpage prices. That one is the national Forest Service itself. If this organization were to adopt a marginal cost pricing mechanism for federally-produced stumpage, this output would not only mesh with that of the nation's privately-owned forest land but would also serve as a barrier to possible future monopolistic developments.

Now let's take a look at the alternative—suppose stumpage were *not* produced under purely competitive conditions. Under this assumption the growth goal established by the suggested method would nevertheless be a logical one toward which public policy objectives might be oriented, though owner profit would no longer be maximized. For achievement of the goal so established would assure production of all the stumpage for which consumers could pay. At the same time production of quantities that might depress market prices below production costs would be avoided.

Such a goal might not maximize welfare for many reasons. Perhaps chief among these is the readily observed fact that not all industries approach purely competitive conditions. Furthermore, the model contains nothing whatever about income distribution—a major consideration in questions of welfare. But maximization of welfare is an objective only slightly less ephemeral than "the greatest good for the greatest number in the long run." A growth goal established under a marginal-cost pricing mechanism will satisfy the needs of consumers as these needs are expressed in the market place, and "consumers" includes almost the entire population.

PROBLEMS OF ESTIMATION

For the sake of simplicity the model was presented as yielding a single set of outputs over time that would be the final answer to the growth goal problem. Actually nothing quite this neat should be anticipated. One should not expect to estimate future costs, prices, technologies, or populations with such accuracy. It should be possible, however, to establish a reasonable range for these variables, and within this range to arrive at production goals that make sense to public and private foresters alike.

Whenever long-range forecasting is involved in a problem, someone will demur at the possibility of achieving any worth-while solution. But forestry by its nature is a longterm proposition and we are only kidding ourselves if we think we can avoid making forecasts. Every time a plantation is established by a private corporation there is an implicit forecast that future market prices will be at least sufficiently high to cover all planting costs and carrying charges. A major advantage of the proposed system lies in the fact that it forces the planning agent (or agency) to be explicit in his estimates.

PERIODIC PLANNING

It might be asked whether a system suited to planning by definite periods is applicable to forestry. As a matter of fact the periodic treatment accorded by the model is a definite advantage, for forestry is by nature a periodic process. Moreover, there is nothing that requires one to be satisfied with a "single stage" approach. Suppose, as an extreme, we plan a 100-year period by 20-year intervals. If more detailed plans are desired, we can use the result of this long-term plan for establishing "boundary conditions," then replan the first twenty years—perhaps by four year intervals. This process could be repeated as often as desired; each time more detail could be incorporated, yet each time the detailed plan would be consistent with the long-term plan.

And this is about the way we do make plans. The greater the time span involved, the more opaque the "veil of the future" becomes. As a consequence we make daily operating plans in far more detail than those for next year, while those for next year are much more certain than our plans for ten years hence. The time-jointness system permits refinement of planning as the time span is reduced, yet insures consistency between short and long range plans. And of course, plans are usually made with the expectation that revisions will be necessary as new information becomes available.

COMPUTATION

The objection might be made that the computations implied by the suggested method are much too involved. No detailed answer to such an objection will be attempted here—it should suffice to point out that modern econometric methods have progressed to the point where ten and fifteen equation systems are rather ordinary. Fifty equation systems have been solved, and the author knows of one project requiring the simultaneous solution of 100 equations. Compared with these latter models, the suggested method requires relatively little in the way of computation.

GOALS OR PREDICTIONS?

What is perhaps the most penetrating question to be raised about the approach runs something as follows: If stumpage production is purely competitive, why worry about growth goals? Won't the competitive system automatically achieve these goals without anyone bothering to "predict" them?

Even though the stated assumptions are granted there seems little reason to suspect the goals would automatically be achieved—that they would, in fact, represent a prediction of what will happen rather than a goal. In making his own plans each entrepreneur necessarily makes his own estimates of costs, prices, etc. For the model to yield a true *prediction,* either each producer would have to have identical expectations, or expectations of the different producers would have to compensate each other so as to yield fortuitously a result identical to that of the model.

Furthermore, the reader should recall that a discounting factor was used to make the different costs and revenues of different time intervals comparable. Nothing, however, was said about the magnitude of this discount factor. Higher interest rates make it profitable to cut present stands at the expense of future harvests, a fact fully reflected by the time-jointness approach. A different cutting pattern will result from every change in the discounting factor unless compensating variables are introduced. It would scarcely make sense to assume that all entrepreneurs would use identical interest rates or, for that matter, that public and private interest rates would be identical. For this reason, then, the goal determined by an agency having the welfare of future generations at heart (and therefore using a low discount rate) would diverge radically from goals determined by private operators who placed a high premium on present profit. Such divergence between goals points the way to needed policy decisions—if private operators are to meet public production goals, credit aids, insurance, etc., may all be required.

CONCLUSIONS

To be realistic any output pattern must be consistent with the production techniques and input schedules one assumes will be used. It must also be consistent with the political and social framework envisaged. If the pattern cannot be achieved within this framework we may wish to change either the output pattern or the political and social institutions themselves; if it is not compatible with technological relationships, we must either change the planned outputs or alter our production techniques; if it proves impossible to achieve a schedule of inputs that will yield a selected output, one or the other must be varied. Lacking this consistency, our production plan is not rational: our output plan becomes only an opinion, a wish, or a dream.

The time-jointness method has been presented as a means of placing national growth goal formulation in a rational production context. The system permits—in fact requires—the integration of both the physical and economic factors of forest production in establishing these goals. While presented as an alternative to current methods, in reality the approach is more than this, for it can be a powerful tool for analyzing the effect of institutional factors, and for determining the probable result of national policy decisions on output patterns over time. It offers a way of determining the realism of any output pattern however determined.

REFERENCES

1. Gregory, G. R. 1955. An analysis of forest production goal methodology. Jour. of Forestry 53:247–252.
2. U.S. Forest Service. 1948. Forests and national prosperity. U.S. Dept. of Agriculture Miscellaneous Publication No. 688. Page 3.
3. Same reference, page 6.

32

TRADITIONAL FOREST REGULATION MODEL: AN ECONOMIC CRITIQUE

Emmett F. Thompson

First published in *Journal of Forestry* 64(11):750–752. November 1966.

BIOGRAPHY

Mr. Thompson is Head, Department of Forestry, Mississippi State University. In the decade to 1973, he served on the forestry faculty at Virginia Polytechnic Institute and State University, where he became nationally recognized for his work on quantitative methods in forest management. A native of Oklahoma, he studied forestry at the State University of Oklahoma and later took graduate studies in forest management and forestry economics at the State Universities of North Carolina and Oregon. He began his professional career with the U.S. Forest Service, on a southern national forest.

EDITORS' SUMMARY

Today's version of the traditional regulation model is a forest with a set rotation and an even distribution of ages up to rotation age. The model fails to conform to the necessities of managerial decision making: (1) specify goals; (2) define alternatives; (3) make a choice. Research is needed on models that will lend themselves to Steps 2 and 3.

QUESTIONS TO CONSIDER

(1) How is Mr. Thompson's criticism of the traditional forest regulation model related to Mr. Marquis's criticism (Item 3) of traditional forestry in general? How do the two differ?

(2) Mr. Thompson explains that the goal of approximately equal yearly harvests of timber (page 256) may be irrational if the costs of reaching it are very high. Suppose that there are no costs: that the forest is already regulated so as to produce annual harvests of equal quantity, age, and other physical characteristics. List the problems that may arise under such a regulatory scheme. What alternative schemes may be preferable? Why?

(3) Mr. Thompson emphasizes the problems of transition to a regulated forest. Such problems may occur in transition to any forest that is time-consuming or costly to achieve. The southern pulp-paper industry, for example, found it difficult to build up timber growing stocks on the depleted lands that the industry acquired in the 1930s and thereafter. Do you know how this particular difficulty was overcome?

(4) In decision making (page 257), there is always the possibility that the preferred alternative for attaining specified objectives may not represent a wise program because superior *objectives* exist which have not been investigated. How does the decision maker solve this problem?

* * *

The objective of this paper is to evaluate the traditional forest regulation model with respect to those forest properties whose management policies are influenced primarily by economic and business considerations.

The traditional forest regulation model derives from the normal forest concept. As originally conceptualized, the model was a forest composed of even-aged, fully (normally) stocked stands, with one age class for each year of the rotation. In addition, each age class was represented by an area of equal productivity. The probable reason for development of the model was a need for long-range planning of forest production. Obviously, attainment and maintenance of the model guaranteed continuous timber production. A number of developments, notably economic and silvicultural research which has indicated that normal stocking is not necessarily a rational management goal, have modified the original concept of the model. A present-day characterization of the model would be a forest property which exhibits:

1. An even distribution of age classes, with each age class occupying an area of equal productivity; and
2. A set rotation which establishes the upper limit on age classes.

The function of a forest regulation model is to fulfill the regulatory objectives of management. Properties, such as those considered in this paper, generally have objectives which reflect a basic profit motivation along with certain managerial or institutional constraints on profit. Maintenance of a minimum level of merchantable wood, stability of a local economy, and job security for employees are examples of such constraints.

A number of foresters have previously criticized or pointed out shortcomings of the traditional model. Fedkiw[1] criticized the model as being inadequate by not specifying the period within which it should be achieved or the area to which it should be applied. Wicks[2] detailed a number of problems encountered by industrial forestry organizations attempting to attain the traditional model. Among the problems were capital limitation, ownership stability and location, mill requirements, forest protection, and economic and fiscal requirements. Gould[3,4] and Dowdle[5] discussed the inconsistencies between traditional approaches to forest management and actual conditions, particularly those associated with economic and business considerations.

Proponents of the traditional forest regulation model have generally presented it as the primary regulation goal or guide for all even-aged forest management. Three generations of forest management textbooks[6,7,8] have stated that attainment of the model produces, among other things, the following three conditions:

1. A yearly cut of approximately equal volume.
2. A yearly cut of about the same age, size, and quality of timber, and hence a yearly income of about the same amount.
3. A current growth and income obtained from a forest capital no larger than is necessary, thus ensuring a maximum rate of return upon this capital.

ECONOMIC EVALUATION

Certain economic implications are inherent in the traditional forest regulation model. The first condition, an approximately equal annual yield, is of paramount concern to the model. However, the concern appears irrational in many cases. The traditional model cannot be attained in less than one rotation; therefore, management planning is oriented toward the second and subsequent rotations. As a result, the costs of attaining the model may be considerably higher than some alternative approach. For example, if regulation of an old-growth forest is initiated according to the traditional model, possible revenue is lost by the delayed harvest of over-mature timber.

The second condition is closely aligned with the first but, in addition, implies that the age, size, and quality of timber which are optimal today will continue to be optimal ad infinitum. Also, stumpage prices will either never vary or will fluctuate around a constant mean, hence an average income of about the same amount. Forest managers who have observed the changes in standards of utilization as well as stumpage and forest products prices do not need to have the fallacies of this condition detailed.

The third condition, while it constitutes an implicit acceptance of timber as capital and forest management as a business activity, is also a clear contradiction of the economic concept of opportunity cost. A forest enterprise should not only strive for internal capital efficiency, as the third condition implies, but, assuming capital efficiency is the enterprise's goal, should be conscious of more attractive investment opportunities. To indicate that a maximum rate of return is assured by attaining the traditional forest regulation model makes rate of return *dependent* upon timber growth within the enterprise. Actually, the appropriate rate of return is *independent* of the particular level of the forest enterprise under consideration. The appropriate rate of return is the alternative rate of return—the opportunity cost. By definition, the opportunity cost is external to a specific level of an enterprise.

The above evaluation of the traditional forest regulation model, particularly that portion concerned with the first two conditions, leads to another general criticism of the model. The model itself can be considered an objective; that is, a physical representation of a forest management goal. However, the model contains no theoretical guides regarding its attainment. The model implies, and is usually associated with certain physical steps for attainment. These steps, however, are divorced from economic considerations and are generally inconsistent with management needs during the period required to attain the model. Changing standards of utilization, ownership patterns, etc., make practical attainment improbable, if not impossible, for the economically motivated forest manager. Such a forest manager needs a regulation model whose attainment is compatible with economic theory. The theoretical void between the specification of an objective and the means for attaining the objective is a primary failing of the traditional forest regulation model. The failing accounts, to a large degree, for the fact that the traditional regulated forest exists only in textbooks.

ALTERNATIVES

As indicated, the objectives a forest manager wants to attain through regulation are prerequisite to a choice of regulation model. The traditional model should be applied only in those cases where management objectives are compatible with the inherent implications of the model. In retrospect, the traditional model seems to be compatible only with the objective of producing an even distribution of age classes, irrespective of cost.

It is impractical to suggest that a single alternative to the traditional model might be appropriate for all forest owners. Unfortunately, alternative regulation models are not abundant in the forestry literature. However, some approaches to forest management, which at least have regulatory implications, are available.

Fedkiw and Yoho[9] have developed what can be considered a regulation model for the stumpage producer willing to assume certainty regarding future yields, costs, and prices, and whose sole management objective is maximization of present net worth. The model is based on the concept of financial maturity; that is, timber harvests are scheduled in accordance with maximizing present net worth. Therefore, annual yields and income may fluctuate quite widely. Chappelle and Nelson[10] have essentially quantified Fedkiw and Yoho's model for loblolly pine.

Curtis,[11] and others,[12,13] have demonstrated uses of linear programming which indicate that this technique may have utility in forest regulation. Gould and O'Regan[14] have demonstrated the same possible utility with respect to simulation.

While the above do not constitute generally recognized substitutes for the traditional forest regulation model, they are compatible with decision making principles. A sound approach to managerial decision making (forest regulation falls within this category) contains three distinct elements:

1. Specification of the objectives.
2. Defining alternative methods for attaining the objectives.
3. Choosing the most efficient alternative.

As indicated, the traditional model does not contain the second element. The possible approaches to forest regulation cited above include all three elements. For example, linear programming is essentially a method for choosing among alternatives to fulfill a stated objective.

Of the elements essential to decision making, the first is most important. Unless the objectives of forest regulation are specified, there will be no direction or continuity to the regulation. Also, the specification of objectives must come from the person or persons responsible for managing the property; outsiders cannot specify objectives. While the first element is important, elements two and three would probably present more difficulty at present; that is, to discretely express alternative regulation procedures and then apply a criterion for choosing among the alternatives. Consequently, research efforts focused on elements two and three should be initiated.

DISCUSSION

The primary purpose of this paper has been to evaluate the traditional forest regulation model with respect to economically motivated forest managers. The importance of a regulation model being compatible with the objectives of management has been emphasized. It is obvious that the traditional model cannot fulfill the objectives of all forest managers.

To merely criticize the traditional model without considering alternatives does little more than create a void. Therefore, some pertinent contributions in the forestry literature were cited. Unfortunately, these contributions are not numerous and do not address themselves directly to regulatory problems. Consequently, the primary conclusion of the paper must be that, while the traditional forest regulation model is an inappropriate guide for many forest managers, completely adequate substitutes are not presently available. Following this conclusion, the primary recommendation is that research efforts be focused on the area of forest regulation. Specifically, research should be directed toward identifying or developing alternatives to the traditional model and establishing criteria for evaluating the alternatives.

Criticism of a model without offering definite alternatives may leave some readers unsatisfied. However, it appears to the author, criticism and development of alternatives are distinct steps in the evolution of more appropriate models. Unless there is criticism of traditional models in forest management, or any other discipline, there will be little impetus for change. If this paper has provoked a few forest managers or forest management researchers to objectively consider the qualifications and implications of the traditional forest regulation model, it will have served its purpose.

REFERENCES

1. Fedkiw, John. 1961. Practical applications of capital budgeting concepts to industrial forest management. *In* Seventeenth Yale Industrial Forestry Seminar. Univ. of Washington, College of Forestry, Seattle, Washington. Pages 1–29.
2. Wicks, W. W. 1962. Regulatory problems of the modest forest ownership. *In* Proc., fifth conference on southern industrial forest management. Duke Univ., School of Forestry, Durham, N.C. Pages 59–65.
3. Gould, Ernest M., Jr. 1960. Fifty years of management at the Harvard Forest. Harvard Forest Bul. No. 29. 29 pages.
4. ————. 1962. Forestry and recreation. *In* Report No. 11, Economics in outdoor recreational policy. Committee on the Economics of Water Resources Development of the Western Agricultural Economics Research Council. Pages 41–50.
5. Dowdle, Barney. 1964. The role of economics in forest management decisions. *In* Proc., 1963, Society of American Foresters. Pages 155–157.
6. Davis, Kenneth P. 1954. American forest management. New York: McGraw-Hill Book Co., Inc. 482 pages.
7. Mathews, Donald M. 1935. Management of American forests. New York: McGraw-Hill Book Co., Inc. 495 pages.
8. Roth Filbert. 1925. Forest regulation. 2nd ed. Ann Arbor, Michigan: George Wahr. 239 pages.
9. Fedkiw, John and James G. Yoho. 1960. Economic models for thinning and reproducing even-aged stands. Jour. of Forestry 58:26–34.
10. Chappelle, Daniel E., and Thomas C. Nelson. 1964. Estimation of optimal stocking levels and rotation ages of loblolly pine. Forest Science 10:471–502.
11. Curtis, Floyd H. 1962. Linear programming the management of a forest property. Jour. of Forestry 60:611–616.
12. Leak, William B. 1964. Estimating maximum allowable timber yields by linear programming. U.S. Forest Service. Northeastern Forest Exp. Sta. Res. Paper NE–17. 9 pages.

13. Loucks, Daniel P. 1964. The development of an optimal program for sustained-yield management. Jour. of Forestry 62:485–490.
14. Gould, Ernest M., Jr., and William G. O'Regan. 1965. Simulation, a step toward better forest planning. Harvard Forest Paper No. 13. 86 pages.

33

PROFIT CRITERIA AND TIMBER MANAGEMENT

Henry H. Webster

First published in *Journal of Forestry* 63(4):260–277. April 1965.

BIOGRAPHY

Mr. Webster is, and has been since 1967, head of the Department of Forestry at Iowa State University. He is known for his active interest in improving forestry education, having served in a variety of roles, including as a member of the professional Accreditation Committee and as co-chairman of a national faculty-workshop series concerned with instruction in integrated resource management. He is known also for his research in the financial aspects of forestry (the present article is illustrative) and in forest-products marketing. He is a past Chairman of the Department of Forestry, University of Wisconsin, and for 10 years was Forest Economist in the Northeastern Forest Experiment Station, Philadelphia.

EDITORS' SUMMARY

Five criteria—internal rate, present net worth at two different rates of interest, and value response per cost dollar at two different rates—gave much the same rankings to 15 investment opportunities in Pennsylvania. Payout period gave different rankings. For mutually exclusive opportunities, the choice of criterion is critical and must be based on the circumstances.

QUESTIONS TO CONSIDER

(1) A firm invested \$1,000 two years ago, took in a net return of \$2,300 one year ago, and is in the red \$1,320 today. To calculate the internal rate of return (p. 261), we must solve

$$-1.0(1 + i)^2 + 2.3(1 + i) - 1.32 = 0$$

i.e.,

$$x^2 - 2.3x + 1.32 = 0.$$

Find the roots of this quadratic equation. Is the rate 10 percent or 20 percent? Explain the ambiguity. What do you think is the likelihood, in general, that the internal rate will be ambiguous?

(2) Evidently the internal rate, present worth, and benefit-cost ratio are measures, often highly correlated, of different aspects of an investment. Instead of recommending one of the three as generally the "best," may it not be possible to identify the purposes or circumstances to which each is "best" suited? Try your hand at such an analysis. One approach is to identify precisely what is measured in each case. Another is to devise a very simple numerical example that will bring out contrasts, then try to spot the principles involved.

(3) Can an analytical approach in the spirit of Question 2 be followed to arrive at a suitable interest rate for use in net-worth and benefit-cost calculations?

(4) Evidently payout period is a criterion fundamentally different from the others. What is this difference?

(5) Explain in your own words the problem of weighing mutually exclusive investment opportunities.

* * *

Relative profitability of timber management opportunities can be determined only after a measure of profitability, a profit criterion, has been selected. Several are available. The *internal rate of return* measures the rate at which an initial investment "grows into" a final value, with due allowance for intermediate costs and incomes. It is calculated by determining the rate of interest which makes the sum of costs (including both the initial investment and the intermediate costs) equal to the sum of values (including both the final value and the intermediate incomes) when all items are discounted back to the present. *Contribution to present net worth* measures the increase in the present net value of an area of forest land (or other asset) that is caused by investment. It is calculated by discounting costs and values associated with the investment to the present at a specified interest rate and then taking the difference, $V-C$. *Value response per cost dollar* relates benefits and costs as a ratio. Costs and values are discounted to the present at a specified interest rate and one is then divided by the other, V/C. For all of these measures, the larger the value, the more favorable a particular timber management opportunity. *Payout period,* also suggested as a profit criterion, measures the number of years required to recover an investment. The smaller the value, the more favorable the opportunity.

CONTROVERSY

The relative merits of various profit criteria for rating investment opportunities have been discussed by both general economists and forest economists. The list of general economists who have advocated one or another of these profit criteria reads like an all-star roster. According to Lutz,[1] Böhm-Bawerk,[2] Wicksell,[3] Knight,[4] Boulding,[5] and Hayek,[6] among others, have favored the internal rate of return. So has J. M. Keynes.[7] A partial list of those favoring contribution to present net worth would include Irving Fisher,[8] Hicks,[9] and Samuelson.[10] R. G. D. Allen[11] has proposed a criterion similar to value response per cost dollar (although not called by that name.)

Discussion among forest economists and other natural resource economists has focused on measures of the maturity of timber, and has paralleled the discussion by general economists to some degree. One group, which has included Duerr, Fedkiw, Guttenberg,[12] and Yoho,[13] has favored "financial maturity." It is a form of contribution to present net worth. Prospective increases in the value of standing timber are expressed as an interest rate based on present value and are then compared with a minimum acceptable rate called the alternative rate of return. Timber is judged to be mature when the two rates are equal. At this point the difference between total revenue and total cost (i.e., the contribution to present net worth) will be at a maximum.

Another group, notably Worrell,[14] has proposed the internal rate of return. Worrell noted that financial maturity considered only the cost of leaving funds tied up in the form of standing timber, neglecting the opportunity cost associated with the land occupied by that timber.

Value response per cost dollar has been used (in somewhat disguised form) in at least one recent study to rank the relative profitability of management opportunities in various kinds of immature stands.[15] Nelson has suggested payout period as a criterion.[16] And Gaffney[17] examined the whole matter of maturity of timber, concluding that all of the preceding methods err and that the Faustman formula, also known as the soil rent approach, is theoretically correct. The Faustman formula is a form of contribution to present net worth reflecting opportunity costs associated with both timber and land. Financial maturity would be equivalent to it if modified to include opportunity cost for land as well as timber.

These various profit criteria measure somewhat different things. Ratios or rates such as value response per cost dollar and the internal rate of return measure the rate of capital growth. Contribution to present net worth measures the absolute size of an investment opportunity. Payout period is a partial measure of the flow of funds during the investment period.

Most discussions of profit criteria have been primarily theoretical. The differences between various criteria have been examined in terms of the algebra used to link cost, revenue, and time. The differences discovered by such an approach are theoretical.

How important are these differences in a practical sense? How much would they change a program of timber management? Under what circumstances would various criteria give different practical results? This article reports a case study directed toward these questions.

PRACTICAL TESTS OF DIFFERENCES: A CASE STUDY

A series of timber management opportunities was ranked by each of seven different criteria. These rankings were then compared.

The management opportunities were those evaluated by the author in an analysis of timber management opportunities in Pennsylvania.[15] These opportunities included three sets of forestry practices each applied under a variety of stand conditions. The three sets of practices included (1) planting softwoods on open and lightly stocked forest land; (2) stand improvement (cleaning and cull-tree removal) in hardwood seedling and sapling stands; and (3) stand improvement (thinning) in hardwood poletimber stands. There were 23 timber management opportunities in all. The cost, response, and value data used in ranking these opportunities were also taken from the Pennsylvania analysis.

All of the timber management opportunities involved an initial investment, intermediate costs and income, and a final harvest. Initial investments ranged from approximately $10 per acre for stand-improvement practices to nearly $60 per acre for planting under difficult circumstances. Intermediate costs at first exceeded intermediate incomes with these incomes gradually overtaking the costs. Major output effects included increased volume with planting, improved species composition and quality, and earlier harvest with stand-improvement practices. Investment periods ranged from 40 to 50 years for stand improvement in hardwood poletimber stands and up to 80 years for softwood planting.

The criteria that were used can be grouped in four categories: internal rate of return, contribution to present net worth, value response per cost dollar, and payout period. All but the first of these categories can be meaningfully subdivided. Both contribution to present net worth and value response per cost dollar require that an interest rate be selected for discounting future values and costs. Many different interest rates could be chosen. Three percent and six percent compound interest were arbitrarily selected for purposes of this case study. Payout period can be used in several ways. It can be used alone as the sole criterion for ranking timber management opportunities, or it can be used in combination with other criteria. Both payout period as a sole criterion, and with site productivity as a secondary criterion, were used in this case study. In summary, the following criteria were used:

Maximize
—internal rates of return.
—contribution to present net worth, future values discounted at 3 percent compound interest.
—contribution to present net worth, future values discounted at 6 percent compound interest.
—value response per cost dollar, future values discounted at 3 percent compound interest.
—value response per cost dollar, future values discounted at 6 percent compound interest.

Miminize
—payout period as the sole criterion.
—payout period as the primary criterion and site productivity as a secondary criterion (opportunities first ranked by payout period, those equal on this score then further ranked in terms of site productivity class).

The stage was set for comparisons once the 23 timber management opportunities had been ranked by each of these criteria. Comparisons could be made in many different ways. Three methods were used.

First, the rankings by various criteria were compared visually. The question was how do these rankings resemble one another and how do they differ? Visual comparison is obviously subjective and crude. Nevertheless, it is a useful first step revealing major similarities and differences.

Second, a correlation analysis was made. For example, the internal rates of return for the 23 timber management opportunities were correlated with the payout periods for these opportunities. Such calculations were made for each 15 pairs of profit criteria. The correlation analysis measured the quantitative association between different criteria. Thus, it went beyond visual comparisons, which considered only the order in which timber management opportunities are ranked.

Third, a net-benefit analysis was made. This was done in three steps. (a) Calculations using area and cost data were made to determine how far down the rankings (or priority scales) given by each criterion each of several alternative budgets would reach. (b) The net benefit to be expected from each of these budgets using each of the criteria was then calculated using response and value data. Contribution to present net worth, with future values discounted at 6 percent compound interest, was used as an arbitrary standard for this purpose. (c) Finally, the net benefit expected from each budget using each criterion was expressed as a percentage of the net benefit expected when this arbitrary standard was used as the criterion. The net-benefit analysis went beyond the correlation analysis in measuring the practical (indeed the dollar and cents) importance of differences between profit criteria.

RESULTS QUITE SIMILAR

The results of these three comparisons are reported in Tables 33–1, 33–2, and 33–3. All point to much the same conclusion. All criteria (with the exception of payout period) gave much the same results in terms of ranking the timber management opportunities examined in this case study. This was particularly true in terms of opportunities that ranked somewhere near the top of the list. Nevertheless, it will be helpful to examine the three comparisons separately.

Visual comparisons show that there are differences in the order in which timber management opportunities appear on a priority scale. For example, thinning in cove hardwood poletimber on productive sites is the most profitable opportunity if several of the profit criteria are used (namely, the internal rate of return, and both contribution to present net worth and value response per cost dollar with future values discounted at 6 percent compound interest). On the other hand, thinning northern hardwoods on productive sites is the most profitable opportunity if other criteria are used (both contribution to present net worth and value response per cost dollar with future values discounted at 3 percent). Similarly, the order differs by an increasing amount further down the priority scale.

However, the differences at the upper end of the priority scale are relatively small. The first 10 timber management opportunities can be separated into three groups (Table 33–1). Differences occur within groups, not between them. The upper end of the scale is, of course, the most relevant end. For example, these ten opportunities alone would absorb a statewide timber management budget of nearly $35 million in Pennsylvania. Visual comparisons are subjective, and the grouping of opportunities was done to minimize differences in ordering between groups. Nevertheless, the order in which high-ranked opportunities appear is quite similar from one profit criterion to another.

The correlation analysis makes the same point in a more objective way (Table 33–2). All criteria are closely and quite consistently related, with the exception of payout period. At a minimum, three-fourths of the variation among timber management opportunities ranked by any one criterion is associated with variation among them ranked by any other criterion. In all but a few cases the relationship is closer, very commonly in the range of 83 to 88 percent, and in one case nearly 95 percent.

Table 33–1. Visual Comparisons of First Fifteen Timber Management Opportunities in Pennsylvania Using Various Profit Criteria

MANAGEMENT OPPORTUNITY	RANK ORDER				
	Contribution to Present Net Worth		Rate of Return	Value Response per Cost Dollar	
	6 Percent	*3 Percent*		*6 Percent*	*3 Percent*
Thin cove hardwood, site 1	1	2	1	1	2
Thin number hardwood, site 1	2	1	2	2	1
Thin number hardwood, site 2	3	3	3	3	3
Thin cove hardwood, site 2	4	7	4	4	6
Clean number hardwood, site 1	5	4	6	6	5
Thin oak, site 1	6	5	5	5	4
Clean number hardwood, site 2	7	4	7	7	7
Thin oak, site 2	8	9	8	8	9
Clean oak, site 1	9	10	9	9	10
Plant bare land, site 1	10	8	10	10	8
Clean oak, site 2	11	16	11	11	14
Thin oak, site 3	12	17	20	17	17
Plant bare land, site 2	13	11	12	12	11
Clean oak, site 3	14	19	22	18	19
Plant bare land, site 3	15	21	17	19	18

Source of data: Calculated from Webster.[15,18]

However, payout period is erratic in its relationship to other profit criteria. The association between payout period and other criteria ranges from more than 80 percent to very little more than 40 percent.

The net benefit analysis makes basically the same comparisons in practical, dollars-and-cents terms. And again, the results are much the same regardless of the specific criterion used to rank timber management opportunities (Table 33–3). Several of the criteria gave identical results in terms of expenditure patterns and benefits. The internal rate of return, and both contribution to present net worth and value response per cost dollar with future values discounted at 6 percent, constitute one set giving

Table 33–2. Coefficients of Determination Relating Various Profit Criteria

PROFIT CRITERION	Rate of Return	Contribution to Present Net Worth		Value Response per Cost Dollar		Payout Period
		3 Percent	*6 Percent*	*3 Percent*	*6 Percent*	
	X_1	X_2	X_3	X_4	X_5	X_6
Rate of return X_1	—	—	—	—	—	—
Contribution to present net worth:						
3 percent X_2	0.881	—	—	—	—	—
6 percent X_3	0.825	0.838	—	—	—	—
Value response per cost dollar:						
3 percent X_4	0.877	0.854	0.792	—	—	—
6 percent X_5	0.831	0.752	0.785	0.931	—	—
Payout period X_6	0.816	0.467	0.738	0.602	0.692	—

Source of data: Calculated from Webster.[15,18]

Table 33–3. Comparisons of Net Benefits at Several Expenditure Levels, Using Various Profit Criteria[*]

	INCREASE IN PRESENT NET WORTH OF PENNSYLVANIA FORESTS, USING AS CRITERION—						
EXPENDITURE LEVEL	Contribution to Present Net Worth, 6 Percent	Value-Response per Cost Dollar, 6 Percent	Rate of Return	Contribution to Present Net Worth, 3 Percent	Value Response per Cost Dollar, 3 Percent	Payout Period Only	Payout with Site as Secondary Criterion
$100 Thousand	547 (100)	547 (100)	547 (100)	488 (89)	488 (89)	351 (64)	547 (100)
$250 Thousand	1,282 (100)	1,282 (100)	1,282 (100)	1,219 (95)	1,219 (95)	878 (68)	960 (74)
$500 Thousand	2,524 (100)	2,524 (100)	2,524 (100)	2,438 (97)	2,438 (97)	1,777 (70)	2,018 (80)
$1 Million	4,962 (100)	4,962 (100)	4,962 (100)	4,876 (98)	4,876 (98)	3,577 (72)	4,456 (90)
$5 Million	20,568 (100)	20,568 (100)	20,568 (100)	20,568 (100)	20,568 (100)	18,044 (88)	20,342 (99)
$10 Million	37,592 (100)	37,592 (100)	37,592 (100)	37,592 (100)	37,592 (100)	36,111 (96)	37,366 (99)
$15 Million	54,584 (100)	54,584 (100)	54,584 (100)	54,584 (100)	54,584 (100)	54,177 (99)	54,290 (99.6)

Source of data: Calculated from Webster.[15,18]

[*First figure in each pair is thousand dollars of net benefit (increase in present net worth at 6 percent). Second (parenthetical) figure is the percent of the corresponding first figure in Column 2.]

identical results at each of seven different budget levels. Contribution to present net worth and value response per cost dollar also give identical results when future values are discounted at 3 percent.

Payout period again gave results which diverge somewhat. This was particularly true when payout period was used as the sole criterion for ranking timber management opportunities. It was much less true when site productivity was used as a secondary criterion.

Expenditure patterns differed more from one criterion to another with relatively small budgets than with large ones. Which opportunity is first on a priority scale is obviously most important if only one can be selected. It is less important if several can be selected. All of the criteria except payout period gave identical results once a statewide timber management budget of $5 million was reached. Indeed there was only a 5-percent difference in net benefit with a budget of $250 thousand, and a 10-percent difference with a budget of $100 thousand. Such differences are relatively small. They may be more apparent than real since the information used to rank timber management opportunities is often quite imperfect.

In summary, all criteria except payout period gave much the same results. Both the close and consistent relationships among 5 of the criteria and the inconsistency of the payout period criterion stem from the same basic cause. The internal rate of return, contribution to present net worth and value response per cost dollar all take account of the same factors. These factors are (a) the costs of various forestry practices; (b) their effects on quantity, quality, and timing of output; and (c) estimated future value of various kinds of timber. Payout period, on the other hand, considers only one factor—timing of output, and then only to the point where the investment is recovered. Output effects are partially considered when site productivity is used as a secondary criterion with payout period. The results are more like those obtained using the other criteria. This illustrates the effects of using more of the same factors.

TWO TYPES OF DECISIONS

This case study makes one major point. It shows that, under some circumstances, many different profit criteria give results that are very similar. Indeed they give results that are virtually identical. To put it another way, under some circumstances there is little use arguing on theoretical grounds about which profit criterion should be used. The preferences of the likely users of a particular analysis or convenience to the analyst may be the most important considerations. If the users understand one criterion better than another or the analyst can use one more easily than another, that is the one to use.

Note that these statements raise an important question. "Under some circumstances" various profit criteria give similar results. But under what circumstances do they do this? And under what circumstances do they give contradictory results?

We can start by reconsidering the case study reported here. What were the important circumstances associated with it? What types of decisions, what types of alternative courses of action, were considered?

An analysis of timber management opportunities in Pennsylvania was used as a source of data for this case study. One fact about that analysis is particularly noteworthy. It dealt entirely with the extensive (as opposed to the intensive) margin. Each of the 23 timber management opportunities was defined in terms of a particular set of

Table 33–4. An Example of Mutually Exclusive Alternatives

SITUATION AND MANAGEMENT ALTERNATIVES:

Low grade oak stand:	
Present age	30 years
Yield without stand improvement	3 M bd. ft. per acre, maturity 75 years
Yield with stand improvement	4 M bd. ft. per acre, maturity 65 years
Cost of stand improvement	\$5 per acre
Stumpage values	Unmanaged \$10 per M; managed \$10.50 per M
Type conversion to pine:	
Pulpwood yield	25 cords per acre, maturity at 30 years
Cost	\$40 per acre
Stumpage value	\$5 per cord

RELATIVE PROFITABILITY:

Profit Measure	Oak Management	Type Conversion
Rate of return	4.4 percent	3.3 percent
Contribution to present net worth (future values discounted at 3 percent)	\$2.00 per acre	\$3.57 per acre

Note: Hypothetical figures.

forestry practices applied under particular stand conditions. Thus the analysis dealt with decisions concerning the relative merits of applying specified forestry practices in various kinds of stands. These alternatives compete with each other only in terms of an overall budget. At a given time the timber management budget may be adequate to cover only the first few opportunities in the rankings (or priority scales) shown in Table 33–1. However, the other opportunities will still be there for future consideration. In short, the opportunities that were considered were not mutually exclusive.

But many timber management opportunities are mutually exclusive. To undertake one opportunity may be to forgo all chance of undertaking another. Stand improvement in a young hardwood stand vs. type conversion of this stand by bulldozing and planting would be one example. Such opportunities are mutually exclusive at any given point in time.

Different profit criteria may give quite different results with mutually exclusive opportunities. Table 33–4 illustrates this using hypothetical, but fairly reasonable, figures. The choice is between improvement and type conversion. Stand improvement involves a relatively small investment and a relatively high rate of return. Type conversion involves a larger investment but a lower rate of return. Therefore, the rate of return criterion points one way; contribution to present net worth, when based on an assumed capital cost of 3 percent, points the other way.

SUMMARY AND CONCLUSIONS

Several different criteria can be used to evaluate the relative profitability of timber management opportunities. Possibilities include the internal rate of return, contribution to present net worth, value response per cost dollar, and payout period. Variations are also possible with most of these basic criteria.

The effects of using various criteria were also briefly tested with opportunities that were mutually exclusive. For example, stand improvement or type conversion could be carried out in a young hardwood stand, but both could not be carried out in the same stand. Here the results were quite different. The internal rate of return favored stand improvement, contribution to present net worth favored type conversion.

These results do not definitively identify all the circumstances under which various criteria will and will not give similar results. However, the case study does not provide rather substantial evidence that many different criteria give results that are much the same when the decision at hand involves opportunities that are not mutually exclusive. Brief investigation of other circumstances suggests that different criteria may give quite different results with mutually exclusive opportunities.

The effects of using different criteria would certainly bear further investigation. For example, opportunities all of which involve relatively the same investment (large or small), might be investigated, so might those involving roughly the same or quite different investment periods might be studied, along with those involving relatively small vs. relatively large differences in inherent productivity. Simulation techniques might be used very effectively to make this investigation.

A practical question remains. What profit criteria should be used by forest managers and forest economists to evaluate the relative profitability of timber management opportunities? The answer depends first upon the alternative courses of action that are open, the resources available, and shifts in opportunities and resources that are likely over time. If the alternative courses of action are not mutually exclusive, there is no particularly serious problem. Any criterion that takes account of cost, output response including timing, and value would appear to give quite similar results.

Under these circumstances, the internal rate of return may have a distinct advantage from the viewpoint of the analyst. A rate of interest for discounting future values need not be specified in advance. Some of the other criteria, notably contribution to present net worth and value response per cost dollar, do require that an interest rate be specified in advance. The appropriate rate can be defined with apparent precision as the marginal cost of capital as determined by alternative opportunities for investment of the same funds. But in many situations it is extremely difficult to determine just what this rate should be. What are the alternative opportunities for investment of the same funds? And what rate of return do they promise? These are often very difficult questions. For example, in public programs, timber management may ultimately compete for funds with public education, highway construction, and mental hospitals. It would certainly be difficult to quantify the rates of return earned on these investments. The tendency for many profit criteria to give quite similar results with alternatives that are not mutually exclusive can greatly simplify analysis. This tendency suggests that difficult questions concerning cost of capital are often not particularly important with alternatives of this type.

With mutually exclusive alternatives there is a problem. This article is not a definitive treatment of the criteria to be used to evaluate mutually exclusive alternatives. Nevertheless, several observations can be made. First, situations where capital is the primary limiting factor differ from those in which land is the primary limiting factor. This observation has been made by several writers, including Worrell.[14] The question can be put in slightly different (and perhaps more explicit) words: which is likely to be used up first, funds or opportunities to use them? Second, likely shifts in opportunities and funds may change this situation. If availability of funds or opportunities is likely to change fairly drastically, this change might be anticipated in selecting a profit criterion.

Capital is frequently the primary factor limiting timber management. Under such circumstances, a profit criterion like the internal rate of return is appropriate. The reasoning is very simple. A high rate of return on a given investment is preferable to a low rate of return, other things equal. And other things are equal if capital is the primary limiting factor. All of the funds available for timber management can be invested in opportunities high-ranked by the internal rate of return.

Suppose, on the other hand, that land and the associated opportunities for investment in timber management are the primary limiting factors. Suppose, in terms of the example shown in Table 33–4, that stand improvement will not expend the funds that have been definitely budgeted for timber management. (The point previously made by the author, that budget determination and budget allocation are frequently two quite different kinds of decisions is relevant here.) Under these circumstances, a criterion that reflects the amount of investment as well as the rate of return will be a better choice. Contribution to present net worth is one such criterion. The reasoning is again quite simple. A relatively low rate of return on a large investment may give a larger absolute profit than would a higher rate of return on a much smaller investment.[19]

Developments over time may blur this distinction between situations in which capital is the primary limiting factor and those in which land occupies this role. Prospect of a large increase in the timber management budget might suggest a shift in criteria before the increase actually occurs. Suppose that a budget increase is clearly foreseen and that it will make land rather than capital the primary limiting factor. With this prospect, there is little use carrying out stand improvement (based on the internal rate of return) if another criterion will soon be more appropriate and type conversion will then be the most profitable course of action. On the other hand, possible but rather improbable changes in the future should not be a reason for doing nothing now!

It may also be useful to use several criteria in combination since they do measure somewhat different things. Depending upon his specific circumstances, a forest landowner or manager might rank opportunities primarily in terms of one criterion with minimum conditions specified for others. For example, he might use the internal rate of return as his primary criterion but also demand that all opportunities promise at least (say) $100 in additional present net worth, and that all opportunities pay out in (say) 35 years or less.

Differences among the profit criteria that have been examined may be important with mutually exclusive alternatives, but unimportant with alternatives that are not mutually exclusive. Where important, the choice of the primary criterion rests on the relative force of capital and land as limiting factors, and upon likely shifts in the foreseeable future.

REFERENCES

1. Lutz, Friederich and Vera. 1951. The theory of investment of the firm. Princeton, N.J.: Princeton Univ. Press. 253 pages.

2. Böhm-Bawerk, E. V. 1921. Positive theorie des Kapitals. 4th ed. Mit einem Geleiwort ven. Jena. G. Pischer. Ff. Wieser.
3. Wicksell, Knut. 1935. Lectures on political economy. London: Macmillan and Co. 253 pages.
4. Knight, F. H. 1934. Capital, time, and the interest rate. Economica 1:265.
5. Boulding, Kenneth E. 1935. The theory of a single investment. Quarterly Jour. of Economics 49:475.
6. Hayek, F. A. 1942. The Ricardo effect. Economics 8:135.
7. Keynes, John Maynard. 1936. The general theory of employment, interest, and money. New York: Harcourt, Brace and Co. 403 pages.
8. Fisher, Irving. 1930. The theory of interest. New York: Macmillan and Co. 566 pages.
9. Hicks, J. R. 1946. Value and capital. 2nd ed. London: Oxford Univ. Press.
10. Samuelson, P. A. 1937. Some aspects of the pure theory of capital. Quarterly Jour. of Economics 51:141–157.
11. Allen, R. G. D. 1938. Mathematical analysis for economists. London: Macmillan and Co. 548 pages.
12. Duerr, W. A., John Fedkiw, and Sam Guttenberg. 1956. Financial maturity: a guide to profitable timber growing. U.S. Dept. of Agriculture, Washington, D.C. Tech. Bul. No. 1146.
13. Fedkiw, John, and J. G. Yoho. 1956. Financial maturity: what is it good for? Jour. of Forestry 54:587–590.
14. Worrell, Albert C. 1953. Financial maturity: a questionable concept. Jour. of Forestry 51:711–714.
15. Webster, Henry H. 1960. Timber management opportunities in Pennsylvania. U.S. Forest Service Northeastern Forest Exp. Sta. Paper 137.
16. Nelson, A. W., Jr. 1961. How forest industries judge investment opportunities. Proc., 1960, Society of American Foresters. Pages 85–88.
17. Gaffney, M. Mason. 1957. Concepts of financial maturity of timber and other assets. Agric. Economics Information Series No. 62, North Carolina State College, Raleigh, N.C.
18. Webster, Henry H. 1963. Timber management and economic analysis: a case study. U.S. Forest Service Northeastern Forest Exp. Sta. Res. Paper NE–14.
19. Solomon, Ezra (Editor). 1959. The management of corporate capital. Glencoe, Illinois: The Free Press. 327 pages.

34

FINANCIAL MATURITY—WHAT'S IT GOOD FOR?

John Fedkiw
James G. Yoho

First published in *Journal of Forestry* 54(9):587–590. September 1956.

BIOGRAPHY

Mr. Fedkiw, Deputy Director of the Office of Planning and Evaluation in the Office of the Secretary, U.S. Department of Agriculture, is a forester with advanced degrees in administration and in agricultural economics. For 12 years following World War II, he served on the forestry faculty of the State University of New York. Thereafter, he joined the Agriculture Department, first in the Forest Service's Branch of Research and later in the Secretary's Office. Among his duties here, he directed supply, demand, and price analyses for the Seaton Panel; and among his honors was a Superior Service Award from the Department for his analytical work. He is best known for his many writings on economic aspects of forest management and for his enthusiastic, diligent, and penetrating studies of public programs and program alternatives in forestry.

Mr. Yoho's biographical sketch is given at Item 8, page 68.

EDITORS' SUMMARY

Financial maturity, a concept useful in designating individual trees for harvest, is the circumstance in which a tree's prospective rate of conversion-surplus increase has fallen below the owner's guiding rate of return. Appropriate judgment modifications of the financial-maturity decision make it applicable to a variety of situations.

QUESTIONS TO CONSIDER

(1) Simple, or unadjusted, financial maturity is the time or condition at which the rate of current value growth (in annual percent of total value) of a tree or timber stand equals or

promises to drop below the firm's guiding rate of return. What, precisely, is the purpose of "adjusting" this figure for "land-use cost"? Does adjusted maturity occur earlier or later than unadjusted?

(2) Why do we work with *total* growth and value to examine financial maturity, but with *marginal* growth and value to examine optimal stocking?

(3) What difference does it make whether we calculate value in terms of conversion surplus, as recommended, or in some other terms, such as a stumpage value based on conversion return?

(4) Why does a tree's rate of value growth tend to decline over time? Would it be a silvical impossibility for the rate to continue unchanged? a social impossibility? If the rate were sustained, what would be an acceptable criterion for scheduling the harvest?

(5) Messrs. Fedkiw and Yoho emphasize that the timber marker uses "judgment" to supplement his estimate of financial maturity. What do they mean by this term? List the points or questions which they recommend be covered by judgment.

* * *

Over the past few decades the planned production of forest crops has been a rapidly expanding business on the American economic scene. Much of the progress can be attributed to foresters' efforts to understand the business considerations underlying timber management. Foremost in this respect has been an increasing realization among foresters that the timber-growing enterprises can not be looked upon as a separate entity, but must be considered as a part of the forest owner's total business or household. Modern forest management planning is giving increasing consideration to the interests of the business as a whole and the functions woodlands serve in the business. This progress in forest management planning is particularly evident in the case of industrial forestry.

For forest crop production to make its maximum contribution to individual firms as well as to the national economy, forest economists must continue to develop sound and practicable methods for appraising business alternatives that will hold foresters accountable for the decisions they make. The development of the financial maturity concept and its application has been one of the principal efforts in this direction.

The financial maturity concept is not new, but was clearly defined in forestry literature as early as 1913 by W. W. Ashe:[1]

> Single trees of yellow poplar can be considered mature financially when their annual rate of increase in value becomes equal to the current rate of interest on money. If timber is held after the rate of increase in value falls below the interest rate, there is a loss, since if the timber had been sold, the proceeds could have been invested as loans at the current rate on money.

While this principle has been more or less generally understood and applied by American foresters insofar as practice and knowledge permitted, it did not receive a great deal of attention in American forestry literature prior to 1950. In 1942 S. O. Heiberg[2] used it in the preparation of economic increment charts for eastern white pine, and in the same year the Lake States regional office of the Forest Service prepared tree value increase tables for sugar maple, yellow birch and basswood.[3]

Beginning in 1951, the preparation of value increment tables for application of the financial maturity concept greatly accelerated. Between 1951 and 1953 the Southern

Forest Experiment Station published tree value increment tables for loblolly and shortleaf pine for the Crossett, Arkansas area,[4] for bottomland red oaks and sweetgum in the Mississippi River Delta and other southern river bottoms,[5] and for oaks and yellow-poplar in the uplands surrounding Birmingham, Alabama.[6] The Northeastern Forest Experiment Station published similar information for yellow-poplar in West Virginia in 1952.[7] In 1954 Bertram Husch[8] prepared value growth percent data for black cherry growing in northwestern Pennsylvania, and in 1955 S. O. Heiberg and Philip G. Haddock[9] published a value increment chart for young Douglas-fir in Lee Forest, Maltby, Washington. Also in 1955, the Southeastern Forest Experiment Station published value increment tables for twelve Appalachian hardwoods.[10]

The foregoing citations are the principal published works concerned with the application of the financial maturity concept to particular species and locations. Although all these references are concerned with the application of the financial maturity concept to the practical problem of choosing between trees to leave and those to cut, they vary considerably in the bases used for defining individual tree values and value increment as well as in details regarding the practical use of the concept. In this paper the writers bring together the essential ideas underlying the application of the financial maturity concept and attempt to present them in a manner and language that the busy practical forester can readily grasp. Although the concept is applicable to both stands and individual trees this article is concerned primarily with the individual tree application.

MEANING AND OBJECTIVES

The meaning of financial maturity is most easily understood by considering the case of the individual tree. The financial maturity concept looks upon the individual tree as a negotiable bond paying a certain rate of interest which is its rate of value increase. It suggests that if the rate of value increase is in excess of the rate the funds from such a bond (or tree) would yield elsewhere, the bond (or tree) is the most profitable use for such funds. Where the yield rate is less, it suggests selling the bond (or tree) and reinvesting the funds in a more remunerative alternative.

Thus, the financial maturity concept makes it possible to better understand and evaluate one of the important business aspects of timber growing. It accomplishes this by increasing the accuracy of judgment in selecting the trees the forester should leave for future income and those he ought to harvest to satisfy current income or wood requirements.

The guiding rate for the financial maturity decision must be determined by the forest owner. As suggested above, the owner may decide upon a guiding rate according to the alternative opportunities for the use of his capital within or outside his forest business. Or he may derive the rate from the cutting budget for his woodland enterprise, in which case it equals roughly the growth rate of the most productive and promising trees that must be harvested to meet the budget. A rate so determined is referred to as an alternative rate and should be adjusted for differences in risk as well as for taxation of earnings.

This is the basic story, but not the whole story. In applying the financial maturity concept, values other than the value increase of the tree itself are involved and should be taken into account. It is, for example, necessary to consider the influence of leaving the tree on: (1) the value increment of other trees; (2) maintenance of site quality, and (3) income from future rotations. These are very difficult, if not practically impossible, to appraise objectively and concretely. Thus, it follows that the personal judgment of the tree marker must still be introduced to determine how these considerations would modify the appraisal of financial maturity based on the value increase of the tree alone.

Appraisal of the intrinsic value increment of a tree provides only the core for the financial maturity decision. Its major utility lies in the fact that it accounts for the largest part of the income from growing timber. Furthermore, quantifying actual value increase in the form of a rate simplifies the decision which ultimately must be made: to harvest or leave the tree in question. It also provides an explicit means for making the forester accountable for his marking decisions to the forest owner. And the matter of accountability is a primary principle of good business management.

MEANS OF APPLICATION

It is impractical to take increment borings to determine the value growth rate of every tree when applying the financial maturity guide or any other guide in selecting trees for harvest. Reference can be had to prepared tables which classify trees by size, quality, and vigor and give average values and rates of value increase for each classification. Such tables have been called economic increment charts, value increment charts, or financial maturity guide tables.[11] These tables, in most cases, are one of the principal parts of the published works cited above. Given such tables and an alternative interest rate as a guide, it is then a simple matter to determine the size, quality, and vigor criteria which will identify trees typically growing below the guiding rate.

The use of the financial maturity guide requires information on tree values in relation to size, grade, and species and the corresponding average growth rates by vigor classes. The simplest basis for tree evaluation, and the only basis which is defensible, is conversion surplus. Conversion surplus is the residual value of the tree based on the price of the end product sold by the forest owner less the direct (variable) costs incurred in converting the tree into its end product. This product could be stumpage, logs at roadside, logs at mill, lumber f.o.b. mill, or any other forest product. Fixed costs incurred by the tree owner for administration and supervision, timber processing, timber management, or land taxes need not be prorated to individual trees since such costs have to be paid whether a particular tree is cut or not. Cutting one tree as compared to another does not change these costs. Conversion surplus, then, is a measure of how much a tree can contribute to payment of fixed costs and profits. It applies to the harvested trees as well as to those left for continued growth.

For the owner who sells stumpage, conversion surplus is practically equivalent to stumpage prices. His principal direct costs are those involved in marking, scaling, and tallying the sale volume. For the owner who sells logs at roadside and does his own logging, conversion surplus is computed as roadside log price less marking and scaling costs as above and all the direct costs of harvesting. The latter are primarily labor costs involved in felling, bucking, skidding, and landing, plus equipment operating costs which consist mostly of the operator's time and fuel expenses. The owner selling logs at the mill should use delivered prices and subtract additional direct costs of loading, hauling, and unloading at the mill to arrive at conversion surplus. The owner who processes his logs into lumber should use lumber sales price as a basis of valuation and subtract the direct costs associated with sawing, grading, piling, drying, and selling which, for him, are in addition to those mentioned for other owners. This same general procedure would apply in the case of products other than lumber and ultimately, whatever the product, all prices and direct costs must be related to individual trees by species, grade, and size.

Financial maturity guide tables give tree values and typical growth rates for trees of different vigor. Rate of value increase is calculated by solving the formula $(1 + r)^n = Vn/Vo$ for r with the use of compound interest tables. (r is the rate expressed as a decimal; n, the time to the next scheduled cut; Vo, present tree value; Vn, expected tree value n years hence.) Such rates, when computed, are averages for a particular tree grade, size, and vigor class and can be used to estimate the expected rate of value increment for similar trees between now and the next anticipated marking date.

In addition to its application in the financial maturity concept, conversion surplus has the added utility of providing a justifiable basis for determining relative profitability of harvesting, converting, and selling lumber or other products from trees of different size and grade. The higher its conversion surplus value the more a tree, log, or other timber product contributes to payment of fixed costs or profits. A tree or timber product with a zero conversion surplus will just pay the direct costs of processing and marketing and contribute nothing to payment of fixed costs or profits. The tree of negative conversion surplus will not permit recovery of the full amount of the direct costs and will, therefore, involve a loss in harvesting and processing. Thus, conversion surplus provides a guide to the relative profitability of processing alternative trees or timber products as well as a basis for estimating financial maturity.

USE OF THE CONCEPT

In the case of individual trees the financial maturity principle is relevant primarily after a decision has been reached to make a partial cut in a particular stand. Then it is useful in making choices between trees to cut and trees to leave so as to assure maximum income from all the resources the investment in the stand represents. This maximum will be realized by leaving, until the next scheduled cut, the maximum value in individual trees per acre growing at or above the alternative rate. Of course, the maximum value of residual stand per acre should meet silvicultural requirements for maintenance of site quality, desired influence on quality development in other trees, and desired composition of regrowth. On the other hand, the trees marked for harvest should satisfy the requirements for an operable cut. In order to meet these requirements it may be necessary to cut some trees growing above the alternative rate or leave some growing below the alternative rate.

It should also be noted that under a system of partial cutting the financial maturity principle can be helpful in determining whether an operable volume is ready for harvest. If the requirements of an operable cut are known, a stand can be examined to determine whether the value growth rate of enough trees is expected to fall below the guiding rate so as to constitute an operable volume.

The utility of the financial maturity concept is not limited to situations where the owner has a specific alternative rate of return in mind. Using one guide or another, decisions are frequently made to remove a certain proportion of growing stock without reference to an explicit alternative rate. Such is likely to be the case when a forest owner harvests part of his stand to meet a need for raw material or capital, e.g., provide logs for a sawmill, finance a debt, or obtain working capital. In such situations the financial maturity principle suggests the removal first of those trees promising the lowest rates of value increase; then trees with successively higher rates of increase. The basis for selecting trees to leave and to harvest would be the value increment of the trees involved and some additional standard such as spacing or desired cut or residual stand per acre. However, if suitable stock and stand tables are available, the cutting budget could be applied to these tables to derive an alternative rate. In either case, the alternative rate will be implicitly determined in the harvested stand. Thus, even though an alternative rate is not explicitly provided, it can be derived approximately from the actual cut for the information of management.

Where there is a choice of cutting one of two or several trees, the alternative rate-growth rate comparison can be used as the exclusive guide for the decision only where the trees have approximately the same conversion surplus value. Such choices occur where the value increments of the several trees involved have fallen, or are expected to fall, below the alternative rate in the next cutting period and removal of one or more would release the remaining sufficiently to bring or maintain their value increment at or above the alternative rate in the next period. If there is a notable difference in value between the alternative trees, but all are expected to grow at the same rate after the cut, the returns to the forest owner will be greater in the next cutting period if the higher value trees are left, other things being equal. This is simply because a twelve dollar tree appreciating at six percent will earn more than a ten dollar tree growing at the same rate.

Now it is true that differences in value which arise from differences in tree species, quality, or size must be taken into account in using the financial maturity principle to discriminate between alternative trees for cutting. However, in favoring the more valuable species, the larger sizes, and the better quality trees, the forester needs to evaluate the effects of this practice upon the current earnings of the business as well as upon the production of the forest in subsequent harvests. A tendency to mark only the poorer species and the smaller size and lower quality trees affects the marketability and selling price of the marked timber adversely, while enhancing the marketability and selling price of prospective future cuts. These effects tend to be multiplied in the case of the vertically integrated operator who usually anticipates a margin for profit at each successive processing stage in which he engages beyond the sale of stumpage. The forester applying the financial maturity principle, therefore, needs to keep one eye on the current market or profits and the other on the future. In other words, he must strike the balance in marking which will maximize the contribution of the forest property to the business of the owner during the period the owner expects to retain an interest in the forest's value and productivity.

SOME ADDITIONAL POINTS

In applying the financial maturity guide to selection of trees to leave for continued growth, risk of decadence and mortality in the time until the next cut, and sometimes beyond, needs to be considered. The principle implicitly requires an appraisal of risk of loss as a part of the decision. Naturally, any tree having positive conversion surplus, which is not expected to survive the next cutting cycle, should be marked now regardless of its value growth rate. Trees which are highly susceptible to disease or insect infestation as a result of injury or other causes should be favored for cutting. These considerations simply acknowledge that some of the costs involved in holding trees over to the next harvest are associated with mortality and decadence which in turn may lower the expected rate of value increase to the point where it becomes more profitable to remove such trees rather than risk losing them.

Anticipated price behavior likewise should enter into the financial maturity decision. When prices are expected to rise in the next cutting period, without corresponding increases in costs of conversion, value increment includes the differential due to the price rise as well as the expected value increase based on the current price level. Thus, trees that might be financially mature at current prices may be held profitably until the anticipated price rise becomes effective. Likewise, trees that are not financially mature may become so in light of an anticipated price fall without a corresponding drop in conversion costs.

It is sometimes observed in the case of the vertically integrated forest business that trees which are not yet financially mature have to be cut or else the firm will have to purchase raw materials in the open market at prices higher than average in order to maintain a given level of production. In such cases the conversion surplus value of trees that would meet the raw material rises by the amount of net income the business would lose by the impending shut down, cut back, or temporarily increased cost of purchased raw material. In these emergencies current tree values are higher than anticipated tree values at some future date under stable conditions of wood supply. This effects a drop in the anticipated rate of value increment in all trees owned by such a firm and results in many trees becoming financially mature.

These additional considerations indicate the further utility of the financial maturity concept in dealing with the practical considerations in the forest business. They demonstrate that the economy of the forest business is dynamic and that judgment needs to be exercised to modify the static analysis which would be given by a less flexible financial maturity guide or any other calculation based on the assumption of stable prices and interest costs.

CONCLUSION

The financial maturity principle has been developed largely for the purpose of providing a guide to timber growing based on individual tree selection. Basically, its application requires two measurements: (1) tree value and the expected value growth percent of the tree between now and the next anticipated cut; (2) some measure, either explicit or implicit, of the alternative rate of return demanded by the forest owner. If the alternative rate is greater than the tree's expected value growth rate the tree is financially mature. The decision may be altered, however, if additional net values or costs associated with market expectations, conversion operations, raw material requirements, growth of other trees, or reproduction of the stand are sufficiently great. These considerations, which are additional to the value increment of the tree itself, are frequently difficult to measure objectively or to standardize; therefore, their evaluation must rest heavily upon the experience and judgment of the forester. These points, once understood, should not deter the practical use of the financial maturity concept.

For all practical purposes the logic underlying the financial maturity principle is identical with that of any other system of choosing trees to cut now and those to leave for future income, where the objective is maximization of net returns to the entire business. Any defensible method of tree marking must consider the additional value which may be earned by leaving the tree for future growth as compared with the value realizable from harvesting it.

The application of judgment in comparing financial alternatives in forest management is inescapable, for the forest business cannot afford a thorough analysis of every tree in the stand. The financial maturity principle and the basic guide tables which are being developed by research foresters provide a practical means for minimizing the vagaries of judgment and thereby are contributing to a more successful choice among alternatives in the business of growing timber.

REFERENCES

1. Ashe, W. W. 1913. Yellow poplar in Tennessee. Tennessee State Geological Survey. Bul. 10-C, 52 pages. Page 36.
2. Heiberg, S. O. 1942. Cutting based on economic increment. Jour. of Forestry 40:645–651.
3. Stott, C. B., et al. 1942. Comparison of investment and earnings of hardwood trees in the northern Lake States (Region 9. U.S. Forest Service, Milwaukee, Wisconsin).
4. Guttenberg, Sam, and R. R. Reynolds. 1953. Cutting financially mature loblolly and shortleaf pine. Southern Forest Exp. Sta. Occasional Paper 129, 18 pages.
5. Guttenberg, Sam, and John A. Putnam. 1951. Financial maturity of bottomland red oaks and sweetgum. Southern Forest Exp. Sta. Occasional Paper 117, 24 pages.
6. Burkle, Joseph L., and Sam Guttenberg. 1952. Marking guides for oaks and yellow-poplar in the southern uplands. Southern Forest Exp. Sta. Occasional Paper 125, 27 pages.
7. Holcomb, Carl J., and C. Allen Bickford. 1952. Growth of yellow-poplar and associated species in West Virginia. Northeastern Forest Exp. Station Paper No. 52, 28 pages.
8. Husch, Bertram. 1954. Some financial aspects of managing black cherry in northwestern Pennsylvania. Jour. of Forestry 52:832–837.
9. Heiberg, S. O., and Philip G. Haddock. 1955. A method of thinning and a forecast of yield in Douglas-fir. Jour. of Forestry 53:10–18.
10. Campbell, Robert A. 1955. Tree grades and economic maturity for some Appalachian hardwoods. Southeastern Forest Exp. Sta., Station Paper No. 53, 22 pages.
11. Duerr, W. A., John Fedkiw, and Sam Guttenberg. 1956. Financial maturity—a guide to profitable timber growing. U.S. Dept. of Agriculture Tech. Bul. 1146.

35

WHO SAYS ACCELERATED ROADBUILDING PAYS?

Con H. Schallau

First published in *Journal of Forestry* 69(5):279–280. May 1971.

BIOGRAPHY

Mr. Schallau has worked upon research for the Forest Service throughout his professional career. Graduating in forestry from Iowa State University in the mid-1950s, he went on to Michigan State University for advanced studies. It was while he was employed in the Pacific Northwest that he produced the research reported here. His current assignment is as Deputy Director of the Intermountain Forest and Range Experiment Station, Ogden, Utah. Active in the professional affairs of the Society of American Foresters, he has served as Chairman of the Society's Division of Forest Economics and Policy and was a charter member of its Forestry Sciences Boards.

EDITORS' SUMMARY

Accelerated roadbuilding has been recommended for public forests in the Douglas-fir region. Studies, however, show that for a wide range of forest site quality, age, and acreage, the timber thinning, prelogging, and salvage benefits of stepped-up road programs fail to pay a satisfactory return on the outlay. And nontimber values are deteriorated.

QUESTIONS TO CONSIDER

(1) Of all the forest-management activities, roadbuilding has about as heavy an impact as any upon all forest values: wildlife, water, developed recreation, wilderness, scenery, range, and timber. Review for yourself the types of impact and their effects upon

value in each case. Think of value as the aggregate of net monetary and other returns, or benefits, to society.

(2) What parts of the value which you have identified has Mr. Schallau attempted to cover in his analysis? With wider coverage, do you think that he would have reached the same or different conclusions?

(3) What bearing do road standards (width, surface, maximum grade and curvature) and spacing have upon one's conclusions about the desirability of road construction?

(4) Generally speaking, what differences in considerations bear upon the roadbuilding question for privately-owned forests as distinguished from public forests?

(5) Some say that in forestry we are in a road-transportation era of limited duration, which will evolve into an air-transportation era. How may we take such a possibility into account in road analysis?

* * *

The Public Land Law Review Commission Report[1] recommends that "timber production units should be managed primarily on the basis of economic factors so as to maximize net returns to the federal treasury" (Recommendation 30). Further, the report recommends that "there should be an accelerated program of timber access road construction" (Recommendation 33), and that "controls to assure that timber harvesting is conducted so as to minimize adverse impacts on the environment on and off the public lands must be imposed" (Recommendation 36). The purpose of this paper is to summarize several research studies which suggest that, for certain areas in the Douglas-fir region, these three recommendations are inconsistent.

The concept of accelerated roadbuilding originated in the Douglas-fir region. But more important, that is where any change in the rate of public road construction would have the greatest impact on timber flows. Fedkiw, for example, estimated that accelerated road construction could lead, on the average, to the utilization of an additional 3.8 billion board feet annually between 1960 and the year 2000[2]—that is, output from the region could be increased by one-fifth. The added production would be in the form of harvests from prelogging small timber and salvaging mortality in old growth, as well as thinning young growth.

Since the publication of Fedkiw's assessment of the physical yield potential of accelerated roadbuilding, three economic analyses have been completed. Collectively, these studies encompass a full range of physiographical and areal conditions—high site to low site, old growth and young growth, and tracts ranging from 50,000 acres to an analysis involving 7.3 million acres of national forest lands in the Douglas-fir region.

ACCELERATED ROADBUILDING UNECONOMICAL

Payne's analysis[3] of the North Umpqua unit of the Umpqua National Forest in southwest Oregon disclosed that, between 1966 and 1995, accelerated road construction could increase total annual harvest by as much as 11 million board feet (a 5.6-percent increase). However, his analysis of the 497,000-acre unit, comprised mainly of old-growth timber, disclosed that none of the examined alternatives to the present program of constructing 75 miles annually would be feasible from an economic standpoint. For example, he compared four alternatives to the current rate of road construction—from an increase of one-third to six times the current rate. The most attractive alternative, which would double the present rate of road construction, earned only a 3.6-percent rate of return—considerably less than the guiding rate now required of federal agencies. The other three alternatives were even less attractive from an economic standpoint.

The *Douglas-fir Supply Study*[4] also considered accelerated roadbuilding. The results of this study, which examined road construction rates on *all* national forests in the Douglas-fir region, also suggest that accelerated roadbuilding is not a particularly lucrative investment alternative. For example, a plan to complete all roads in Region 6 in 20 years rather than 40 years would not earn a positive rate of return.

The merits of accelerated roadbuilding have also been considered for a smaller geographical area. In an attempt to answer the question—"Will increased revenues from thinning justify quicker access to young-growth stands?"—I studied the feasibility of accelerated roadbuilding for the Bureau of Land Management. This study involved their 50,000-acre Tillamook Resource Area located in the Coast Ranges of western Oregon.[5] At the time the study began, less than one-half of the permanent road system had been completed—160 miles of the planned system of 500 miles. But even at this stage of development, the road system was supporting an active thinning program. We knew accelerated roadbuilding would generate additional thinning volume but did not know if this volume would be enough to compensate for additional maintenance, timber sale administration, and interest costs resulting from constructing roads sooner.

The results of this study were quite conclusive. Although doubling the current rate of construction would increase thinning yields, added stumpage revenues would not cover additional interest, timber sale administration, and maintenance charges. Investment in advance roadbuilding for such a plan cannot be justified at any positive discount rate.

SLOWER CONSTRUCTION RATE DESIRABLE

A subsequent analysis of the Tillamook situation has disclosed that it might be economical to slow down the current rate of construction. The tabulation below compares the economic returns of alternatives to a 10-mile rate of road construction. Currently 15 miles of timber access roads are being constructed annually.

MILES OF ROAD CONSTRUCTION IN EXCESS OF 10 MILES PER YEAR	RATE OF RETURN *(PERCENT)*
5 *(current rate)*	0.8
10	0.2
15	0.3
20	−0.1

It shows that, although several alternatives to a slower rate of road construction would earn a positive rate of return, investments in such rates would not be lucrative given the current guiding rate of 5½ percent for federal agencies.

ACCELERATED ROADBUILDING CONFLICTS WITH NONTIMBER USES OF FOREST LAND

None of the above studies included costs and revenues associated with nontimber uses of forest land. Likewise, the possible economic gain associated with being able to salvage catastrophic losses due to disease and insect attacks and fire was not assessed. However, road networks are sufficiently completed to allow for relatively quick salvage of catastrophic losses. For example, the 50,000-acre Ox Bow Burn in Oregon was completely salvaged within 3 years. Besides, roads can increase the risk of fire. In the case of the Ox Bow Burn, blame was attributed to sparks from roadbuilding equipment.

. . . The *Douglas-fir Supply Study* disclosed that accelerated road construction could have a detrimental impact on fisheries. Similarly, speaking of nontimber values, Payne[3] asserted that

> the implications for advance roading do not look particularly promising. Soil and watershed, and fisheries and wildlife are undoubtedly affected adversely by a faster roadbuilding schedule. Recreation and fire protection may be positively affected on balance, but this result is highly questionable.

Evidence suggests, therefore, that an accelerated roadbuilding program would contradict a Public Land Law Review Commission recommendation which asks that controls be imposed to assure that harvesting will have "minimum adverse impacts on the environment."

In summary, managing public timber-production units so as to maximize the net return to the federal treasury would seemingly favor the adoption of practices other than accelerated roadbuilding.[4,6] Although constructing timber access roads faster than the current rate might lead to significant increases in timber output, we know of no situations involving public lands where it would be economically feasible. In addition, available evidence suggests that accelerated roadbuilding could have a deteriorating impact on nontimber uses of forest land. Consequently, one would have to conclude that Recommendations 30, 33, and 36 of the Public Land Law Review Commission's Report are inconsistent.

REFERENCES

1. Public Land Law Review Commission. 1970. One-third of the nation's land, a report to the President and to the Congress by the Public Land Law Review commission. Washington, D.C.: U.S. Government Printing Office.
2. Fedkiw, John. 1960. Advance roading for increased utilization in the Douglas-fir region—Oregon and Washington. Western Forest and Conservation Association Proc. 51:64–69.
3. Payne, Brian R. 1969. An economic analysis of alternative rates of investment in national forest road construction: the North Umpqua case. 190 pages. Illus. (Unpublished Ph.D. thesis on file at Univ. of California, Berkeley.)
4. U.S. Forest Service. 1969. Douglas-fir supply study. 53 pages. Illus. Cooperatively prepared by Regions 5 and 6 and the Pacific Northwest Forest and Range Exp. Sta.

5. Shallau, Con H. 1970. An economic analysis of accelerating road construction on the Bureau of Land Management's Tillamook Resource Area. U.S. Forest Service Pacific Northwest Forest and Range Exp. Sta. Res. Paper PNW-98, 29 pages.
6. Marty, Robert, and Walker Newman. 1969. Opportunities for timber management intensification on the national forests. Jour. of Forestry 67:482–485.

36

ECONOMIC GUIDES FOR ALLOCATING FOREST FIRE PROTECTION BUDGETS IN WISCONSIN

Stephen S. Sackett
Henry H. Webster
William B. Lord

First published in *Journal of Forestry* 65(9):636–641. September 1967.

BIOGRAPHY

Mr. Sackett is known professionally for his interest in prescribed-fire and fuels-management research. His first work as a research forester was with the Forest Service in the mid-1960s. After a period of advanced training in agricultural economics, he was assigned to the Southern Forest Fire Laboratory at Macon, Georgia. He headed a team of foresters and meteorologists studying the transport and dispersion of smoke from prescribed fires. Now he is part of the Fire Management Project at Tempe, Arizona, headquarters for fuels-management and fire research in the Southwest.

Mr. Lord is a forester and agricultural economist of wide-ranging experience. He served for 3 years as research forester with the Forest Service, and for 2 years as economic advisor to the Secretary of the Army. His principal association has been with the University of Wisconsin, where, toward the end of a 13-year tour of duty, he was Professor of Agricultural Economics, Forestry, and Environmental Studies. Most recently, he has worked for Resources for the Future in Mexico, as Visiting Professor in the Postgraduate College of the National School of Agriculture. His interest there is in the design and evaluation of rural-development programs.

Mr. Webster's biographical sketch is given at Item 33, page 260.

EDITORS' SUMMARY

Data from two regions of Wisconsin which differ in the intensity of their fire-control programs permit an empirical study of intensification benefits and costs. Three vegetative cover types found in both regions are examined. Conclusions are drawn about priorities in intensification and, more decisively, about promising avenues of research.

QUESTIONS TO CONSIDER

(1) Messrs. Sackett, Webster, and Lord had the interesting opportunity to compare areas which were similar in most respects except intensity of fire control. If such areas had not been available for study, what alternative procedures could they have followed for guiding decisions about the intensity of control?

(2) How is the method used in this study of control intensity related to the traditional rule: Find the level of control at which the sum of costs and losses is minimized?

(3) Does the traditional "minimum-cost-plus-loss" rule give the same answer about the optimal intensity of fire control as the economist's rule for finding the financially best combination where marginal unit costs and revenues are equal? Propose a proof that the two rules are not the same.

(4) What considerations not taken into account by the authors might significantly affect their conclusions? What would the effects be?

(5) What do you see as the principal defect in this fascinating and highly useful analysis? How might this defect be remedied?

* * *

An unintentional experiment has been carried out in Wisconsin's forest fire protection program. Various areas in the state, quite alike in many ways, have been given different intensities of forest fire protection. Most of heavily forested northern Wisconsin has had quite intensive fire protection for 30 or 35 years. A large forested area in the central part of the state has had significantly less protection. Areas which are otherwise quite similar can be found in both the low-intensity and high-intensity protection districts. As a result, valuable information concerning the relationship between intensity of protection and level of loss is available.

There are three potential forest fire situations to be considered . . . (1) oak forests, (2) upland grass and brush, and (3) pine forests. Each consists of a particular blend of timber type and stand conditions, climatic conditions, and conditions of human use. Two protection units illustrating each situation have been selected, one with high- and the other with low-intensity protection. The additional costs and benefits of high-intensity fire protection in each of the three situations have been measured by comparing costs and fire losses for high versus low intensity. (Nonmarket-determined values which may result from investments in forest fire protection are not considered in this analysis. Therefore this analysis is only a partial guide to forest fire protection decisions.)

* * *

The three potential fire situations differ significantly in terms of potential fire losses and protection costs. . . . Disastrous crown fires are a threat in pine; they do not occur in oak or grass. Surface fires are the major threat in these latter situations. Both losses and protection costs are likely to be highest in pine, intermediate in oak, and lowest in grass. Physical interrelationships between fire situations may also be important, as when grass and valuable timber are adjacent to each other. Fires may then spread from grass to timber.

Two sample areas representing each potential fire situation have been selected. One of each pair of sample areas has had fairly low-intensity fire protection for at least the last ten years. The other has had fairly high-intensity protection for at least ten years. . . .

Low-intensity and high-intensity fire protection differ from one another in several respects. High-intensity units have more equipment and personnel, greater detection facilities, and more prevention activities. Lookout towers are used for fire detection. Low-intensity units have less equipment, personnel, and prevention activities. They use fixed-wing aircraft, together with reports of local residents, for fire detection. Consequently fires are often larger on arrival in low- as compared with high-intensity protection units.

The sample areas were chosen to be as nearly alike as possible in every important respect except intensity of fire protection. Important respects include density and size of timber, and factors related to fire risk such as weather and woods use by visitors. Similarities and differences are summarized in Table 36–1. Woods use has been measured in terms of number of campgrounds. Number of visitors would be a more appropriate measure, but unfortunately requisite data were not readily available. . . .

Table 36–1. Selected Characteristics of Sample Areas

FIRE SITUATION	INTENSITY OF FIRE PROTECTION	
	Low	High
Oak forest		
Stand size distribution by area		
Sawtimber	26.5%	19.4%
Poletimber	45.8% } 73.5%	33.4% } 80.6%
Saplings	27.7%	47.2%
Average timber volume	6.7 cords/acre	5.6 cords/acre
Average number of high-risk fire days [per year]	119 days	98 days
Number of campgrounds	12 campgrounds	16 campgrounds
Upland grass and brush		
Proportion of area occupied by upland grass and brush	14.5%	14.5%
Average number of high-risk fire days [per year]	119 days	110 days
Railroad mileage	203 miles	224 miles
Pine forest		
Stand size distribution by area		
Sawtimber	20.4%	9.2%
Poletimber	36.0% } 79.6%	19.0% } 90.8%
Saplings	43.6%	71.8%
Average timber volume	7.7 cords/acre	4.4 cords/acre
Average number of high-risk fire days [per year]	119 days	110 days
Number of campgrounds	3 campgrounds	3 campgrounds

Approximately three-quarters of both the oak forest low-intensity and oak forest high-intensity areas support trees of pole-timber size or smaller. These sizes are most hazardous and most easily damaged. The low-intensity area contains slightly larger timber, hence a greater timber volume per acre. Both areas have a very large number of high-risk fire days. Recreational campground areas are also similar. These sample areas are part of the same general oak forest region in Wisconsin.

The sample areas illustrating the upland grass and brush situation are contiguous and alike. The area covered by brush and grass in both is almost identical; the number of high-risk days is quite comparable. Railroad mileage (a useful index of ignition probability in grass and brush) is about the same in each sample area.

The two pine forest sample areas are contiguous. Pole-size or smaller timber covers more than four-fifths of both samples areas. Such young stands are most vulnerable to crown fires. Average timber volume per acre in the low-intensity area is greater due to a higher proportion of sawtimber stands. Both the number of high-risk days and the number of campgrounds are about the same in the two pine forest areas.

MORE INTENSIVE FIRE PROTECTION: BENEFITS AND COSTS

Benefits and costs at various intensities of fire protection can now be compared. Benefits have been measured by reduction in losses including cost of fire suppression activities. Fire losses recognized include suppression costs, current value of destroyed merchantable timber, future value of destroyed immature timber discounted back to the present, and appraised value of destroyed or damaged improvements, equipment, and forest or farm products. Timber losses are valued in terms of reduction in stumpage values resulting from fire damage. This method of measuring losses is simplified and incomplete, but useful when applied in a consistent manner over all sample areas. Costs recognized include prevention, presuppression, personnel training, and capital outlays for fire fighting equipment and detection systems.

Loss and cost data were obtained from records of the Forest Protection Division, Wisconsin Conservation Department. Comprehensive reports covering all forest fires in the sample areas during an 11-year period, 1953–1963, were used. Only suppression costs could be measured directly for sample areas illustrating particular potential fire situations. District totals for the other cost elements were allocated to sample areas on the following arbitrary, but presumably reasonable, bases: Prevention [was] distributed in proportion to protected areas. Stand-by time and maintenance of equipment and tools [were] distributed in proportion to the number of fires. Capital outlay and training [were] distributed in proportion to protected area. Detection, communication, and miscellaneous [were] distributed in proportion to protected area.

Tables 36–2 and 36–3 summarize average annual protection costs and fire damages for all sample areas. Table 36–2 contains the basic information. Table 36–3 measures the incremental benefit-cost ratios for additional expenditures to achieve high-intensity (as opposed to low-intensity) protection. Some care must be taken in interpreting these ratios. Low-intensity and high-intensity protection involve quite different patterns of fire prevention and control. Therefore, a continuous range of fire protection intensities does not exist. Nevertheless, the benefit-cost ratios measure the relative cost-effectiveness of moving from low-intensity to high-intensity protection in these three fire situations.

Table 36–2. Protection Cost and Fire Loss for Three Potential Fire Situations—in Dollars per Year per Protected Acre

FIRE SITUATION		INTENSITY OF FIRE PROTECTION	
		Low	High
Oak forest	Loss	0.0290	0.0022
	Cost	0.0390	0.1167
Upland grass and brush	Loss	0.0377	0.0061
	Cost	0.1625	0.2461
Pine forest	Loss	0.2184	0.0350
	Cost	0.0626	0.1654

Source: These cost and loss figures were calculated from accounting and other records of the Forest Protection Division, Wisconsin Conservation Department.

In the case of pine, a small addition to protection costs brings a large reduction in losses. A similar addition to costs brings only a very small reduction in losses in upland grass-brush. Oak falls slightly below upland grass and brush in terms of the reduction in losses associated with a given increase in protection costs. Stumpage values and the damage done by fires cause these relative positions. Pine has a relatively high value for a variety of products including lumber, pulp, and Christmas trees. Upland grass and brush, on the other hand, has a very low value. Increased protection reduces losses in oak stands comparatively little because many fires do little damage to these stands even though value per acre is fairly high.

Table 36–3. Loss Reduction in Three Potential Fire Situations

FIRE SITUATION	DOLLAR VALUE OF LOSS REDUCTION PER ADDITIONAL DOLLAR INVESTED IN HIGH-INTENSITY PROTECTION
Oak forest	0.34
Upland grass and brush	0.38
Pine forest	1.78

These incremental benefit-cost ratios were calculated from Table 36–2. High-intensity protection was compared with low-intensity protection for each fire situation. The reduction in loss per protected acre caused by moving from low- to high-intensity was divided by the associated increase in cost. The figures for pine illustrate this:

$$\frac{0.2184 - 0.0350}{0.1654 - 0.0626} = \frac{0.1834}{0.1028} = 1.78$$

It is not possible to provide high-intensity protection only for the area in a particular forest type. Therefore, it is *not* legitimate to conclude from this table that no areas in oak or upland grass and brush should be given high-intensity protection.

IMPLICATIONS FOR FIRE PROTECTION

The purpose of this study is to compare differing intensities of forest fire protection in several potential fire situations. Nearly all forest fire protection organizations protect areas in several land uses and timber types. Protection units within a large area, such as a state, have different needs according to their respective situations. A method for allocating fire protection budgets among units has been developed, and will now be illustrated.

Suppose that a forest fire protection agency is formulating its budget request for next fiscal year. One use of additional funds would be to raise one or more of three forest protection units from low-intensity to high-intensity protection. Acreages by timber type for each protection unit are shown in the upper portion in Table 36–4. These types conform to the corresponding potential fire situations in terms of stand-size distribution, timber volume, number of high-risk fire days, and relevant human use characteristics.

Benefit-cost ratios for intensification in each protection unit are required to evaluate the priority and desirability of raising the three units to high-intensity protection. These ratios have been calculated by multiplying the ratio of damage reduction to increased protection costs for each type (Table 36–3) by the percentage of each protection unit occupied by that type, and aggregating for all types in each unit. The result is a weighted average benefit-cost ratio for the entire unit. Thus, for Unit A,

$$1.78 \times 0.25 + 0.34 \times 0.15 + 0.38 \times 0.60 = 0.724.$$

Similarly, average per acre additional cost of intensified protection for each unit has been calculated using the type percentages as weights. Average benefit-cost ratio of intensification, and average additional cost per acre for intensification, total additional cost of intensification for each protection unit are shown in the lower portion of Table 36–4.

Table 36–4. Calculation of Benefits and Costs of Intensified Protection: An Example

UNIT CHARACTERISTICS	UNIT A	UNIT B	UNIT C
Total acreage	285,000	295,000	275,000
Timber type composition, percent:			
Pine	25	20	50
Oak	15	50	30
Upland grass and brush	60	30	20
Cost data:			
Average cost per acre to intensify protection	$0.0783	$0.0780	$0.0851
Average benefit-cost ratio for intensified protection	0.724	0.640	1.068
Total cost to intensify	$24,942	$24,925	$25,143

Unit C clearly should have priority over Units A and B, and Unit A should be favored (although by a narrower margin) over Unit B, since the benefit-cost ratios are 1.068, 0.724, and 0.650, respectively. Furthermore, the additional benefits of high-intensity protection in Unit C exceed the additional costs, making this a desirable, if marginal, project. (Marginal because benefits only slightly exceed costs.) Raising either Unit A or Unit B from low- to high-intensity protection is undesirable on strict economic grounds; the additional benefits do not match the additional costs.

The relative extent of the three types—pine, oak, and upland grass and brush—on two of these three hypothetical protection units exemplify conditions in parts of Wisconsin. Protection Unit A represents average conditions in northeastern Wisconsin and Unit B represents average conditions in Central Wisconsin.[1] Unit C has no parallel that we know of on a protection district level. Its composition was chosen to illustrate the very high proportion of pine types necessary to qualify a particular area for high-intensity protection. Any district which is not at least 50 percent pine probably should not receive high-intensity protection if the decision is based on timber values alone, and if the protection district is the minimum unit for which high-intensity protection is feasible. Most, if not all, Wisconsin forest fire protection districts are in the less-than-50-percent category.

The situation may be quite different if high-intensity protection is feasible for smaller areas. Two notable examples are sizable pine areas in extreme northwestern Wisconsin and along the Wisconsin River in the central part of the state. Indeed, a redrawing of district boundaries and a reallocation of protection efforts and funds in favor of these areas may be in order.

This analysis is a crude economic guide for allocating forest fire protection funds. It should not be used in mechanical fashion for at least two reasons. First, high-value and low-value types may occur next to each other, but wild fires do not respect type boundaries. A part of the cost of protecting particular low-value stands may be justified by the threat to adjacent high-value stands. Second, values other than those for timber must be considered. Some areas may yield larger nonmarket-determined values than do other areas. Recreational and wildlife values are prime examples. A simple economic analysis may underestimate the intensity of protection which is appropriate for areas yielding such values. In short, this analysis must be used cautiously and with due regard for its limitations. Nevertheless, it is a useful starting point for allocating forest fire protection funds. It compares alternative ways of allocating funds and thereby focuses on relative effectiveness. Such explicit consideration of alternatives is the method most likely to identify the best one.

RESEARCH FOR MORE COMPLETE ANSWERS

This study was made during a relatively short period using existing fire protection records. More detailed and extensive studies in the economics of forest fire protection would be less dependent upon existing records and arbitrary assumptions.

More detailed and extensive studies could go further in at least three ways. First, intensity of fire protection could be defined more precisely and more levels of intensity of fire protection, "high" and "low." The definitions of these intensities are simply the patterns of fire prevention and control carried out by the Wisconsin Conservation Department in two large geographic areas of the state. These patterns make sense from an operational viewpoint. However, more precise definition of intensity of fire protection, and more levels of intensity, would improve the precision and expand the utility of research. Intensity might be defined more precisely in terms of particular kinds or levels of protection expenditures. More levels of intensity would disclose the form of the relationship between protection expenditures and fire losses. Something approaching a continuous relationship might be derived.

Second, cost accounting systems permitting direct measurement of all relevant cost elements could be devised. In this study only suppression costs could be directly measured for sample areas illustrating particular fire situations. Other costs including prevention, presuppression, etc., had to be allocated from protection districts to sample areas on arbitrary bases. Presuppression costs are a major element of forest fire protection costs—approximately three-fourths of the total in the protection units selected in this study. Therefore, two different methods of allocating presuppression costs from protection districts to sample areas illustrating particular fire situations were tested. One allocated presuppression costs by number of fires alone, the other by a combination of number of fires and area. The results were nearly identical. This sensitivity analysis suggests that the method of allocation is not overly critical in this particular instance. Nevertheless arbitrary allocation of a cost element which constitutes three-fourths of the total cost is not theoretically satisfactory even though it appears to work fairly well in this particular instance. Direct measurement would be distinctly better. Revised, and probably more elaborate, cost accounting systems in forest fire protection agencies would be required.

Third, the relative efficiency of alternative patterns of forest fire protection could be studied. This study was limited to the effects of two different patterns, each applied at a particular intensity. Several different paths or patterns could be selected and compared at each of several different intensities. Some comparisons might involve different methods for carrying out a given aspect of fire protection. For example, fire detection by aircraft versus fire detection using towers or other fixed observation points might be compared. Other comparisons might involve quite different aspects of fire protection. These latter comparisons might determine the relative cost-effectiveness of control of crown fire danger in pine stands by means of major firebreaks and other presuppression devices versus control by means of faster initial attack on fires in pine stands. Still more widespread comparisons might eventually rate the cost-effectiveness of hazard reduction in particular timber types versus intensified railroad locomotive inspections versus more fire prevention advertising. All these comparisons of alternative patterns of fire protection would help protection agencies to more systematically choose the most effective pattern for each fire situation they face.

The economics of forest fire protection is one important area for research. Research in physical aspects of fire behavior, control, and use is another broad and extremely important area. The two are potentially very closely related. Physical aspects of fire behavior, control, and use must be defined with some precision before economic considerations can be effectively introduced. This will be increasingly true as progressively more comprehensive studies are undertaken in the economics of forest fire protection. These studies may compare the relative effectiveness of systems of fire protection, not all of which have been tried on an operational basis. Such comparisons may contribute to major advances in fire protection. Direct observation of the relationships between protection expenditures and fire losses of the type carried out in this study would be impossible in studies comparing quite different and untried fire protection systems.

Synthesis of physical and economic information from a wide variety of independent studies will become increasingly important.

REFERENCE

1. Stone, Robert N., and Harry W. Thorne. 1961. Wisconsin's forest resources. U.S. Forest Service Lake States Forest Exp. Sta., Station Paper No. 90, 52 pages.

Part Four

MANUFACTURING

37

CONVERTING FOREST RESOURCE STATISTICS TO TIMBER SUPPLY

Sam Guttenberg

First published in *Proceedings Seventh Conference on Southern Industrial Forest Management,* Duke University, Durham, N.C.; pages 46–51. 1967.

BIOGRAPHY

Mr. Guttenberg's biographical sketch appears at Item 24, page 184.

EDITORS' SUMMARY

The firm that uses *Forest Survey* timber statistics as a measure of supply to guide its plant-location decisions must reduce the reported figures to allow for such factors as physical accessibility, owners' timber-sale policies, supply elasticity and competition among buyers of stumpage, forest availability for purchase or leasing, standards of wood merchantability, and the plans of competitors and landowners.

QUESTIONS TO CONSIDER

(1) What does Mr. Guttenberg mean by the term, *supply?*

(2) Under what circumstances may several wood-using firms be drawing their raw material from the same group of counties and yet not be in competition for that raw material?

(3) What considerations would lead a forest owner to sell more timber at a high price than at a low price? What considerations would lead him to sell less at a high price?

(4) Why was the timber harvest from public and industrial forests (page 298) limited to less than half of the growth?

(5) Mr. Guttenberg opens his discussion by referring to "statistics on the forest re-

sources." Imagine for a moment that he is speaking not of timber resources alone but of all forest resources. Picture a survey designed to gather such statistics. How would the field work be carried out? In what form would the results be presented? With respect to each of the resources other than timber, would there be a question of *availability* analogous to the question that Mr. Guttenberg discusses for timber?

* * *

The *Forest Survey* provides the only statistics on the forest resources of each state. In industrial application these data must be converted to timber supply—the portion of the forest resource economically available for processing in the short run. Let's discuss the conversion problem from the standpoint of a firm estimating the prospects for profitable recovery of investments in major wood-using plants.

Inventory volumes reported by *Survey* typically exceed timber supply by a wide margin. The *Survey* task is to determine the volumes of timber resources regardless of operability for commercial purposes. Hence, the basic concepts underlying data collection and compilation are more mensurational than economic. To be sure, economic concepts are also present to the extent that they bear on national appraisals.

While the economic principles for analyzing timber supply are universally applicable, the details necessarily vary with such factors as type of forest industry and size of firm. The men in *Survey* are keenly aware of the problem. Their continuing efforts to improve the compilation of resource statistics reflect the numerous requests for information by those concerned with timber supply. . . .

Appraisal methods may range from the rudimentary to the exhaustive. For example, twenty years ago two hurried entrepreneurs came to the Southern Forest Experiment Station with this query: How much timber can be grown on a 100,000 acre tract located anywhere in southern Alabama? The forester's answer was that the specific tract must be examined, data collected, and a growth analysis made. "Never mind all that," they said. "Just tell us from your professional experience and knowledge of the general area if we can count on at least 50 board feet per acre annually." The answer was, "yes!" That quick appraisal launched a sizable forest investment. Today, however, corporations planning plant modernizations and expansions or seeking locations for new multimillion dollar pulp mills, sawmills, or plywood plants would demand a more thorough analysis of timber supply.

There is an economics of timber supply appraisal as of anything else. How much time priority should be assigned, and how much money should one spend? Priorities may be determined largely by the activities of other firms interested in an area. One rough yardstick of reasonable cost is the size of the proposed investment. But perhaps of greatest significance is the proportion of wood costs to product value. It's axiomatic that as this ratio rises the need for thorough analysis of timber supply heightens. In lumbering, for instance, where wood costs typically exceed 50 percent of product value, small changes in log values greatly affect net returns. In contrast, the price of bourbon staves is a trivial fraction of the value of branch water additive, and hence distillers can afford to be indifferent to wood costs.

The industrial questions raised will largely set the economic interpretation of the data in *Survey* reports. Variations in financial structure, in need for timber, and in forest landownership patterns may easily cause different industries—or even firms within an industry—to draw contrary conclusions from the same set of statistics. Timber is a bulky raw material that gains considerable value per unit of weight between stump and ultimate consumer. The relative costs of transporting timber to mills and products to markets favor setting mills in timbersheds.

In principle, a timbershed is the minimum area that can economically supply the needs of a particular conversion plant. In practice, the timbersheds being appraised usually cannot be delineated merely by drawing circles of fixed radii from likely centers of transportation. The numerous pulpmills operating along the eastern seaboard and the Gulf of Mexico, for example, are neither drawing much timber from the seas nor growing it there either. Neither are rail rates respecters of simple geometric forms. Is it a one-line haul within the area in question? Will a state line slice through the area? These considerations materially affect rates and the shape of the timbershed. Major rivers may be an asset if river logging is contemplated or a hindrance if rail and road bridges are few. Similarly, major water impoundments in the Coastal Plain can materially add to the cost of hauling timber. In effect, a carefully drawn timbershed will rarely resemble a perfect circle. Interactions between the selected timbershed and adjacent ones must also be considered. Economic boundaries are apt to change with time, and some of them are even predictable at the time of analysis.

We are ultimately concerned with the cost of wood at the plant. In addition to transportation, there are the costs of stumpage and logging to consider. What determines stumpage price in the timbershed? Is it largely the degree of competition? Is it accessibility? Is it the size, grade, and concentration of timber available for cutting? What changes in the volumes of timber and in prices will the proposed capital investment be likely to cause? What are the prospects that landowners will change their forest investment policies? These are major questions for those concerned with converting resource statistics to economic timber supplies.

What does the *Survey* offer the analyst who is faced with these questions? The main items are county data on forest areas, timber volumes, and cut. There are also broad data on growth and ownership.

County data are basic for roughly approximating timbersheds. Volumes are reported by species group and diameter classes. These breakdowns offer considerable advantage in partial analysis. For example, pine plywood plants or sawmills are primarily in need of timber at least 15 inches in diameter. Pulpmills generally rely on smaller trees.

Hardwood lumbermen are looking for the volume in trees at least 15 inches in diameter. The Survey eyes volume to an 8-inch top diameter in logs meeting minimum local-use standards, but the lumbermen are after the volume in logs that are at least factory grade 3, and 10 inches in top diameter. Such adjustments can be deduced from the reported statistics, and can quickly and cheaply eliminate unpromising timbersheds. The next step in shrinking resource totals down to timber supply requires some attention to conceptual devices developed by economists. In particular, the notion of derived demand and elasticity is germane to our problem. Demand for timber is a derived, not a direct, demand. Because it stems from consumers' demand for end products, it can fluctuate sharply—with consequent effect on price. This effect can be readily seen in the prices paid for pine sawtimber stumpage from the national forests. When the lumber market weakens, stumpage prices sag and when the market strengthens, prices jump. Effects on pulpwood stumpage are less evident, partly because of the aggregative year-to-year increases in demand and partly because the industry has built into its procurement practices the alternatives of shifting from high- to low-cost sources of wood. An organization's relative ability to compete for the available stumpage can offset unfavorable price effects. Firms that have a substantial land base or enjoy the economics of integrated operations can exercise a degree of monopsonistic power.

The price effects are due partly to the nature of the demand for timber and partly to the inelasticity of supply. That is to say, for each percent rise in stumpage prices there is less than a percent gain in volume offered for sale. And the volume gains tend to diminish progressively. Thus, from the standpoint of processors, sharply rising stumpage prices bring forth disappointingly low gains in timber supplies. This phenomenon is largely due to widespread adoption of forest management—especially on public and industrial holdings. Management programs have sharply reduced the timber available for cutting and stimulated a substantial demand for timber and timberlands for investment purposes. These forces contribute to the inelasticity of timber supply. In the South, at least, incremental increases in reserved growing stock are likely to occur for several decades.

Even if the South's timber lands were fully stocked, the managers would require a reserve growing stock of perhaps 10 to 15 units of volume for each unit harvested. And with our typically understocked stands, the managers must increase the ratio of reserves before they can achieve their goal. In the interim, as more and more forest owners rebuild their stands, timber supplies become progressively more inelastic. Stumpage prices rise, especially for sawtimber, and some of our market shifts to the forest industries in the West, British Columbia, and overseas. We have all seen this happen in both softwood and hardwood markets. The process is, of course, reversible. Meanwhile, the analyst needs to recognize the effects on timber supply.

Forest resource statistics are reported by broad ownership classes. The public holdings and forest industry lands are essentially alike in regard to stocking and management policy. These forests have above-average stocking, but management considerations hold the cut—that is, the volume going into the area's timber supply—to less than half the current growth. Overall comparisons of growth to cut or of cut to inventory can suggest how the current timber supply is bearing on the area's resources. But the basis for more refined answers on timber supply is to be found at the county seats.

The SCS offices there often have county ownership maps and usually complete aerial photograph coverage on which individual ownerships are delineated. The number of ownerships with meaningful forest acreages, say at least 500 acres, can be quickly, cheaply, and accurately determined. These are the ownerships from which the nucleii for supporting expanded or new forest industries can be created. In some cases, outright purchase of the assets of family-held lumber firms is the route to take. While many such firms have been bought out in the past two decades, a goodly number still remain. There are also the extra-market devices for accomplishing the same end. Leasing, tree farm family systems, and other means of obtaining long-term stumpage commitments on the larger holdings will suffice for example. If the major landowners are under obligation to provide timber to one firm it is difficult for other firms to enter the timbershed.

Let's return to the aerial photographs for a moment. They can be used to develop a refined timbershed, since they depict the transportation scheme and the net economically meaningful forest acreage. And they can be used to answer at least one more question: How much acreage can sustain highly mechanized logging operations?

A number of projections have been made recently to point up the impending crisis in woods labor. Typically the projections take the current output per man-day of pulpwood and multiply it by the expected increase in pulp production. The result is an alarming additional demand for woods labor. I would submit that such calculations are not very likely to prove true. If the expansion in pulp output is to occur, as I believe it will, man-hour requirements per cord will decline. But there is a hidden stinger.

Substitution of capital for labor both in the plants and woods has been going on for so many decades that we may be inclined to take it for granted. In large measure, capital has made up for the substantial decline in quality of our forest resources. By quality I mean operable volume per acre, tree diameter, tree grade, and volume per tract taken together. We have probably reached bottom, and from here on I would expect capital improvements to be geared to an improving rather than a declining resource. The photographs lend themselves to a determination of how much acreage can meet the needs of highly capitalized logging operations.

There is another group of timber supply problems that deserves brief mention. The question is, "What to do about demand for highly specialized portions of the forest resource?" The reference is to such items as persimmon, high-quality ash, hardwoods that meet the exacting standards for face veneer, and pines suitable for conversion to long poles and piling. Volumes of the desired timber as reported by the *Survey* are usually very small and widely dispersed. Effective demand is small, but obtaining the timber may be difficult. Survey reports provide clues to areas wherein suitable timber is likely to be concentrated. The problem is mainly one of choosing among alternative procurement strategies. It is generally infeasible to locate among the multiplicity of woodlands those few that have the desired timber and then attempt to deal with the owners. The two principal alternatives are: (1) to arrange with the leading wood processors in the area to harvest the special class of timber as an incidental part of their logging activities, or (2) to deal with a few forest managers of extensive holdings for the required volume.

Finally, no matter how sharply the analyst calculates, he cannot accurately predict the investment decisions of the timbershed's landowners and of the competing forest industries. That is unfortunate, for gross misguesses in this key area can raise hob with estimates of the economically available timber supply. For short-run investment planning, a measure of protection can be gained by ranking the relative attractiveness of alternative timbersheds. But for those who view their investments as commitments to stay in the forest products business indefinitely, there is probably no satisfactory substitute for intensive management of fee-simple timberlands or their near equivalent.

38

A POSITIVE PROPOSAL TO STRENGTHEN THE LUMBER INDUSTRY

Walter J. Mead

First published in *Land Economics* 40(2):141–152. May 1964.

BIOGRAPHY

Mr. Mead is Professor of Economics in the University of California at Santa Barbara. During his academic career as an economist, which included doctoral study at the University of Oregon and teaching at Lewis and Clark College, Portland, he served as a consultant in the lumber and plywood industries in the West and grew interested in their problems. A result was this and other articles and a book, *Competition and Oligopsony in the Douglas-Fir Lumber Industry* (1966). He has consulted also with public agencies and served as president of the Western Economic Association and Senior Economist for the Energy Policy Project of the Ford Foundation. His recent research interest has been in oil shale and petroleum economics.

EDITORS' SUMMARY

The lumber industry, notably in the Douglas-fir region, has been experiencing problems that center upon the industry's relative loss of efficiency. If federal timber were sold as peeled and sorted logs in public markets, sawmill efficiency would improve through increased specialization, the use of bark-free logs, fuller wood utilization, reduction in uncertainty and buyer-seller conflict, and lowering of capital requirements.

QUESTIONS TO CONSIDER

(1) Mr. Mead depicts for us a lumber industry beset by problems, just as did Mr. Bond (Item 23) and Mr. Duerr (Item 27). Evidently the industry has difficulties in many aspects of its life. What lies at the root of these difficulties? What do its difficulties portend for the industry?

(2) It has been argued that the sort of competition that Mr. Mead sees as socially beneficial (page 303) is one of the sources of the lumber industry's problems. Do you agree or disagree? Why?

(3) Suppose that the lumber industry had access to a limited but sufficient supply of skilled labor willing to work for very low wages—because, say, of an inordinate fondness for the smell of freshly cut wood. Would this solve the industry's problems?

(4) How would the situation of the United States lumber industry be changed if the Jones Act (page 302) were repealed?

(5) Why, do you suppose, has Mr. Mead's proposal not been followed or even tried out? To what degree can this be explained in terms of what Mr. Mead calls the "disadvantages and problems" of his proposal? Then again, do you think Mr. Mead is mistaken about the "anticipated benefits" of his system?

* * *

The lumber industry of the United States, and of the Douglas-fir Region in particular, is in serious economic difficulty: (1) More Americans are consuming less lumber. (2) A net surplus of lumber exports abroad has become net imports. (3) Profit from lumber production has declined. (4) Employment in lumber production has declined sharply. (5) And there is a mass exodus from the lumber industry by small independent producers, unmatched in other segments of United States manufacturing industries.

* * *

REASONS FOR LUMBER INDUSTRY DIFFICULTIES

Before listing in detail some of the principal causes of the economic problems identified above, a general comment is in order. Most of the lumber industry's difficulties are the result of normal economic forces at work in a mixed free enterprise system. The heavy demand for lumber in the post–World War II years induced entry into the industry. Barriers to entry into lumber production are at a minimum, given the ability to bid freely for federal timber. The lush United States market in the post-war period also caused a reallocation of lumber in world markets toward the United States market. These forces brought an end to lumber price inflation by 1951. Competition in lumber production carries penalties for the least enterprising producers but involves corresponding benefits for price conscious consumers.

In more detail,the reasons for present difficulties of the lumber industry must include, first, the penalties of past bonanzas. From depression levels in the mid-thirties to 1950, the demand for and price of lumber increased at unsustainable rates. The price of lumber over this period increased at approximately twice the rate of all commodity prices. This was the result of rapid expansion in residential construction and a relatively inelastic supply of lumber. By 1950 the ability of lumber to compete with substitute building materials was substantially impaired. Continued high-level demand for building materials included a shift from lumber to other construction materials. Under the impact again of a relatively inelastic supply, the price of lumber declined in the 1950s relative to all other commodity prices.

Second, the mass exodus of small producers from the lumber industry may in part be accounted for by the exhaustion of some private timber resources. However, there was a nearly compensating increase in annual log supply from public timber resources. The fact that small mills all but disappeared from production while medium-large mills expanded sharply in both number and output suggests that inefficiency may have contributed to the mass exodus of small mills. Similarly, the demise of some of the large mills suggests that they failed to maintain efficiency relative to the medium-large plants.

Third, the rapid increase in number of operating sawmills (in the Douglas-fir Region) from 1940 to 1947, followed by a matching dropout rate, has created an overhang of excess capacity at present output levels. Firms without timber reserves and facing the alternative of closure make the perfectly rational short-run choice of bidding away all their profit and sunk costs when such action is necessary to obtain timber. This excess capacity condition with its harsh alternative, undoubtedly contributes to the apparently excessive bidding for federal timber in certain national forests within the Douglas-fir Region.

Fourth, the cost of producing lumber increased relative to lumber prices in the 1950s. While relative price-cost inflation favored the mill operation during the 1940s, from 1950 to 1960 the ratio of cost of producing lumber in the Douglas-fir Region to "average realization" increased from 76% to 93%.[1]

Under present conditions, effective competition requires a relatively high degree of utilization at the processing level. In the late 1940s a mill could profitably convert logs into one product, lumber, and burn the residual. Under presently prevailing conditions in the Douglas-fir Region this practice is becoming economically hazardous. Competitive pressures increasingly require that a mill convert its residual wood into chips for pulping, that appropriate logs be converted into veneer, and that other logs be converted into lumber.

Fifth, technological progress in the lumber industry has been both slow and uneven among producers. Over the period 1899 to 1954 the increase in output per man-hour of labor employed in lumber mills has been only 1.1 percent per year. Comparable data are available for 79 other United States industries. Only six of them have productivity growth rates lower than lumber mills. The average rate for all 80 industries is exactly twice that of lumber.[2]

Sixth, in addition to being buffeted between rags and riches by the interplay of economic forces as noted above, government interference, often at industry request, has also taken its toll. Under pressure of the maritime industry, Congress passed the Merchant Marine Act of 1920 (called the Jones Act). Section 27 required the use of American flag vessels in United States inter-coastal shipping. British Columbia cargo mills are not similarly restricted and can therefore use foreign bottoms for lumber shipment from Pacific ports to Atlantic and Gulf ports.

From mid-1950 to early 1952, and from January 1955 to September 1957, the Jones Act was not restrictive for United States compared to Canadian producers because charter rates were usually $3.00 to $7.50 per M board feet higher than United States conference rates. However, since September 1957, United States cargo mills have been burdened by a Jones Act penalty of $5.50 to $12.00 per M board feet compared to competing British Columbia cargo mills.[3]

The foregoing sections have outlined the nature and causes of the economic problems facing the lumber industry. If the nature of the problems has been correctly diagnosed as, in essence, an inability to compete with foreign lumber suppliers and domestic producers of competing products, in turn the result of a tardy reaction to dynamic economic forces, then corrective measures should concentrate on (1) efforts to increase efficiency, (2) expanded research in new product development, and (3) the introduction of more effective marketing techniques. While efforts in all three areas would be highly desirable the concern of this proposal is entirely with the first area.

One prerequisite of sharp increases in productivity (or, in the extreme, the application of automation) is a more uniform raw material input. Historically, lumber mills have been designed and constructed to process a wide range of log diameters and lengths in addition to a variety of species. But equipment able to process the largest possible log sizes and lengths will be ineffective and under-utilized when processing small logs. Engineers involved in sawmill design indicate that mills designed to process logs of more uniform length, diameter, and species would be capable of producing lumber at substantial reductions in cost.

The cost of attaining full utilization at the processing level includes investments ranging from $150,000 to $250,000 for log barking and chipping equipment, $750,000 to $1,000,000 for a highly efficient 75–100 M-board-foot (per 8-hour shift) sawmill, and somewhere between $750,000 and $2,500,000 for an efficient veneer and plywood plant. A typical pulp mill requires a minimum intitial investment of not less than $12,000,000. A fully integrated firm may achieve relatively full utilization and relatively uniform input into each processing facility by having not only the processing methods illustrated above but several sawmills to process sawlogs of different sizes, two or more veneer lathes to peel logs of different sizes, and two or more pulping processes. All of these facilities are illustrated in the Longview, Washington operations of Weyerhaeuser Company.

If present trends continue, the lumber industry in the Douglas-fir Region by 1975 will consist of a few very large and fully integrated firms accounting for most of the Douglas-fir lumber production, plus a fringe of efficient medium-size lumber mills able to operate profitably on isolated sawlog supplies. A broad wholesale log market which once served the needs of independent mills has nearly disappeared as timber companies became integrated producers of various wood products and as other private holdings were liquidated. While data on free market log volumes are not as complete as might be desirable, the best information available indicates an 81 percent decline in sales volume over the period 1947–1960. According to the Pacific Northwest Loggers Association, Douglas-fir log sales in the Columbia River (Grays Harbor Area) declined from 531 to 103 million feet over this fourteen-year period.[4]

Some benefits for the economy of the Pacific Northwest would follow from the timber industry becoming concentrated in the hands of a few large integrated firms. Studies by the United States Forest Service indicate that timber resource management improves with the size of ownerships. At the same time hazards must be recognized. If a competitive free enterprise system is to operate effectively, economic control must be diffused rather than concentrated. If economic control becomes concentrated (oligopoly replacing competition) a free price system becomes replaced by "administered prices." Prices of log input and product output would no longer be determined by free market forces (supply and demand) but rather established through control over output. The normal result of oligopoly behavior is that output is reduced to obtain higher prices and profits than would be possible under free competition. Further, there are political hazards in a concentration of economic power in a few hands.

A multitude of solutions will be offered to compensate for the economic problems of the lumber industry. One solution that may be proposed during the decade of the 1960s to ameliorate the lumber industry's problems is that the small firms be given a farm-type subsidy. The subsidy may be direct in the form of a cash payment for smaller firms—or indirect as under the present federal timber set-aside legislation. The latter is designed to restrict competition for federal timber and place the small producer in a favored competitive position. A direct subsidy raises other difficult problems. Ultimately a direct subsidy requires production controls and thorough policing. Such measures are inconsistent with a competitive free enterprise system.

Included among the proposals of the Lumberman's Economic Survival Committee are two types of restrictions on free economic activity. The Committee has asked that a quota be imposed against lumber imports from Canada and that the presently very modest tariff on lumber be increased. These two measures represent another form of indirect subsidy, one that consumers would pay in the form of higher retail prices for lumber. They are short-run steps which would create more problems than they alleviate, while reducing pressure to become more efficient.

A POSITIVE PROPOSAL

This proposal calls for a change in federal timber sales policy, particularly in the Douglas-fir Region, to reestablish a vigorous and active wholesale log market. The proposal suggests that logs be sold not as a mixture of species and grades on the stump as at present, but rather from relatively uniform log decks at one or two concentration yards near the exit(s) of a national forest. Under this proposal timber would be logged by a contract logger operating under contract with the United States Forest Service rather than with the successful bidder. The contract would include delivery of logs to a dry yard. The logs would be processed through barkers appropriate to their size, then bucked and sorted into as many separate sorts as the market required. They would be sorted by species and grade. At this point buyers would make their needs known by the customary market procedures. A buyer having a small log sawmill would be able to buy relatively uniform bark-free logs to meet his specialized needs.

Price would be determined by the usual principles of supply and demand. A price would be established that would "clear the market." If the price established was too high, inventories would accumulate beyond normal needs. If excess inventories of a given species and grade were persistently below normal requirements, demand would be brought into balance with supply by means of a price increase. Thus the method of pricing would be similar to the method of pricing turkeys in a supermarket, for example. To avoid a favoritism charge, market prices should be reexamined and posted at established intervals, perhaps monthly or quarterly.

The supply of all species and grades of timber is determined initially by allowable-cut determinations, not by market conditions. Further, action to adjust supply in order to influence price must be specifically disavowed in principle and in practice in order to avoid the charge of monopolistic behavior. Concerning the so-called "sub-marginal" log, a particular species and grade should be logged, transported to the wholesale log market, and offered for sale when its long-run prospective market value exceeds the marginal cost of placing it in the market.

The demand for timber would be substantially altered under this proposal. At present a refusal price (the appraised price) is necessary. In practice, working arrangements occasionally become established which regard federal timber located near or intermixed with a private ownership as the exclusive preserve of the nearby operator. A concept of "my operating area" occasionally develops where other potential bidders respect the claim of an operator to obtain the timber within his "normal sphere of operations." Thus the market demand is often limited to a single buyer, and the timber is sold at appraised price. The record contains sufficient documentation of this practice to require that refusal price be established via the appraisal procedure.

The wholesale log market would eliminate the limitation in demand resulting from the fixed location of stumpage in particular spheres of operation. With prices posted periodically that reflect a balance between supply and demand, the very intricate and hazardous appraisal procedure would become unnecessary.

The proposal offered in this document is designed to strengthen the position of independent firms in the heart of the lumber industry by creating conditions under which non-integrated firms may operate more efficiently. The proposal is thus intended to maintain the industry as a competitive industry within the free enterprise context and avoid the political necessity of a subsidy in either the direct or indirect form. As of 1953, approximately 57 percent of all remaining sawtimber in the Douglas-fir Region was in federal, state, county, or municipal government ownership. Privately owned timberland in the Douglas-fir Region is becoming concentrated in the ownership of a few firms. In 1960 the largest private timberland owner held 16.1 percent of all privately owned commercial forest land. The eight largest private timberland owners in the Region held 34.1 percent of all privately owned commercial forest land. As recently as 1953 the eight largest firms owned only 27.4 percent of all such land.[5] Since most of the remaining timber is in government ownership and private timberland is becoming increasingly concentrated, a wholesale log market that is reliable must rest on publicly owned timber, principally that of the federal government.

Under present federal timber sales procedures, bidders compete for the timber standing on an identified piece of ground. The timber will invariably consist of a mixture of species and grades. Old-growth timber particularly will have a wide range of species and grade. The successful bidder, therefore, obtains a "hodge podge" of logs for processing. He will either log the timber with his own crews and equipment or he will contract with a third party to do the logging. All logs in the sale may be delivered to a single milling facility which may be either a veneer plant or a lumber mill. Alternatively, the buyer may separate logs into two or more sorts to be processed by different facilities. Interviews indicate that two or three sorts are common but some buyers made no separation while a few made as many as seven sorts. The buyer who makes no separation but rather processes all of his logs in his own mill would likely be a high cost producer and an ineffective competitor. It is a truism in production management that the flexibility needed to process a wide variety of input is attainable only at increasing cost. Making many sorts in the woods often at restricted landings is a costly procedure. Small quantities of minor species and sizes involve expensive sorts. Trading arrangements made with other processors to handle small quantities involve high unit trading cost.

The essential difference between this proposal and the existing method of selling federal timber is that a given buyer would be able to purchase bark-free logs, highly uniform in species and quality, in contrast with the present procedure in which he, as a successful bidder, obtains a hodge-podge of timber species and grades. By obtaining a more uniform log input, an independent mill would achieve one of the great advantages of the fully integrated firm without the necessity of being fully integrated and incurring the very high capital cost. This proposal offers the possibility of a high degree of efficiency through regional integration of independent firms in a reinvigorated free market mechanism, rather than integration within the context of a large single ownership.

Many sawmills still operating in the Douglas-fir Region, and most of the small sawmills, still do not have barkers and chippers. Under present conditions many operators feel that the volume of production is too small and uncertain to merit large investments in barking and chipping equipment.

In certain areas there are milling facilities having far greater capacity than can be supplied by timber currently available. This proposal would not solve the problem of over-capacity. The over-capacity problem will continue to be present in certain areas regardless of which method is employed to sell federal timber.

However, certain log classes might well be increased in supply as a result of the adoption of this proposal. At present the quantities of a particular species and grade available as a result of separate stumpage sales are often so small that it is unprofitable to utilize such timber. Under this proposal odd species and grades would be concentrated at the wholesale log market with larger quantities of such odd species and grades available. Reliable markets would be expected to develop sooner than otherwise would be the case. For example, in the areas relatively distant from a pulp mill, second-growth thinning and old-growth prelogging have not advanced. The principal reason for this is that reliable markets are not yet developed for the small-diameter logs and some of the species produced in thinning and prelogging.

* * *

BENEFITS ANTICIPATED FROM ESTABLISHMENT OF WHOLESALE LOG MARKET

First in importance among the anticipated benefits would be that small, independent lumber operators would be able to establish efficient specialized milling operations enabling them to become more profitable than presently possible. Thus, they would achieve one of the benefits of the large integrated operation without the necessity of being integrated under single ownership. An operator owning a gang mill would be able to buy the type of logs best suited for gang mill processing. The operator having a small veneer lathe would be able to buy logs best suited for his operation. The fully integrated producer presently enjoys this advantage. He makes many sorts when his timber comes from federal sources or from timber owned by his company. Logs may be sent to the processing facilities yielding the greatest possible economic value.

Second, all mills buying logs through the wholesale log market would obtain bark-free logs without the necessity of a large investment in barking equipment. With bark-free logs even the small producers would be in a position to realize revenue from chip sales. The cost of chipping equipment is approximately $50,000. In addition to obtaining economic value from chip sales, additional benefits follow from sawing bark-free logs: for example, (1) saw maintenance costs will be reduced as rocks and grit embedded in bark are removed; (2) the probability of spotting spikes will be greater when bark has been removed; (3) milling efficiency will be improved because bark-free logs can be handled with greater speed; (4) mill clean-up costs will be reduced since bark will not accumulate on the mill floor; and (5) more profitable cuts can be made from a log because the sawyer can see "under the bark."

Third, utilization of the area's timber resources should be improved by a wholesale log market. All sawmills obtaining logs through the wholesale log market could in turn become suppliers of chips for pulp mills. The supply of such chips would be significantly increased. This would allow a shift of resources from the traditional waste burner to a positive economic value as a result of pulp and paper manufacture. A high community cost can similarly be avoided by the elimination of the air-polluting burners. Utilization will further be improved by stimulating such advanced timber management practices as thinning and prelogging. Such intensive practices would be encouraged by a very large increase in log volume from such sources and the probable development of reliable markets based upon such production.

Fourth, as small mills cease operations and large mills (sustained in part by timber ownership) are combined through the merger process, the number of independent firms competing for federal timber must be expected to decline. A continuation of this process into the 1960s places critical importance on the appraisal process which establishes a "refusal price" for federal timber. The appraisal process is laden with subjective judgment and controversy-spawning uncertainty. The establishment of an active wholesale log market based on vast supplies of federal timber would allow logs to be sold at prices determined by the interaction of supply and demand. The appraisal process could thus be avoided.

Fifth, capital required to purchase federal timber would be reduced. At present buyers of federal timber must make substantial cash deposits against future timber removal in order to safeguard the public interest. In addition, if road construction is required, large sums must be committed in advance of log production. Under the present proposal, the time period between log purchase and delivery to the mill for processing would be reduced to a matter of days rather than months. Further, road construction would be financed by the federal government. Construction would be by contract with private firms much as at present.

Sixth, conflict between the Forest Service and timber buyers would be substantially reduced. The present degree of discord may be accounted for in part by different objectives between the Forest Service and timber purchasers as they meet in the market for logs. Forest Service personnel are primarily responsible for carrying out silvicultural operations and timber harvesting in keeping with multiple-use objectives. On the other hand, the first interest of the timber buyer is to obtain logs to supply his mill, to operate profitably, to keep his men employed, and to meet his customers' needs for wood products. These different objectives are not at all conducive to the harmonious execution of the timber sale contract. The result is an unnecessary and undesirable degree of conflict between the two parties.

A separation of timber management from log sales should eliminate some of the present discord. The Forest Service would be able to pursue advanced forest-management (including harvest) procedures as deemed desirable and appropriate by arranging with contract loggers to thin, prelog, clear-cut, etc. The wholesale log market could be managed by personnel trained in marketing and administration. As the allowable cut is delivered to the wholesale log market, the ensuing objective would be limited to selling logs to those who wish to buy logs.

DISADVANTAGES AND PROBLEMS ASSOCIATED WITH LOG MARKET PROPOSAL

First among the disadvantages of this proposal is the fact that it would extend an operation of the federal government from selling timber as stumpage to selling timber in log form. This disadvantage must be weighed against the advantages listed above. Further, if the wholesale log market enables the independent mill operator to become an effective competitor, and enables the nation to avoid political judgments leading to subsidies to producers as a solution to the industry's economic problems, then it is possible that the competitive free-enterprise system may actually be strengthened by the wholesale log market operation. Second, this proposal is not a solution to all of the economic problems of the lumber industry. Lumber marketing problems, for example, are very great and will persist independent of which method is used to sell federal timber. Third, Congress must be sufficiently convinced of the merits of the wholesale log market proposal to appropriate a revolving fund to meet road construction costs, contract logging costs, and wholesale log market expenses. All such expenses would be recouped as logs are sold. Fourth, while a wholesale log market would avoid the economic necessity of investing in a whole-log barker for those mills not presently having such equipment, it would also render unnecessary a barker now installed in a mill able to supply all of its log requirements from a wholesale log market. This fact would involve a capital loss, which may, however, be offset by new operating efficiency resulting from increasingly specialized log input. Fifth, a wholesale log market will not provide an assured supply of raw material to any one operator as a basis for obtaining loans and amortizing long-term investments in better utilization facilities. Nor does the present system meet this pressing problem. The proposed wholesale log market would be more conducive to eventual development of a futures market for logs which would at least extend the period of future certainty.

* * *

REFERENCES

1. West Coast Lumberman's Association. 1959–1960 Statistical Yearbook. Portland, Oregon. 9 pages.
2. Kendrick, John W. 1962. Productivity trends in the United States. Princeton, N.J.: Princeton Univ. Press. 161 pages. (The reader should be cautioned concerning the accuracy of the productivity gain quoted above. The author of the study admonishes that "such measures are not precision tools of analysis, but are subject to unknown and probably not inconsequential margins of error." Same reference. Page 17.

   The Kendrick study ends with 1954. The "weeding out" process taking place since the early 1950s may have substantially changed the productivity record for lumber.
3. United States Tariff Commission. 1963. Softwood lumber. Washington, D.C.: Tariff Commission, Publication No. 79, 59 pages.
4. The Pacific Northwest Loggers Association cautions that their listing of log sales is not complete since not all firms report to the Association.
5. Mead, Walter J. February 1964. Mergers and competition in the Douglas-fir lumber industry. U.S. Forest Service, Res. Paper PNW–9.

39

ECONOMICS OF PLYWOOD PRODUCTION IN THE SOUTHERN PINE REGION

Sam Guttenberg
Clyde Fasick

First published in *Forest Products Journal* 18(5):43–47. May 1968.

BIOGRAPHY

Mr. Fasick's research on southern forestry economics and marketing is widely known through his many writings. A southerner by birth, he began his higher education at the University of the South and continued at Duke University with a Master of Forestry degree and a Ph.D. in economics. He joined the Forest Service in 1956, at the Southern Forest Experiment Station, commencing a career not only in research but also involving teaching and participation in professional affairs, notably those of the Forest Products Research Society. His Forest Service stations have included Athens, Georgia, where he served incidentally on the University faculty; Asheville, North Carolina; and, most recently, St. Paul, Minnesota, where he is an assistant director of the Experiment Station.

Mr. Guttenberg's biographical sketch appears at Item 24, page 184.

EDITORS' SUMMARY

Abstract analysis confirms that southern pine plywood manufacture is most profitable when supplied with large, high-grade logs and that profit falls sharply on poorer logs. The industry, which has expanded fast on favorable log supplies of the West Gulf region, will soon face poorer supplies, lower profits, and greater incentives to integrate with lumber manufacture.

QUESTIONS TO CONSIDER

(1) How does one explain the sudden appearance and subsequent spectacular growth of an industry in a region? In the case of softwood plywood, the South always had an advantage in terms of wage rates, proximity to principal markets, and some other influential determinants of industry location. But before the 1960s, something was lacking. Can you explain the South's breakthrough in terms of developments in the dominant softwood-plywood region, the Douglas-fir region?

(2) The principle of comparative advantage states that each person or region does what it can do best—not necessarily what it can do better than others. Yet we find the rise of the South's softwood plywood industry being attributed to the larger profitability of plywood manufacture in the South than in the West. At first glance, this view of the matter appears to contradict the comparative-advantage principle. Explain why it is not a contradiction.

(3) How do you characterize the method used by Mr. Guttenberg and Mr. Fasick for explaining and predicting trends in development of the southern softwood plywood industry? What alternative approaches might be taken to analyzing industry trends? What are the advantages and disadvantages of these approaches, considering their sensitivity to questions of comparative advantage, resource trends, technological change, and so on?

* * *

The first southern pine plywood plant became operational in December 1963. Other mills soon followed, and by 1967 the capacity of the new industry reached 2.6 billion square feet, ⅜-inch basis. Plants that operated yearlong in 1966 achieved production equal to 80 percent of their rated capacity.

A combination of factors appears to be responsible for the rapid growth of plywood production in the South. First, the softwood plywood market is now predominantly for sheathing, underlayment, and plyform, and southern pine is eminently suitable for making these grades. Second, machine manufacturers have developed equipment for processing small-diameter bolts at high speed and low cost.[1] Third, more than 70 percent of the nation's plywood market is east of the 100th meridian. The South alone takes more than a fourth of the total output.[2] Southern mills thus enjoy a transportation advantage. Fourth, labor is a major item in plywood manufacture, and southern wage rates are about half those paid in the West. Southern plants with access to natural gas also enjoy cost benefits in drying veneer.

These factors give southern plywood manufacturers a competitive advantage. Presently they cannot even satisfy their own region's demand. What will happen to the margin of advantage as southern capacity and output continue to climb? What are the economics of production? What are the possibilities in various strategies for economically obtaining that portion of the timber supply suitable for veneering? How can plant managers allocate their supply of trees or portions of trees to the best use—plywood, lumber or pulp? To answer these questions data were gathered on production possibilities of southern pine bolts, on various forest resource situations, and on values for veneer and byproducts. Analysis of the data was by linear programming. Sensitivity analysis was relied on to draw broad inferences about the industry. Simulation provided insights on the relative effectiveness of alternative methods of procuring a timber supply, the item most likely to affect the development of the industry.

THE LINEAR PROGRAMMING MODELS

This basic model was developed to maximize the objective function:

$$\text{Net revenue} = \sum_{i=1}^{n} P_i Y_i - \sum_{j=1}^{m} C_j L_j + \sum_{j=1}^{m} P_j L_j$$

This objective function has 23 variables or activities, 2 inequality constraints, and 52 equality constraints. All variables are nonnegative.

The variables and coefficients are:

P_i = prices for veneers, cores, and chips
Y_i = yields of veneer by grade, and of cores and chips by weight
$i = (1, 2, \ldots, n)$ $n = 5 = 3$ veneer grades + chips + cores
C_j = costs for logs, by input class
L_j = log volume, by input class
$j = (1, 2, \ldots, m)$ $m = 9$ two-inch diameter groupings from 8 to 24 inches, inclusive. The diameter groupings were programmed to be converted to whatever individual log classes became available.
P_j = prices for logs of classes not peeled (resold)

Prices per thousand (M) square feet of B-, C-, and D-grade veneer were varied from a mean expected price to simulate relative price changes. Values of chips and cores were also varied. Costs for logs of each input class were varied for several timber-availability situations. Three basic log distributions were simulated: (1) A full range of logs from 8 to 24 inches; (2) 8- to 18-inch diameters only; and (3) 8- to 16-inch diameters with the average diameter much less than in the second distribution.

The prices for reselling logs of classes not peeled were varied under three major assumptions: (1) Resale price was equated to cost of log input, to simulate a situation where only those logs peeled are brought to the mill; (2) resale price was set less than initial log costs to simulate buying woods-run timber, bringing it to the mill, and absorbing the costs of handling and marketing logs not peeled; and (3) the basic model was extended to place veneer and lumber production into competition for timber. To maximize the joint returns from veneer and lumber production, the marginal dollar values from lumber production were the resale prices to the veneer plant.

The constraints to the basic model are on lathe time, proportion of D-grade veneer, log input distributions, volume of veneer by grade, and byproduct recovery for each log class. The two inequality constraints are: That total lathe time per hour $\leqq$ 49 minutes of net available processing time; that yield of grade-D veneer must be less than two-thirds the total output, giving the equation,

$$Y_D \leqq 0.67\,(Y_B + Y_C + Y_D) \text{ or}$$

$$0.67Y_B + 0.67Y_C - 0.33Y_D \geqq 0$$

The 52 equality constraints fix the log input volume available to the mill and the output of veneer and byproducts for each log class. The right-hand side of the constraint matrix was varied to assess the effects of management alternatives on optimum net revenues. Prices for veneer, byproducts, and excess log supplies were modified through multiple objective functions. In these functions the prices were multiplied by variable quantities of output, input, and alternatives for inputs not veneered.

INPUT DATA

Veneer and byproduct yields in Table 39–1 are from a recent study of southern pine production possibilities.[3] For practical purposes, the 32 bolt classes include the categories of veneer timber to be found in the South. Bolts were graded by the Forest Service rules for southern pine lumber logs.[4] Bolt density conforms to the Southern

Table 39–1 Unadjusted Yields per M bd. ft. Doyle of Southern Pine Veneer Bolts, by 2-inch Diameter Groups

	VENEER YIELD BY GRADE			BYPRODUCT YIELDS	
BOLT CLASS	B	C	D	Chips	Cores
	M Square Feet			*M Pounds*	
Grade 1, all densities					
18	0.97	6.31	2.43	1.82	0.56
20	1.33	5.33	2.86	1.62	0.43
22	1.85	4.43	2.96	1.46	0.34
24	2.37	3.73	3.00	1.36	0.27
Grade 2, dense grain					
10	—	4.50	5.28	5.26	3.03
12	0.52	4.13	5.68	3.48	1.70
14	0.62	3.80	5.86	2.62	1.09
16	0.59	3.36	5.94	2.12	0.76
18	0.49	2.91	6.31	1.82	0.56
20	0.38	2.38	6.76	1.62	0.43
22	0.09	1.85	7.30	1.46	0.34
Open grain					
10	—	4.40	5.38	5.26	3.03
12	0.21	4.23	5.88	3.48	1.70
14	0.51	3.39	6.37	2.62	1.09
16	0.59	2.57	6.73	2.12	0.76
18	0.49	1.94	7.28	1.82	0.56
20	0.19	1.43	7.90	1.62	0.43
22	—	0.83	8.40	1.46	0.34
Grade 3, dense grain					
8	—	—	12.23	13.36	6.81
10	—	0.51	9.64	6.35	3.03
12	—	0.39	8.02	3.86	1.70
14	—	1.23	6.94	2.69	1.09
16	—	1.43	6.09	2.03	0.76
18	—	1.64	5.50	1.64	0.56
20	—	1.85	5.01	1.38	0.43
Open grain					
8	—	—	12.23	10.55	6.81
10	—	0.30	9.84	5.66	3.03
12	—	0.35	8.56	3.85	1.70
14	—	0.41	7.76	2.97	1.09
16	—	0.38	7.14	2.44	0.76
18	—	0.36	6.79	2.13	0.56
20	—	0.34	6.52	1.92	0.43

Pine Inspection Bureau's rules[5] for dimension lumber—bolts with at least 6 rings per inch are termed dense; those with fewer than 6 rings are considered open-grained.

The data are for yields of dry 1/10-inch veneer from bolts peeled to a 5.2-inch core diameter. Yields are unadjusted, that is, no allowance is made for upgrading by plugging and patching. The chip yields are the green weight of roll-down waste, fishtails, and other forms of veneer clippings. Core weights are also for green material. While yields will vary with timber type and milling practice, the pattern is probably representative of southern pine.

In the analysis it was assumed that the principal physical constraints on the production of plywood are availability of bolts by class, the proportion of D veneer that can be utilized, and lathe time. In general, if bolts are continuously available, the limit on output is set by the lathe, and lathe time to peel an M bd. ft. Doyle varies with bolt size:[3]

Diameter Inches	Minutes	Diameter Inches	Minutes
8	30.34	18	5.32
10	15.63	20	4.71
12	10.27	22	4.28
14	7.69	24	3.96
16	6.23		

The time requirements apply to an operating hour of 49 minutes. Unavoidable delays such as for cleaning the chuck and touching up the knife take 11 minutes per hour. For convenience, all analyses are presented in terms of an operating hour.

All costs other than the price of timber delivered to the plant were considered fixed. Prices of bolts, veneer, and byproducts were varied to determine the effect of price changes on management decisions. Differences between the costs of timber and the values of dry veneer plus byproducts were maximized. The results are presented as a series of problems, beginning with the most atypical timber supply situations and proceeding to the typical.

WIDE AVAILABILITY OF TIMBER

There are a limited number of situations in the South where quality timber is available for a full range of sanded and unsanded items of plywood. Such a situation may occur where an industrial holding has been managed for several decades. The firm may either build a plywood plant itself or contract to furnish timber to a company proposing to build a plant. The hypothetical firm, analyzing its forest inventory data, management alternatives, and outside sources of timber, arrives at an allowable cut that can be sustained for at least 15 years—the time needed to depreciate fully investment in a plywood plant. The analysis indicates that the firm can offer the plywood plant up to 50 million board feet of timber annually. The timber is comprised of 32 classes of bolts that can be delivered at the average rates per hour following [see table at the top of page 314].

This delivery potential is far in excess of the needs of the plywood plant, which is designed to process 50 M bd. ft. Doyle in an 8-hour shift. As the timber-supplying firm is in position to market the excess, however, the plywood plant has first choice on bolt classes.

Regardless of bolt size and grade, timber is priced at $60 per M delivered to the

Bolt Class by Diameter, Grade, and Density (O = open grained, D = dense); M bd. ft. Doyle							
24-1-D&O	0.1	18-1-D&O	0.4	14-2-D	1.6	10-2-D	1.6
		18-2-D	0.3	14-2-O	1.6	10-2-O	1.6
22-1-D&O	0.2	18-2-O	0.2	14-3-D	1.6	10-3-D	1.6
22-2-D	0.2	18-3-D	0.2	14-3-O	1.6	10-3-O	1.6
22-2-O	0.1	18-3-O	0.1				
				12-2-D	1.6	8-3-D	3.2
20-1-D&O	0.3	16-2-D	0.5	12-2-O	1.6	8-3-O	3.2
20-2-D	0.3	16-2-O	0.5	12-3-D	1.6		
20-2-O	0.2	16-3-D	0.3	12-3-O	1.6		
20-3-D	0.1	16-3-O	0.3				
20-3-O	0.1						

plant. The volume of D veneer cannot exceed two-thirds of the total output. Given average dry-veneer prices of $24 per M square feet for B grade, $12 for C, and $9 for D, plus $2.50 per M pounds for chips and cores, which bolt classes or portions of classes should be taken?

The solution by linear programming is to process all the available 14-inch and larger bolts in grades 1 and 2 and 650 board feet of bolts in the 12-inch grade-2 dense class. This intake pattern would yield nearly 72 M square feet of veneer: 6 percent grade B, 36 percent C, and 58 percent D. Byproducts would be 16,700 pounds of chips and 6,500 pounds of cores. The margin above cost of the timber would come to $417.76 per hour, the net revenue. This margin would be reduced substantially by the bill for wages, other operating costs, and capital. If the cores are split-sawn, a thousand pounds will yield 150 board feet of 2 by 4s meeting the requirements of the Southern Pine Inspection Bureau's stud grade. At a net of $50 per M bd. ft. for studs, value per M pounds of cores rises to $7.50, and the margin of $417.76 increases to $440.26. This margin justified converting cores to studs in all subsequent calculations.

Modifications of the example were investigated. What should be done when veneer values fluctuate and what are the possibilities in sharply reducing the output of D veneer?

To represent strong markets, prices of $26 for B, $14 for C, and $11 for D were applied to the model; and for weak markets, $23, $11, and $8, respectively. These prices affected the returns but had no effect on choice of bolt classes to peel. Changes in veneer prices were evaluated in all subsequent appraisals of alternatives, and for practical purposes they had no effect on bolt intake patterns. Hence, all net revenues reported hereafter are based on average veneer prices.

When output of D veneer was held to 40 percent, the restrictions on timber supply limited production to 1 M bd. ft. of grade-1 bolts (which yield 30 percent D) plus only 800 board feet of the very best 10-inch bolts, that is, grade 2 dense. The lathe consequently remained idle for 65 percent of the time. Setting the D restriction to 50 percent of output did permit the lathe to operate to capacity. All of the 18-inch and larger grade-1 bolts were then processed, together with the dense 14-, 12-, and 10-inch bolts of grade 2. Net revenue shrank from $417.76 per hour to $331.39. An analysis extending to finished panels instead of dry veneer would probably not have shown a gain offsetting this difference in revenue. The 72 M square feet of the first solution shrank to 49 M. Even the yield of C and better veneer declined from 30 M square feet to 24.5 M. In effect, the restriction on D veneer forced a substitution of small, high-grade bolts for large ones of lower grade. And the extra lathe time required to process small bolts reduced total product output.

Some land owning firms with an allowable cut of the kind just described also operate a sawmill. The question is how to allocate timber among sawmill, plywood plant, and outside markets. A partial analysis was made by allocating all 20-inch and larger logs to the sawmill. Such logs permit manufacture of bill stock and other high value products, and also compensate for the disproportionate machining time required for the small timber rejected by the veneer plant.

If all 20-inch and larger material goes to the sawmill, what will the intake pattern for the veneer plant be? All the grade-2 and better bolts in the 14-, 16-, and 18-inch diameter groups will be peeled, plus 1.3 M of the 12-inch grade-2 dense bolts. In effect, 600 board feet of 12-inch bolts will be substituted for the 1.4 M of large logs released to the sawmill. The sawmill will also receive all grade-3 logs of any size.

This pattern will yield 65 M square feet of veneer: 6 percent B grade, 36 percent C, and 58 percent D. Chip recovery will total 16,700 pounds and cores 7,000 pounds. The net revenue to the veneer plant will be $421.49. The $18.77 reduction in net revenue resulting from the allocation of big timber to the sawmill is reasonably small.

An evaluation of the effect on joint returns was obtained by using the marginal values of logs to the sawmill. These values were available from a linear programming analysis of sawmill operations.[6] The method of marginal pricing permits log inputs to be allocated as they would by the forces of competition. The joint evaluation resulted in the following distribution of bolt classes:

BOLT CLASS (DIAMETER-GRADE-DENSITY)	TO VENEER	TO LUMBER
	M BD. FT. DOYLE	
24-1-D&O	0.1	
22-1-D&O	0.2	
22-2-D		0.2
22-2-O	0.1	
20-1-D&O		0.3
20-2-D		0.3
20-2-O	0.2	
20-3-D		0.1
20-3-O		0.1
18-1-D&O		0.4
18-2-D		0.3
18-2-O	0.2	
18-3-D		0.2
18-3-O		0.1
16-2-D	0.5	
16-3-D		0.3
14-2-D	1.6	
14-2-O	1.6	
14-3-D		1.6
12-2-D	1.4	0.3
12-2-O		1.6
12-3-D		1.6

The net revenue to the veneer plant under this allocation increased to $473.30 without affecting the returns to the sawmill. The classes of bolts not in the above listing were neither peeled nor sawn. To determine the possibilities from plugging and patching it was assumed that 5 percent of the D veneer could be raised to C and 10 percent of C to B. When this was done, net revenues rose 7 percent, to $507.28. The bolt input patterns were essentially unchanged, except that the 20-inch grade-1 bolts shifted from lumber to veneer.

Not many firms have access to the full range of bolt diameters. To simulate for a more representative group of industrial ownerships, an analysis was run by limiting log input to diameters 18 inches and under. The input patterns given above for 18-inch and smaller bolts remained unchanged. Net revenue was $433.49 for unadjusted yields, and $465.66 when veneer were plugged and patched.

In these analyses all bolts were considered straight and sound. In a feasibility study an estimate of timber supply should include logs that contain defects. Some classes of logs, notably those with sweep and crook, or with large knots or knot clusters on one or two faces, can be handled much more profitably in a sawmill than on a lathe.

TYPICAL TIMBER SUPPLY

Throughout the South there are numerous situations in which quality timber of large diameter is unavailable. There is, however, a sufficient source of C-grade veneer in the butt portions of small sawtimber trees. The veneering problems anticipated with butt cuts have not developed,[7] and the resource base for the plywood industry has thereby been considerably enlarged. Plywood plants designed to produce sheathing only could be located in these settings. Purchase of stumpage in a typical timbershed might generate an hourly pattern of bolt classes as follows:

BOLT CLASS (DIAMETER-GRADE-DENSITY)	M BD. FT. DOYLE	BOLT CLASS (DIAMETER-GRADE-DENSITY)	M BD. FT. DOYLE
16-2-D	0.1	14-2-D	0.2
16-2-O	0.1	14-2-O	0.2
16-3-D	0.1	14-3-D	0.2
16-3-O	0.1	14-3-O	0.2
12-2-D	0.5	10-2-D	0.4
12-2-O	0.5	10-2-O	0.4
12-3-D	0.5	10-3-D	0.4
12-3-O	0.5	10-3-O	0.4
		8-3-D&O	1.2
		6-3-D&O	0.2

What are the possibilities if the bolts in these classes can be purchased for $60 per M bd. ft. and the unused portions or classes sold for the same price? The intake pattern selected by the linear programming model is almost entirely based on bolt grade. All grade-2 bolts are peeled plus a portion of the dense 8- and 10-inch bolts of grade 3, which for veneering purposes are essentially the same as grade 2. Returns per hour over timber costs are $266.75. The net revenue includes the gain from raising 5 percent of D-grade veneer to C. Hourly costs of a plant designed to produce sheathing only might be about $250.[8] The net revenues would increase somewhat if the gain from the manufacturing of sheathing were added, but even then the overall margin would be small. Furthermore, it is very doubtful that the unused bolts, which are essentially of low grade, could be sold for the price at which they were purchased.

Relaxing the assumptions permits examination of more feasible alternatives for obtaining a log supply. For example, what would be the outcome of buying only the optimized pattern of bolts at a premium price of $70 per M? For this assumption, the net revenue shrinks to $242.75, and the likelihood of being able to buy even on this basis is still not great. It would be realistic to purchase open-market timber at $60 and sell excess logs at $50 to a sawmill. Under this option, net revenue declines to $236.75 and 3 M of logs per hour must be resold. It will enhance the feasibility of plywood production to establish a sawmill to convert unpeelable logs to lumber and chips. All 8-inch timber can be profitably diverted to pulpwood, because in logs of this size 1 M bd. ft. Doyle yields more than 4 cords of pulpwood, currently worth about $70.

IMPLICATIONS FOR THE INDUSTRY

Transportation and other sources of favorable cost differentials currently provide a price umbrella for the southern producers. This margin favors further expansion of plant capacity. How large the industry will get depends partly on the national demand for softwood plywood and the degree of interregional competition, but the availability of suitable timber will be a major determinant.

Developments so far have been in areas with the most favorable timber supplies. The West Gulf region, which includes Louisiana, Texas, Arkansas, and Oklahoma, has nearly 60 percent of the southern industry's capacity in plants either operating or under construction. Decades of sustained forest management have provided the West Gulf with pine timber of a superior diameter distribution for industrial development. Forty-two percent of the South's total inventory of pine 15 inches and larger is found there. And half of the volume in trees of all sizes is on industrial and public ownerships comprising 30 percent of the forest land. In the rest of the South, which has three-fourths of the acreage, 70 percent of the timber volume is on nonindustrial ownerships—principally small tracts. It is understandable then to find the West Gulf dominant in softwood plywood manufacture. Even outside the West Gulf the new plants are largely located where the timber is well above average in diameter distribution and is also primarily from industrial holdings. While similar opportunities for plant locations still remain, they are not numerous.

The analysis suggests that, as the industry continues to expand, it will have to adapt to progressively more typical timber-supply situations. Among other things, timber will have to be obtained from a multiplicity of small ownerships. Average log diameter is apt to be about 12 inches, and the veneer is likely to be suitable only for sheathing grades of plywood. Hence expected profit margins will be less than those enjoyed by firms operating in above-average timber with minimal procurement problems. With plywood from typical timber limited to sheathing grades, opportunities for modifying output in response to market changes are also limited. In effect, the lack of flexibility in product alternatives and the need to compete with the lumber and pulp industries for timber may encourage prospective firms to combine plywood and lumber manufacture.

REFERENCES

1. Dorsett, Ronald. 1966. Central Rockies provide plywood manufacturing potential. Forest Products Jour. 16(12):27–30.
2. Sherman, D. F. 1967. Plywood weathers rough '66 yet scores notable gains. Forest Industry 94(1):42–45.

3. Guttenberg, Sam. 1967. Veneer yields from southern pine bolts. Forest Products Jour. 17(12):30–32.
4. U.S. Forest Service. 1962. Forest Service standard grading system for southern pine yard lumber logs. Southeastern Forest Exp. Sta., Asheville, N.C.
5. Southern Pine Inspection Bureau. 1963. 1963 standard grading rules for southern pine lumber. New Orleans, Louisiana.
6. Row, Clark, Clyde Fasick, and Sam Guttenberg. 1965. Improving sawmill profits through operations research. U.S. Forest Service Southern Forest Exp. Sta. Res. Paper SO–20, 26 pages.
7. Haskell, H. H., W. M. Bair, and William Donaldson. 1966. Progress and problems in the southern pine plywood industry. Forest Products Jour. 16(4):19–24.
8. Chiang, T. I. 1963. The feasibility of producing southern pine plywood in Georgia. Atlanta, Georgia: Georgia Institute of Technology.

40

CONDUCT IN THE PULP AND PAPER INDUSTRIES

George R. Armstrong

First published as Chapter III in *An Economic Study of New York's Pulp and Paper Industry.* State University of New York, College of Forestry. 1968.

BIOGRAPHY

Mr. Armstrong is Professor of Forestry Economics and Chairman of the Department of Managerial and Social Sciences in the State University of New York at Syracuse. Since his first appointment at Syracuse, in 1949, he has spent some 7 years off campus in research and consulting work and has produced several major writings, including one book (with J. A. Guthrie) on the western forest industry and one (excerpted here) on New York's pulp-paper industry. On campus, he developed one of the earliest undergraduate-graduate programs in forest-industry economics, centered around his own courses in the economics of the wood-using industry and the economics of pulp and paper.

EDITORS' SUMMARY

A model describing the different size classes of pulp and paper mills in terms of production cost, product demand and price, competitive retaliation, and product differentiation helps explain management's behavior—e.g., the leadership of large mills (firms) in raising prices and of small firms in lowering them; the greater failure rate of small than of large firms.

QUESTIONS TO CONSIDER

(1) Mr. Armstrong describes the long-run average-unit-cost function for a pulp and paper mill (cost related to mill capacity in tons per day) as down-sloping over a very wide range. What are some of the factors which produce these scale economies? What factors might reverse the cost trend beyond a certain point?

(2) How does the short-run average-unit-cost function differ from the long-run in respect to (a) the identity of the independent variable and (b) the shape of the function? Explain both differences.

(3) Why is the demand for "personal-use" paper (e.g., writing paper) more sensitive to price (page 324) than is the demand for "industrial-use" paper, such as building paper?

(4) Explain why the demand curve for a paper mill's product may have a positive slope.

(5) What puts the kink into a kinky demand curve?

(6) Intuitively, one would perhaps expect large firms, with their greater financial strength, to cut prices more readily than small firms. Mr. Armstrong describes a somewhat opposite situation. Explain in your own terms why a decline in demand may lead the smaller paper firm to cut prices, but the larger firm not.

* * *

Modern price theory provides the ingredients for describing the forces acting on the pulp, paper and board mills of New York State and elsewhere, and in interpreting the conduct of mill managements. Among them are the concepts of demand and cost, and models describing price and output determination under conditions of oligopoly. Each of these elements is considered separately in the sections that follow.

COST

Writers who have treated the economics of the paper industry have been appropriately careful in making assumptions about the shape of the cost curve of the typical mill. Stevenson,[1] in 1940, gave several reasons why, for a given mill, unit costs might either increase or decrease as output rose. At the same time, he developed from empirical evidence gathered in two paper mills a long-run curve showing decreasing cost per unit of output. He felt that "paper mills . . . are subject to the law of increasing costs as the units get above 500 tons daily capacity." But he added, "The constantly improving technique of paper-making has tended toward lower costs and will probably affect the law of increasing costs to some extent at least."[2]

Guthrie[3] addressed himself to the question whether the short- and long-run cost curves for individual firms in the paper industry should be assumed to be U-shaped as in the economic literature. He pointed out that the newsprint industry might be expected to show decreasing costs in the long run because of scale economies in production, but did recognize certain limitations on the ease of aggregating raw material for pulp mills of constantly increasing size. He saw no clear indications that costs ever rose within the range of possible output to produce the traditional U-shaped short-run curve.

One important observation by Guthrie was that, "Newer mills have in general been somewhat larger than the old, reflecting some efficiency in greater size, which apparently was not offset by the increased cost of pulpwood assembly."[4] This, together with Stevenson's optimism about the opportunity to hold down costs through use of improved papermaking techniques, perhaps explains the proclivity of manufacturers in the pulp and paper industries to increase average mill capacity through expansion or new construction year after year (see Figure 40–1) apparently without

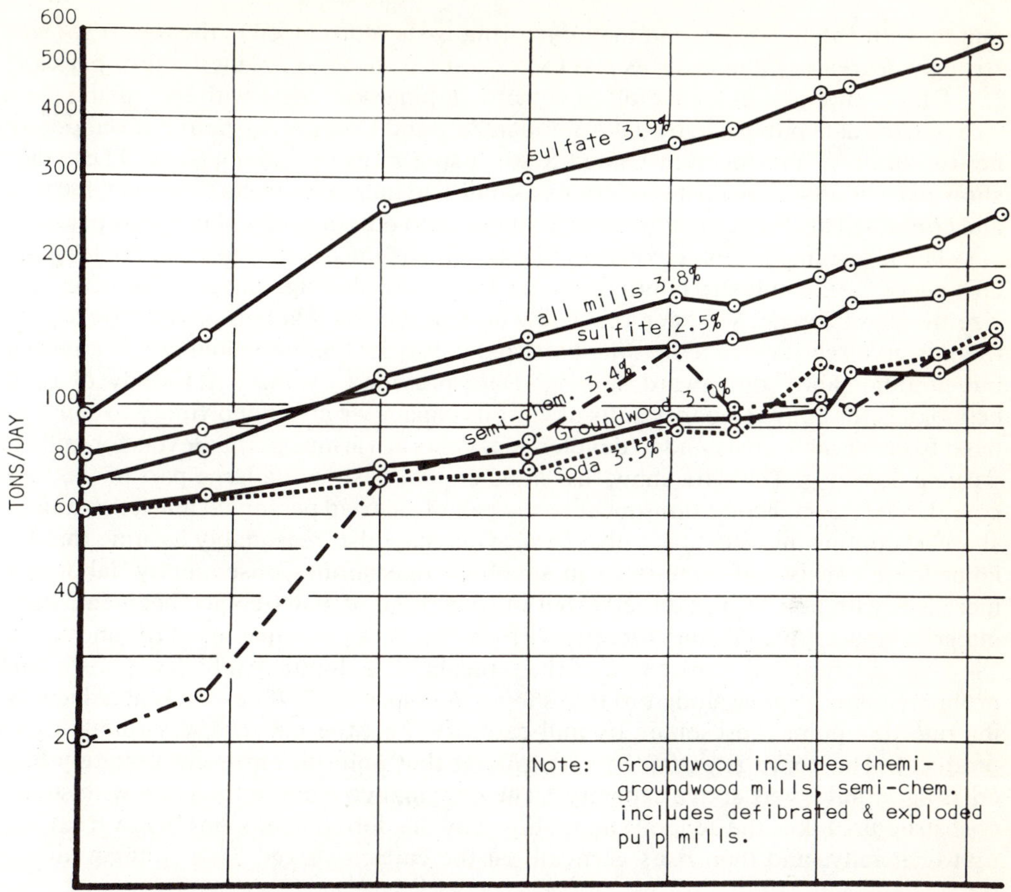

Figure 40–1. Average capacity of United States pulp mills, by type of mill for selected years, showing percent rate of increase in size, 1930–1961. (Source: U.S. Department of Agriculture, Forest Service, "Woodpulp Mills in the United States" various years.)

causing unit costs to rise. For that matter, the tendency for new pulp and paper mills to be larger than their forerunners suggests pretty strongly that unit costs (at or near capacity) decrease as mill size increases.

In 1950, Markham[5] made a careful analysis of the United States rayon industry, which was then consuming close to three percent of the Nation's wood pulp. Although the rayon mills employed a process quite different from that used by paper plants, there were important similarities. In both cases the amount of capital required for mill construction was large, rapid advances in a complex technology tended to restrict entry and therefore kept concentration (for given products) relatively high, and large and medium-size producers tended to increase mill size over time. Consequently, a study prepared by the staff of a large rayon producer and used by Markham to show a decline in unit costs both with increasing size of plant and with percent of installed capacity operated, and the average-unit-cost and marginal-cost curves derived by Markham,[6] provide some support for the notion that similar cost conditions exist in the paper industry.

More recently, in 1964, the FAO Pulp and Paper Advisory Group for Latin America[7] published a paper on the economics of pulp and paper manufacture specifically aimed to show the effects of economies of scale in the industry. Schedules of unit investment and direct unit manufacturing costs reproduced in the report showed strongly decreasing unit costs as plant size varied from 50 to 200 metric tons per day.

Other empirical evidence of downward-sloping unit costs with increasing plant size is in the data published by the American Pulp and Paper Association[8] on labor cost per ton of paper produced in United States paper mills of various sizes. These data show particularly large cost reductions as mill size increases from 25 to over 100 tons and continued decreases in unit labor costs even to output levels of 600 tons per day.

The foregoing data and expert opinions support the assumption that the long-run cost curve for the industry continues to be downward sloping; that as mills increase in size the point has not yet been reached where unit costs level off or rise. Moreover, there is no direct evidence that the unit-cost curve for the individual mill is anything but predominantly downward sloping. Data provided by the FAO study of scale economies for pulp and paper mills of various capacities give opportunity to separate fixed from variable costs and to develop unit costs at various output levels for mills of varying capacity. The data show that fixed capital costs are a large part of the total cost. Under normal conditions most of the pay of salaried personnel and a substantial share of contingency costs are also fixed. One may also reasonably assume that the labor force consists of three tours in a 24-hour operation. Consequently, labor cost increases with output, not directly, but in some kind of stairstep arrangement. Raw-material inputs typically vary directly with output. As a consequence of the above, the short-run average-unit-cost curve of the typical mill is dominated by fixed costs, and probably takes a shape similar to that shown in Figure 40–2. It descends at a decreasing rate to a point represented by mill capacity (i.e., the mill's maximum ability to produce). The curve is extended vertically at that point to represent infinitely high costs at output levels above capacity.[9] The marginal-cost curve for a mill with such a cost structure takes the general shape shown by the dotted line. It lies below total cost up to capacity, and then rises vertically at the capacity level. This configuration is

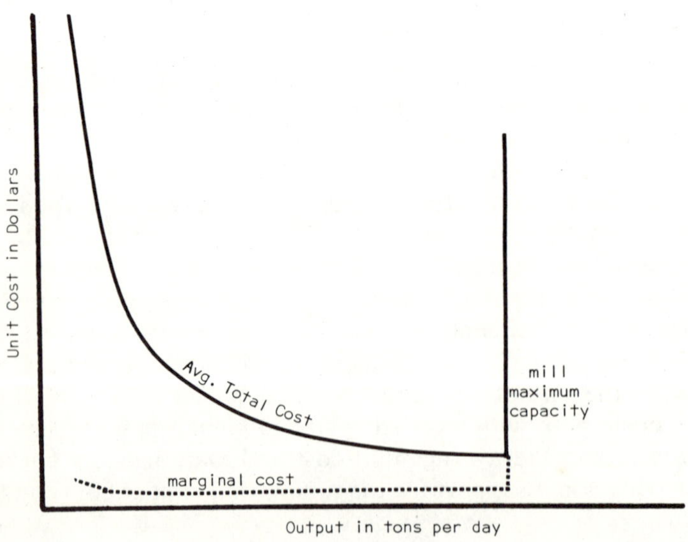

Figure 40–2. Average cost in the short run for a hypothetical integrated mill, showing decreasing cost with increasing output.

similar to that arrived at by Markham,[10] although in this study no evidence was found to justify a significant slope of the marginal cost curve, either up or down.

In line with the further assumption that unit costs tend to decrease as mill size increases, it is possible to generate a set of short-run cost curves representing mills of various sizes, as one might find them in the industry today (see Figure 40–3). Because such mills often employ different technologies, and because their managements differ as to capacity, these curves cannot be used to generate the long-run average-cost curve acceptable to theorists. In order to move from one point to another in this series, the mill owner would normally have to make adjustments in technology as well as size, and sometimes in the management force as well.

He would, in effect, move through a series of planes representing various levels of technical achievement while at the same time increasing mill size.

PRICE

A second important addition to a workable model illustrating profit opportunities of pulp and paper mills of various sizes is price itself. In a general way, pulp and paper approximates the industry structure of oligopoly, where many individual firms are large enough that price and output decisions by any one of them affect the kinds of decisions that the others make. Oligopoly theory stresses the development of price leadership under such conditions, especially in an economy which prohibits collusive pricing. Such leadership acts to curb strong price competition which has, in the past, plunged industries like pulp and paper into a dwindling spiral of prices, thus eroding profits and imperiling the normal profitability of even the strongest firms.

There is a long history of price leadership in the paper industry. Evidences of this form of pricing practice are noted by Guthrie[11] in his study of the newsprint industry, and by the same writer (along with other less widely used price-determining methods)

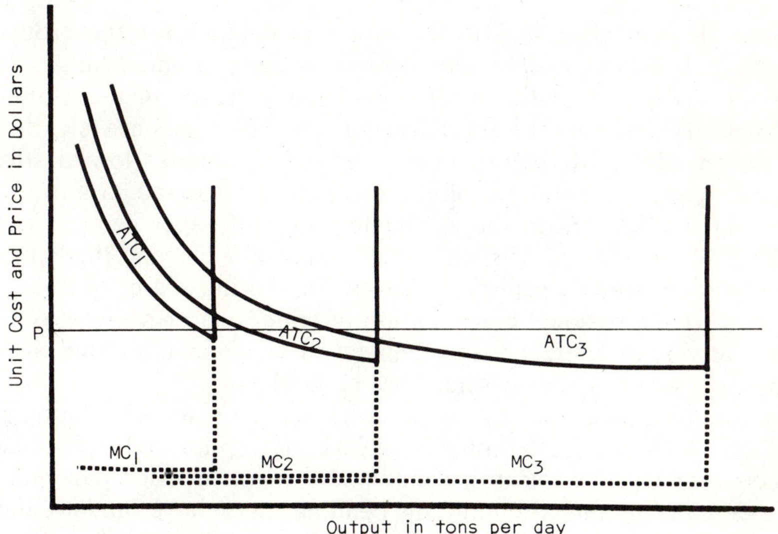

Figure 40–3. Probable configuration of short run cost curves for pulp and paper mills of increasing size.

in the paper-producing industry in general. Markham[12] found evidence of price leadership in the allied rayon industry, and recent investigations in the printing paper markets by Kolhonen[13] demonstrate industry acceptance of a form of price leadership.

To the economic theorist, price leadership connotes a going price for a commodity of a given type and quality which is generally set by one firm and adhered to by most, if not all, of the participants in a given market. In other words, the price is to all intents and purposes "given" by the leader. The price line is horizontal, and no matter what quantity of product is sold, the quoted price is the same.

In the paper industry, although there is price leadership in the sense that certain large firms typically initiate price changes, the list price is not necessarily closely followed. For one thing, discounts based on such variables as size of the purchase, loyalty of purchaser, or variations in the quality of the product (e.g., by virtue of damage or defect) tend to drop the market price below the list price. But another factor underlying deviations from the list price is the existence of a wide range of mill sizes and technical capabilities such that the competitive effects of price and output decisions of individual firms are not always of the same dimension, nor do they elicit the same retaliative action by competitors. Moreover, these same size differences call forth different price responses by the mills when faced with identical demand shifts. In effect, the leader's announced price becomes a ceiling price, rarely exceeded by other firms, but many times undercut.

The notion of price has been introduced into Figure 40–3 as a horizontal line representing the going market price (*P*). This will tend to be below the announced (or list) price, and although not introduced simultaneously by all sizes of firms (empirical evidence shows that smaller firms tend to be the first to cut prices and among the last to raise them), will be representative of a narrow band of prices at about that level.

DEMAND

The third element necessary to the model is demand for the product of the individual mill. It is well known that the demand for paper products in the Nation as a whole has been rising since the inception of the industry in the United States. Although empirical studies of the elasticity of demand for paper are few, Stevenson[14] notes that industrial use (derived demand) generally connotes low elasticity while "personal use" papers generally exhibit a more elastic demand. Guthrie[15] supports this notion, but implies that in the aggregate pulp and paper demand is inelastic, although the gradual invasion of such substitutes as plastics and the light metals in fields formerly dominated by paper promises an increasing degree of elasticity. Conceptually, then, rising national consumption of paper can be viewed as a series of points on a relatively inelastic industry demand curve which over time has tended to move to the right and to become somewhat more elastic.

But the rate of increase in consumption has varied markedly among paper and board products. With respect to both per-capita consumption and total consumption, such product groups as newsprint, groundwood and construction papers, and board products other than container, building and bending board, have shown—and promise to show—a low rate of growth compared to other grades.[16] So, although paper mills generally have faced rising national consumption of their product, the rate of advance for individual firms has varied substantially with the paper grade group produced.

Another factor which significantly affects the demand for the products of a single plant is the geographic concentration of the markets. In the nonspatial models of early economic theorists, all market transactions took place instantaneously at a single point in space. Cost of transportation and elapsed time were not considered. For the New York mill these are real and valid concerns and have a great deal to do with mill profitability. Since freight absorption to distant markets is a common practice in the industry, the rate of market growth at various distances from the mill is a prime variable. The differences in market growth rates derive primarily from two factors. One is the amount of competition from other producers in each major marketing area. The other is the rate of population growth and the rise in the index of industrial production. Both of these latter factors have been found to correlate well with the rate of change in demand for various paper products.

Given markets of various sizes (e.g., cities) at increasing distances from the mill, a price reduction (assuming no change in costs) will decrease the distance that a plant can profitably ship its product. Thus, it may be, in effect, "shut off" from certain markets and the quantities of product that those markets would absorb at the prevailing price may no longer be looked upon as part of the firm's new demand schedule. Thus, the effect of a price reduction may sometimes be to reduce the market area so that the amount demanded within the area falls below that realized at a higher price.

Even if we overlook the effect of distance, however, the demand schedule for the individual firm is also strongly affected by two other factors: the degree of product differentiation and the reactions of rivals. Consider first the demand schedule for the price leader, typically a large, strong firm with large and efficient mills. Once the market price (*P*) has been arrived at, the leader will reason that a premature price rise announced by him (assuming that cost increases reach various mills at various times and that industry demand has at least some elasticity) might not be followed by others. Empirical evidence in some paper fields (newsprint, printing papers and bag papers among them) suggests the reality of these fears, for announced price raises have indeed not always been followed when demand has dropped off unexpectedly for one reason or another. Thus, above the going price, given a static demand curve, the demand for his product would be somewhat more elastic than industry demand. The degree of inelasticity will be due to the ability of the firm to differentiate its product from that of others by nonprice means like service or brand name. Another logical assumption would be that competitors will follow any price reduction in order to protect themselves against loss of any significant part of their market share to a big producer.

The consequence of this thinking is the so-called kinked demand curve of economic theory which often provides, as shown in Figure 40–4, a wide margin for change in cost of production without changing either the optimum output (i.e., at which marginal cost and revenue intersect) or the going price.

But fundamental to further analysis is the notion that the "kinkiness" of the demand curve varies significantly for firms of various sizes. Consider first the extreme case of a firm (plant) so small that total cost per unit of product at capacity is high (i.e., at or close to the market price). Such a firm typically busies itself with cost-cutting and, partly to reduce expenditures and partly because of limited capabilities, normally does not try to differentiate its product on a nonprice basis. As long as its product is relatively undifferentiated it cannot consider a price rise at all, because its customers will defect to larger mills which are constantly affording "excess" capacity in the industry. And it will not drop its price if it is already operating at capacity, for a price

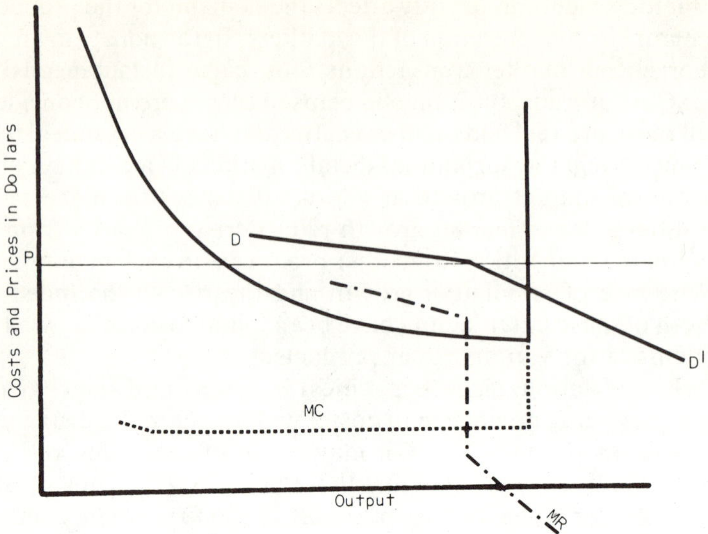

Figure 40–4. Prospective kinked demand for the product of an integrated pulp and paper mill, assuming product differentiation and competitive reprisal.

reduction will only diminish current revenues without a corresponding or offsetting cost benefit. Only if we postulate operation at less than capacity (as in Figure 40–5) may the mill consider price reduction, but in this extreme case the result will be only to reduce losses and to postpone a more serious decision. Thus, the least efficient mill we can conceive of (and by way of generalization, the smallest) will tend to have a perfectly horizontal demand curve to the left of the point marking current sales at the current market price, and a relatively elastic demand curve at lower prices but of which it cannot ordinarily advantage itself.

This is not the case for the medium-size mill. For such a mill it may be assumed that unit costs are moderately below the going price. The ability to raise price above the given level depends heavily upon the degree of differentiation of its product, which in turn is affected not by technical quality of the paper, which is assumed here to be constant, but by sales services and advertising, all of which loom very large as marketing determinants in the industry today. Since expenditures on these attributes are typically keyed to volume of sales and to profits, it is reasonable to assume that large, integrated, profitable firms will generally provide them in greater measure than smaller, less broadly integrated firms. In short, the medium-size mill will tend to have a relatively elastic but not perfectly elastic demand above the going market price because it will generally enjoy some nonprice differentiation of product. And the larger the mill, the less elastic the demand above the kink will tend to become (see Figure 40–5).

In the medium-size mill, the elasticity of demand below the going price will also be different from that of either smaller or larger mills. The medium mill can think in terms of reducing the sales price below the going rate without sustaining a loss, because its costs at high rates of output are somewhat lower than price. Unlike the large mill, it stands the chance of reducing sales price without causing widespread competitive reaction by the industry. This is to say, it will often be small enough that the sales it takes away from other mills will not be so substantial as to call forth immediate

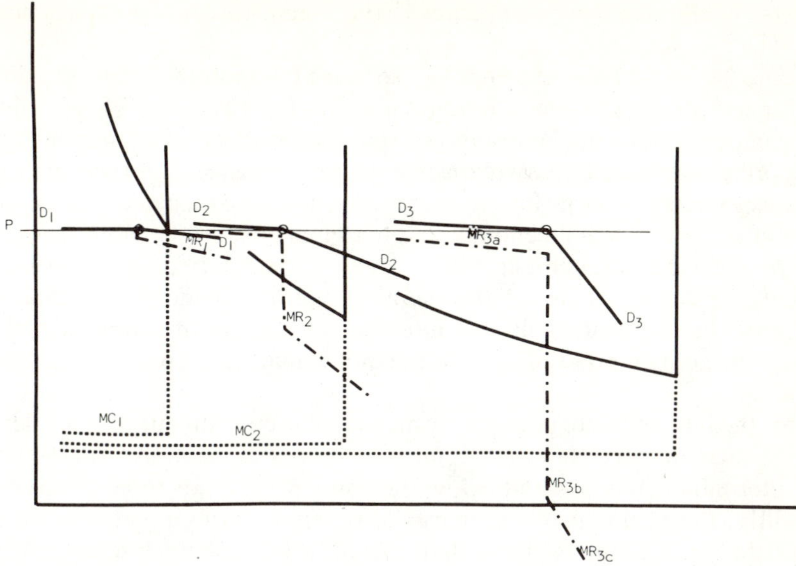

Figure 40–5. Demand considerations for integrated pulp and paper mills of various sizes demonstrating difference in "kinkiness."

reprisal. So the elasticity of demand below the going price will tend to be greater than that of large mills.

In the foregoing section on demand, reference has been made, at times, to the demand for the product of the mill, and at other times to the demand for the product of the firm. The relationship between the mill and the firm is an important one and calls for more than cursory attention.

For purposes of simplicity, the assumption has been made in this chapter that small mills are typically owned by small firms, medium mills by firms of medium competitive influence, and large mills by large firms. Thus, the shape of the demand curve for the mill is visualized as determined by the way the management of the firm looks at the market situation. In the real world, smaller mills are sometimes owned by large firms. In such cases their price and quantity decisions do not necessarily correspond with those shown in the model. Both price and quantity in such cases will be dominated by the aggregate of productive facilities available to the firm and by the aggregate sales that it seems to command. In times of declining demand, therefore, the appropriate decision may be to close the small mill, or to operate it at capacity, or at some intermediate level, depending on the manner in which the marginal cost curves for the firm's several mills are related. The simplifying assumption made here is not unrealistic, and for New York holds reasonably well to the actual situation. Most of the small mills until recently have been owned by small firms.

APPLYING THE MODEL

The kind of price-cost-demand relationship described above provides an explanation for such statements as the following: "We have never had so much competition as we have now—strong pulp and paper companies have been unable to raise prices to meet costs, while weaker companies have been forced to make some price cuts. . . ."[17]

We can assume that the reference to "so much competition" refers to diminishing demand for the product of the individual mill (perhaps because of recession, extra-industry competition, or new entrants) so that the quantity which any mill can sell at the going price is reduced below capacity, as in Figure 40–5. Given the kink in the demand curve which the large firm is likely to assume, it can logically seek to increase price only after costs rise high enough that marginal cost (MC_3) intersects the sloping marginal-revenue curve segment MR_{3a}. Even if demand for some reason moves substantially to the left, causing the plant to operate at much less than capacity, marginal cost and revenue will still intersect at a level of output which justifies maintaining the going market price, and the mill management will see no advantage in a price change.

But for the intermediate (weaker) firm, a similar curtailment of sales (let us say of comparable size in percent so that market shares remain the same) provides a different alternative. His demand below the going price tends to be relatively elastic. Consequently, the segment of the marginal-revenue curve developed from the demand curve below the kink is higher than that for the big plant. It tends to lie close to, or may even intersect, the vertical portion of the marginal-cost curve. Since profits decline with the leftward shift of the demand curve, the mill management normally reacts by paying attention to cost-cutting procedures. But if costs do drop, the marginal-cost and -revenue curves are just as likely to intersect at an output level to the right of the kink as they were before. The firm is therefore strongly and logically impelled to adopt a lower price (P_1) which will maximize its revenues at the indicated new output level. As more and more of the smaller firms take this course of action, their aggregate effect on market shares held by big firms is to attract enough trade that the large firms must retaliate. So the observed sequence of smaller (though not necessarily the smallest) firms cutting prices and big firms grumbling about the consequences and delaying, but finally following is substantiated by this simple kind of model.

In the real world, several types of activity make this model dynamic. New, large and more efficient technological units are constantly being added at the right-hand end of the diagram, and small mills at the left-hand end tend to drop out of the picture as their costs rise irretrievably above market price. In addition, aggregate costs for mills of all sizes tend to rise, particularly because of increases in costs of variable inputs.[18] These cost increases occur infrequently, but vary in intensity and come at odd intervals because of the broad mixture of labor, manufactured products and raw materials typically used in production, and because of differences in input prices due to location of the mill or size of purchase. Product prices tend eventually to rise and to offset these cost increases, but only infrequently and not always commensurately. The arrival in the industry of highly efficient new mills with a relatively low unit cost of production, the suspicion by price leaders that a price rise may not be followed by others—thus leading to a reduction of the leader's share of the market, and fear by such firms of an unanticipated recession or a cutback in demand following a newly announced price rise all militate against prompt reaction to increasing costs. Consequently, announced price rises may come only at intervals of several years. And even when they do come, they may only represent a "ceiling," with current market prices often falling substantially below the published level because of discounting and price competition by growing firms.

The result is for small mills to experience the familiar "cost-price squeeze" to a greater extent than do larger mills, the duration or cycle of the squeezes being principally the resultant of rate of entry and exit, rate of technological innovation and rate of shift of demand for the product.

In summary, the profit position of a paper mill at any moment in time is dominated by four effects. One is the height of the cost curve, which over protracted periods of time tends to rise as costs of inputs increase. The second is the level of market price of the product produced, which also tends to rise over long periods of time, but not in unison with costs. The third is level of demand at the going market price. And the fourth is the degree of elasticity of demand for the product of the firm both above and below the going market price.

Mills of various sizes differ in their ability to control or offset changes in these variables. The small mill considers itself to be extremely weak in its ability to raise the price. In fact, even the professed price leader can raise the "ceiling" only infrequently. The small mill also has little control over rising input prices. Over the longer run, labor costs, costs of stock capital goods, utilities and the like tend to rise independent of the small buyer and even local raw material prices are keyed to regional and national trends. The firm is forced generally to rely on internal controls (i.e., increasing its efficiency in the use of resources).

The arrival of new capacity in the industry, either due to introduction of new mills or the expansion of established ones, tends to cut back the amount demanded from each mill if the market price remains at a level and the size of the market does not increase commensurately. The degree of cutback depends upon the individual mill's ability to hold its customers in spite of the competition of other mills. This ability is due to nonprice product differentiation.

Although the degree of elasticity above and below the kink in the firm's demand curve is to some extent dependent on the actions of competitors and is therefore a function of size of the mill, it is also true that increasing product differentiation tends to change the elasticity of the demand schedule above the going price and below it. So product differentiation may be used to tilt the demand curve as well as to shift it from left to right.

Since so many of New York's mills are small, it is instructive to review the several means available to the small mill owner faced by the "cost-price squeeze" to escape from his predicament. There appear to be seven major means available. One, noted above, is to differentiate the product of the mill so effectively that control is maintained, or increased, over both the height and elasticity of the firm's demand curve. This also reduces the freight-absorption problem, because low-profit sales in distant markets may to some extent be traded for added sales in local markets where transport costs are low. Rich[19] touched on this solution in his characterization of the "successful" nonintegrated mill for which two of the required attributes which he stipulated were (1) the manufacture of specialty papers of a fairly technical nature; and (2) a continual upgrading of the total product mix. In effect, it adds to the latitude of the firm in product pricing and gives the small producer an opportunity to raise prices as well as to lower them. The drawback is that there are costs attached to effective product differentiation, for example in increased advertising, service, and quality control, and these costs erode the rising revenues accruing from larger local sales at chosen price levels.

A second alternative for escape is to continue to produce the same grade of paper but to expand the capacity of the mill through installation of new, modern equipment.

In effect, the mill shifts to a new aggregate-cost curve which lies below and to the right of the original.

A third alternative, good for the very short run, is to attempt cost-cutting measures which will temporarily increase the margin between unit costs and revenues. This measure is represented by a downward shift of the firm's cost curve.

A fourth alternative, and another short-run advantage, may be to reduce the price of the product in order to take advantage of demand elasticity below the prevailing price.

A fifth viable alternative is to shift to manufacture of a different paper grade; one which promises to yield a higher return in the foreseeable future. Here, the apparent costs of production at the newly attained level of capacity and the cost of making the transition have to be balanced off against a different market price which is typically higher, though it could be lower, than that of the original product of the mill. This kind of adjustment may be diagrammed as a shift of the mill to a new cost curve in a new market.

A sixth alternative is to sell out to another operator, usually because the next purchaser is optimistic as to his relative efficiency, feels that he has marketing advantages, has a highly integrated operation which boasts additional economies of scale (particularly in purchasing or distribution), or the like.

A seventh, and final, alternative is to close down the mill and get out of the business because alternative sources of investment appear to be superior. This entails the dismantling and sale of the mill or its parts, and conversion of assets into a more liquid form for reinvestment elsewhere.

* * *

REFERENCES

1. Stevenson, Louis T. 1940. The background and economics of American papermaking. New York: Harper and Brothers. Pages 112–122.
2. Same reference. Page 128.
3. Guthrie, John A. 1950. The economics of pulp and paper. Pullman, Washington: State College of Washington Press. Page 164.
4. Same reference. Page 164.
5. Markham, Jesse W. 1952. Competition in the rayon industry. Cambridge, Massachusetts: Harvard Univ. Press. 245 pages.
6. Same reference. Page 103.
7. FAO. December 1964. Pulp and Paper Advisory Group for Latin America, Economics of Pulp and Paper Manufacture under Average Latin American Conditions.
8. Annual report on labor and operating statistics. New York: American Paper and Pulp Association.
9. These two assumptions are made for reasons of simplicity. In the real mill it may be true that the slope is variable and may even rise over short portions of the curve, as when expanded output calls for a substantial increase in the labor force. And as maximum output is closely approached, the unit cost may rise gradually rather than instantly.
10. Markham, Jesse W. Work cited. Page 103.
11. Guthrie, John A. 1941. The newsprint paper industry, an economic analysis. Cambridge, Massachusetts: Harvard Univ. Press.
12. Markham, Jesse W. Work cited. Pages 69–70, 95–96.
13. Kolhonen, Jukka A. Treating structure and conduct in the printing paper markets of New York City. Unpublished Ph.D. thesis, State Univ. of New York College of Forestry, Syracuse, New York.
14. Stevenson, Louis T. Work cited. Pages 139–140.
15. Guthrie, John A. 1949. The economics of pulp and paper. Pullman, Washington: State College of Washington Press. Page 109.

16. U.S. Forest Service. 1965. Timber trends in the United States. Washington, D.C. Page 48.
17. Pulp and Paper. 1961. 35(5), page 5 (March 6).
18. There are occasional decreases, too, but over the long run they tend to be offset by the rising costs of most inputs.
19. Rich, Stuart V. 1961. Product policies of non-integrated New England paper companies. Res. Rep. No. 13 to the Federal Reserve Bank of Boston. Page 142.

41

CONVERSION SURPLUS AS A MEASURE OF VALUE

William A. Duerr
Sam Guttenberg

First published in *Forestry Chronicle* 44(4):24–27. August 1968.

BIOGRAPHY

Mr. Duerr's biographical sketch is given at Item 4, page 30.

Mr. Guttenberg's biographical sketch is given at Item 24, page 184.

EDITORS' SUMMARY

Conversion surplus is a measure of the value of goods-in-process, calculated as the difference between end-product sales value and variable costs of converting the goods to the end products. The calculation is useful in guiding the firm's wood-utilization decisions. It is not ordinarily useful in forecasting either the wood's market value or the firm's profits.

QUESTIONS TO CONSIDER

(1) A question commonly raised by skeptics concerning conversion surplus runs as follows: "Would you be willing to buy a log or tree for its conversion surplus?" (The answer: "Ordinarily not; I couldn't afford to pay so much.") "Well then, what sort of 'value' do you call that?" How would you reply to this line of questioning?

(2) Under what circumstances would conversion surplus represent a reasonable approximation of market price?

(3) Conversion surplus is said (page 333) to measure the excess of marginal revenue over marginal cost. One thinks of the production process as being continued, rationally, to the

point at which marginal revenue equals marginal cost. Does this mean that material utilization is rationally carried to the level at which conversion surplus falls to zero? Why or why not?

(4) Explain in your own terms, using a diagrammatic or numerical example if convenient, that the firm which is guided by conversion surplus in its wood utilization is being consistent with the "unity principle" (page 334). For which of the firm's enterprises is "unity" being respected in this case? May there be other enterprises which are unaffected?

(5) Why does the use of conversion surplus as a guide to material utilization not ensure a profit for the firm?

* * *

In the nearly twenty years since we introduced, or at least christened, conversion surplus,[1] the forestry profession has been putting the idea to work. At the same time, there has always been a certain amount of misunderstanding about how to use or derive conversion surplus. . . . To clear the record and to help familiarize the profession with a value measure which is serviceable in forestry, let us try an explanation.

MEANING OF CONVERSION SURPLUS

Conversion surplus is a measure of the value of goods-in-process, including raw materials. It is designed to answer the specific question: *What difference will it make financially if the producer converts these goods into end products?*

The conversion surplus of any good-in-process is calculated in three steps: First, the value of some end product into which the good may be converted is forecast. The end product in question is that form of the good in which it leaves the hands of the producing firm. Second, the variable cost of converting the good into the end product is forecast. Here variable cost is any expense which the producer will incur if he converts the good but will not incur if he doesn't convert it, other things remaining the same. Third, variable conversion cost is subtracted from end-product value; the remainder is conversion surplus.

If the conversion surplus of a good-in-process is, say, $10, then the answer to the production question is that the producer stands to gain $10 more by converting the good than by discarding it. If, on the other hand, conversion surplus is negative $5, then the production question is answered in this vein: that to convert the good will entail a $5 loss.

As an illustration, consider a *sawlog* lying in the woods. The log is a good-in-process, and the question concerns whether the lumber company that owns the log will send it to the mill or abandon it: What difference will this make financially? The estimated variable cost of converting the log into marketed lumber is $14. That is to say, if the log is hauled and milled and the lumber sold, the firm's total outlays will be $14 more than if the log is left untouched—other things equal. If the prospective sales value of the end product—lumber—is any amount greater than $14, the log has a positive conversion surplus which expresses how much the firm stands to gain by converting the log to lumber instead of leaving it to decay.

Thus conversion surplus is a measure of the excess, if any, of marginal revenue over marginal cost for a given unit of production. A positive conversion surplus identifies the opportunity to cover direct outlays and estimates how much revenue will be left over to apply toward the payment of fixed costs and, conceivably, profits. A negative conversion surplus measures the extent to which revenue promises to fall short of direct outlays and thus subtract from profit or add to loss.

USE OF CONVERSION SURPLUS

There are many questions other than the one for which it was designed, that conversion surplus will *not* answer. One, for instance, is the sales value of the log: How much would a buyer be willing to pay for it? A buyer might well expect the log to carry a share of his fixed costs and hoped-for profits of conversion, and in this event he would not be willing to offer the seller as high a price as the log's conversion surplus: An allowance for the fixed costs and profits would first have to be deducted. Again, there is the question whether, if the owner gathers up the log and puts it through his sawmill, he will maximize his net financial return. The conversion surplus of this log based on lumber is not enough data for answering the question: The producer may have alternative outlets for the log that are financially superior to lumber. Or perhaps his processing capacity is limited and he has alternative logs that are superior to this one. Still another question that conversion surplus will not answer is whether the producer will make a profit. Even if he arrives at that combination of logs and end products which will absolutely maximize the total conversion surplus of logs for the year, he will make no profit if the total surplus is less than fixed costs of conversion.

It does not follow that because conversion surplus is designed for a highly specific question, its usefulness therefore is highly limited. On the contrary, as a measure of the difference between marginal revenue and marginal cost, conversion surplus finds use in all problems where marginal analysis is applicable: where the producer's aim is to maximize net financial return per unit of time and his choices are an ordered series of resource combinations whose costs and revenues are single-valued.

Since conversion surplus takes account of all those costs and revenues to which materials will become subject as they move through a given path to the end of the firm's production line, it follows that this measure of value is notably useful for the vertically integrated firm. Conversion surplus recognizes the interactions among the enterprises of such a firm and gives answers to the production question that are consistent with the "principle of unity": the principle that the financially motivated producer seeks high net revenue, not necessarily from any particular enterprise, but from all enterprises together. Thus the timber-owning lumber company, under this principle, would wish to increase its woods activity only up to the point where an extra dollar of net outlay in the woods resulted in at least an extra dollar of current net revenue in the other enterprises of the business. For example, the company would move a log out of the woods only if, for every extra dollar of cost involved therein, there were at least an extra dollar of revenue in prospect. This is the original proposition about the log which was taken earlier for illustration: that it must have positive conversion surplus to be worth moving.

EXAMPLE: STANDING TIMBER

The conversion surplus of a *tree* or of a *stand of timber* is a useful measure for certain purposes. The details of its calculation depend on what the timber owner's end product is. If he sells stumpage, then the conversion surplus of the timber is the sales value of the stumpage minus the variable expense of sale. Where there is no such expense, conversion surplus approximates "stumpage value" in the ordinary sense: something which is net of allowances for fixed as well as variable cost and profit in further conversion. But note that the further conversion, in this case, is not something which will be performed by the timber-owning firm: The buyer of this firm's product will perform it. For this buyer, the cost item in question is fixed, but for the seller, the item is revenue, and it is variable—i.e., marginal.

If the timber owner sells, not stumpage, but logs, then the prospective sales value of the logs is the starting point, and the total variable cost of producing and selling the logs is subtracted therefrom, to derive the timber's conversion surplus. And so with any end product as far forward as the firm's integration extends: Its sales value is the starting point, and the resulting timber conversion surplus, being net of only the firm's prospective variable costs of conversion, contains an allowance or counterpart, such as it may be, for the firm's prospective fixed costs and profits in all its enterprises: the woods enterprise and the others extending forward from the woods. The higher the value of the conversion surplus in a given case, the greater the likelihood that it will contain, not only a contribution to fixed cost, but a contribution to profit as well.

Take as an example a single tree[2] in the forest owned by a manufacturer of plywood, dimension, and pulp chips. The sales value of these end products which the tree will produce is estimated at—		$364.50
Estimates are also made of the variable costs (mostly labor) of converting the tree to end products. These variable costs are—		
For felling the tree:	$ 0.62	
For making the stem into as many logs as will have conversion surplus larger than their bucking cost, thus maximizing the conversion surplus of the tree:	$ 2.50	
For yarding these logs:	$ 15.69	
For trucking the logs to the mill, including loading and unloading:	$ 20.97	
For processing the logs into end products at the mill:	$126.00	
For selling the end products:	$ 1.40	
So that the total variable cost of conversion is—		$167.18
AND THE TREE'S CONVERSION SURPLUS IS—		$197.32
This conversion surplus value is markedly larger than "stumpage value" as ordinarily estimated. To derive stumpage value, normally one would subtract—		
First, an arbitrary fraction of the firm's prospective fixed costs of felling ($0.12), bucking ($0.24), yarding ($2.49), trucking ($11.00), manufacturing ($42.00), and selling ($0.30) its products—i.e., a fraction "allocated" to this tree:	$ 56.15	
Second, an arbitrary fraction of an allowance for the firm's "profit and risk", again an allocation to the tree:	$ 30.98	
Thus the total subtraction is—		$ 87.13
And the tree's "stumpage value" is—		$110.19

This value is distinctive in being gross of overhead, profit, and risk allowances (such as are contained in conversion surplus) only for the timber-production enterprise. It is net of these allowances for all other enterprises of the firm.

QUESTIONS ANSWERED AND UNANSWERED

Now let's consider the possible usefulness of such a figure as $197.32 conversion surplus of the tree, keeping in mind all the while the corresponding figure of $110.19 stumpage value.

First, there is the question about the tree's merchantability: How much can the firm gain by taking the tree instead of leaving it? The answer is $197.32, not $110.19: Only if the tree is cut will it be able to bear any part of fixed cost or make any contribution toward profit. If the tree is not cut, then any charge for overhead and profit will have to be borne by other trees. Thus stumpage value is unhelpful in answering the question whether to utilize this tree, because stumpage value is derived with the use of costs which are fixed with reference to this decision and thus irrelevant.

Indeed, the use of stumpage value may actually stand in the way of optimum timber utilization. The stumpage value of the tree may be negative, and yet if its conversion surplus is greater than zero, the tree is merchantable so long as it forms part of an acceptable logging chance.

Or suppose for a moment that it is not the forest owner who will process the tree: Suppose the tree is part of a group being sold as stumpage. Here again, though the tree may have no positive stumpage value, the buyer can increase his net return by harvesting the tree if it has positive conversion surplus and by utilizing all portions of it which have positive conversion surplus.

The foregoing line of thought can be pursued in order to contrast pay-as-cut with lump-sum timber sales. If the buyer has purchased timber on a pay-as-cut basis, the payment is a variable cost of his production and thus a subtraction in computing conversion surplus. But if the purchase basis is a lump sum payment, this payment is a fixed cost, conversion surplus is that much higher, and timber utilization is correspondingly closer. Both buyer and seller can benefit from the extra quantity of material thus made economically available. And both parties can still retain whatever timber-volume symbols they like for purposes of record keeping, taxation, and analysis.

Second—and returning now to the case where the forest owner will process the tree—even if the tree has some conversion surplus and thus is merchantable, the question may still arise whether to cut the tree now or later: Does its value promise to increase fast enough so that the tree is best held as an investment? An answer is reached by predicting the tree's conversion surplus at the alternative future time of logging, then finding the present worth of this future value, discounted at a rate appropriate for the firm and the question. Whatever course maximizes the present value of conversion surplus is the course to follow: If, for example, the present worth of any future value is equal to or less than $197.32, then it is best to cut the tree now: The tree is "financially mature."[3]

The financial maturity calculation based upon conversion surplus value identifies the time for harvest which fulfills the financial aims of the entire firm in all its interacting enterprises: The financial maturity guide based on conversion surplus is consistent with the firm's whole profit maximization or loss minimization. This would not generally be true of a guide based on "stumpage value." For it to be true, the allowances for overhead, profit, and risk would have to be allocated to the tree in proportion to its conversion surplus. Actually, in most cases the allowances are allocated in proportion to the quantity of wood measured in the tree. The two methods of allocation produce the same result only where conversion surplus per unit quantity of wood is a constant. But conversion surplus per unit is precisely what varies from one tree to another or from one time to another, because of variation in tree quality or in labor or product markets.

Third, there are still other questions about the tree or the stand of trees which can be answered with the help of conversion surplus values, but generally not with stumpage values, for reasons already given. Examples are questions about the financially best forest rotation, the financially best timber stocking, the financially best spacing of trees, and the financially best programs of site, stand, or tree improvement. Conversion surplus serves well in these cases because it permits viewing the woods enterprise, not in isolation, but in the context of the firm.

Fourth, there is at least one question about the tree or the stand of trees which can be answered with the help of stumpage value, but generally not with conversion surplus: How much might a buyer be willing to pay?

Fifth and last, there is a host of questions about the tree and the stand that cannot, in general, be answered with the help of either stumpage value or conversion surplus. Perhaps the most vital of these is how large the operator's profit or loss will be.

REFERENCES

1. Guttenberg, Sam, and William A. Duerr. November 1949. A guide to profitable tree utilization. Southern Forest Exp. Sta. Occasional Paper 114, New Orleans, Louisiana.
2. The tree in the example is a Douglas-fir, 40 inches d.b.h., 2,700 bd. ft. net volume by Scribner Dec. C rule. For the revenue and cost estimates, we are indebted to Karl Bergsvik, of the Portland, Oregon, Service Center, Bureau of Land Management, U.S. Dept. of the Interior.
3. To cut now may be the best course even if discounted value somewhat exceeds $197.32. This is because of "land-use cost." See Duerr, William A., and Neils B. Christiansen: Exercises in the managerial economics and forestry. Wm. C. Brown Co., 1964. Pages 71–77.

Part Five

MARKETING, TRADE, DEMAND FOR FOREST OUTPUT

42

FOREST PRODUCTION GOALS: A CRITICAL ANALYSIS

Henry J. Vaux
John A. Zivnuska

First published in *Land Economics* 28(4):318–327. November 1952.

BIOGRAPHY

Mr. Vaux, one of the founders of American forestry economics in the modern sense, is Professor in the College of Natural Resources, University of California, Berkeley. A member of the faculty for more than 25 years, he was for 10 years Dean of the School of Forestry and Conservation. His earlier professional experience was with Oregon State University, the Forest Service Branch of Research, the Louisiana Agricultural Experiment Station, and the Crown Willamette Paper Company. He has been a leader in the Society of American Foresters, the American Forestry Association, and the Forest History Society. A knowledgeability in natural-resource affairs and a gift for orderly thinking, penetrating analysis, and succinct expression mark his many writings.

Mr. Zivnuska's biographical sketch appears at Item 2, page 16.

EDITORS' SUMMARY

To replace the production-goal ("potential-requirements") model currently in use by the Forest Service, a model is proposed which (1) accounts for production costs by accepting a less than perfectly elastic supply function; (2) recognizes intertemporal relations of supply and of demand and thus admits of a time series of goals; (3) allows for change in the relative price of wood products.

QUESTIONS TO CONSIDER

(1) Why are foresters so concerned about setting national goals? Do we have national goals for peanut butter? razor blades?

(2) Why, do you suppose, do foresters, when faced with the prospect of resource "shortage," so commonly think of greater production as the remedy—not lower or more efficient consumption?

(3) What is the matter with a timber-growth goal equal to the quantity of removals ("drain")? Is this same fault shared by a goal equal to the quantity demanded (at the cost for supplying it)? That is to say, what reason have we to suppose that demand (or supply either) represents a social norm? Refer again to Mr. Reidel's remarks on the subject (page 57) and to Mr. Worrell's (page 93).

(4) What is meant by the term *requirements?* How do "requirements" differ from "demand"? from "wants"? "needs"? What is the matter with a timber-growth goal set equal to wants?

(5) If past producer-consumer behavior provides us with poor norms, what are the opportunities in broad national planning in which forest (including timber) goals are defined in terms of socially rational resource use, irrespective of past producer-consumer habits?

(6) What is the relation of the proposal by Mr. Vaux and Mr. Zivnuska to that by Mr. Weyerhaeuser for a "target forest" (Item 13)?

(7) What do *you* think is the best approach to estimating a timber-growth goal?

* * *

"To furnish a continuous supply of timber for the use and necessities of citizens of the United States"[1] became an objective of federal policy in 1897. Since then, one of the central problems of a public forest policy has been to determine how large a supply of timber should be grown to meet these necessities. Although it has not ordinarily been identified as such, the emergence of this problem represented one of the earliest excursions of government into the economics of production goals, a function which has since been extended to much broader fields.

The development of production goals as guides to public forest policy is thus of peculiar interest because it involved a pioneer effort on the part of government in an area of economic policy which has become of increasing political and economic importance. The purpose of this article is to review the development of production goal concepts by the Forest Service, to criticize these developments from the standpoint of the logical function of such goals in economic planning, and to indicate the conditions which must be satisfied by a more complete model of a forest production goal. The analysis may be useful, not only in an evaluation of certain current issues in forest policy, but also in relation to comparable issues concerning other natural resources.

I

Early recognition of the problem of production goals in forestry was in general rather than in specific terms. Beginning in the 1870's a few farsighted individuals saw clearly that the then existing physical surplus of timber represented a relatively transient state of affairs. They understood that, if the opposite sort of economic maladjustment were to be avoided in the future, immediate effort was needed to place forest land in productive condition. The problem was to raise forest production[2] from practically zero to some substantial quantity. So long as no timber forestry was being practiced the question of what level of productivity should be achieved was largely academic. Accordingly, the early emphasis by leaders in forest conservation thought was on stopping forest devastation and on getting some forest production under way. A goal of "more production" was sufficiently precise for the needs of the time.

As progress was made in this direction, the need for a more accurately defined production goal arose. Rough data on long-run timber supply (in terms of current growth) were becoming available, and such quantitative information had to be evaluated in the light of the policy of furnishing a "supply of timber for the use and necessities of the citizens of the United States." One of the earliest quantitative bases for such evaluation involved the comparison of current growth with current drain. Thus, Secretary Meredith reported to the President in 1920 that "we are using timber four times as fast as we are growing it."[3] During the ensuing thirty years comparisons of current growth with current drain have frequently been used to diagnose the general health of the forest economy.[4] "Drain ratios" (ratio of current drain to current growth) in excess of unity have been cited as evidence that increased forest production was needed.

When rigidly applied, the drain ratio concept implies a production goal equal to the sum of current wood use and losses to forest enemies. The difficulties inherent in such rigid application of the drain ratio have long been recognized. For example, a substantial proportion of the growth in any stand takes place on trees of less than merchantable size. The volume of such growth cannot be directly compared with the volume of harvested trees of merchantable size. If board foot units of measure are used, growth on the small trees may be omitted entirely. If cubic foot units are used, we ignore the fact that 100 cubic feet of new wood distributed over 100 small trees is not the precise equivalent—silviculturally or economically—of 100 cubic feet of wood harvested from a single large tree.

A second difficulty with the drain ratio is that it indicates what is happening to the total volume of growing stock, not what is happening to the growth capabilities of that stock. Thus, wherever virgin stands are being cut at even moderate rates a large excess of drain over growth *must* appear initially. Reduction in the total volume of stock (excess of drain over growth) is a necessary condition to increasing the growth under these circumstances.

Finally, the drain ratio is ineffective as a diagnostic tool because it ignores the question of the appropriate *level* of yield. Growth and drain could conceivably be in balance with annual growth either at 1 billion board feet per year or at 100 billion. It is vitally important to have some idea of where within this range the balance should be struck. Indeed, this is the fundamental question at which production goal analyses are directed. But on this matter drain ratios throw no light whatsoever. Because of such problems as these there has been an increasing tendency to use the drain ratio as only one of a number of indicators of the forest situation.[5]

The McSweeney-McNary Act (1928) clarified considerably what Congress meant by its earlier policy of furnishing supplies for "the necessities of the citizens," and at the same time provided the means for more intensive study of the subject. Section 9 of the Act charged the Secretary of Agriculture with the responsibility of making "a comprehensive survey of the present and prospective requirements for timber and other forest products in the United States, and of timber supplies, including a determination of the present and potential productivity of forest lands therein, and of such other facts as may be necessary in the determination of ways and means to balance the timber budget of the United States."[6] Here was a clear statement of the goal of timber production policy, defined in terms of present and prospective balances between demand for timber and forest productivity.

Since 1928 the Forest Service has periodically published appraisals of the forest situation, based in part on the Forest Survey authorized by the Act, which analyze the long-run aspects of balancing the timber budget.[7] Each of these has presented figures for the drain ratio as of diagnostic value. But the Service has continually tried to develop a more critical type of analysis and more satisfactory long-run quantitative goals for forest production. In these reports increasing emphasis has been laid on long-run future "requirements"[8] as the basis for determining the growth objectives which are fundamental to program planning. The programs recommended in each case were those which, in the opinion of the Forest Service, would lead to future production of timber in the amount of estimated future "requirements," due allowance being made for a safety margin. Thus the Forest Service has regarded its requirements concept as the major ingredient of a forest production goal.

II

Before turning to an analysis of requirements, a clear distinction should be drawn between two different applications of the term which have appeared in practice. World War II gave substantial impetus to study by the Forest Service of current and short-run requirements for forest products.[9] The purposes and orientation of this sort of requirements analysis are very different from those of requirements studies underlying appraisals of the long-run forest situation.[10] During any brief period of years, and particularly in a war economy, the facilities available for converting stumpage into lumber and other products are sharply limited, and the total supply of standing timber is relatively fixed. Under pressure of wartime demand, and with the need to allocate these inelastic supplies to the most pressing uses, it is necessary to know something about the quantity of short-run demands for wood, and the urgency of such demands relative to the military program. The objective of wartime and other short-run requirements studies is thus to provide information on how to divide up the existing forest-products pie. In contrast, long-run requirements studies, if they are to be of use in determining production goals, must be oriented to the question of how big a pie we ought to bake for the future.

In order to avoid confusion on this point, this paper will follow the distinction made by Rettie[11] and will use the term "potential requirements" to identify the concept as used in production goal determination.

III

We now consider the adequacy of the potential requirements concept as the basic determinant of production goals in forestry. It should be emphasized that this discussion is limited to the conceptual level and does not consider the operational problems of implementing the model. Thus the *validity* of recent forest requirements estimates is not here in question, but their *significance* will be closely scrutinized.

Rettie and Hallauer[12] define "potential timber requirements" as "the quantity of timber products that might be used by consumers afforded reasonable latitude in choice of readily available materials, including timber products, in a national economy functioning at a high level of employment and output." They further stress that "these estimates are not forecasts of consumption" since actual future consumption may differ from requirements due to deficiencies in supplies.

Rettie, in a further expansion of the requirements concept, provides an alternative definition of "potential timber requirements" as "the quantity of timber and timber products which we would probably use at some future date under conditions of reasonably adequate supply."[13] He emphasizes that estimates of requirements differ from forecasts of consumption, since future consumption will be a function of actual future supply, and from statements of "needs," since such statements are subjective value judgments.

As has been discussed elsewhere,[14] due to the long period required for timber production and the peculiar nature of forest capital, prolonged periods of disequilibrium between the rates of production and demand may occur. In such a field of economic activity, the difference between a forecast of consumption and an estimate of potential requirements may be very large. The forecast of consumption is the amount that probably will be consumed with full allowance for the anticipated probable supply situation, while the estimate of potential requirements is the amount that probably would be consumed if supplies were adequate. The difference between the two estimates, then, is based on the difference between probable actual supplies and "adequate" supplies.

However, if "requirements" is to be a precise economic concept and is to be distinguished from the subjective statement of "needs," it is necessary to develop an objective measure of "adequacy" of supply. For the purposes of economic analysis the logical measure of "adequacy" is readily apparent. When the rate of production is in such a state of balance with the rate of consumption that the average costs of production equal the average selling price and the marginal costs of production equal the marginal selling price, supplies may be described as "adequate" since there is no economic incentive either to increase or to decrease production.

In short, the estimate of potential requirements, if it is to be economically significant, must be an estimate of the rate of production and consumption which would be characteristic of the long-run competitive equilibrium situation. While it may seem obvious when so stated that the goal of economic policy in production should be the equilibrium situation, the growth goal analyses which have been made to date give no explicit recognition to the implications of this fact for such work. This is surprising in view of the fact that the legislative charter for production goal analysis which was quoted above clearly gives equal coordinate emphasis to the influence of *both* "prospective requirements" and "potential productivity."

IV

While this identity of potential requirements and the production goal with the equilibrium situation appears to follow logically both from Rettie's and Hallauer's definitions and, more generally, from the language of the McSweeney-McNary Act, the actual pattern of the Forest Service requirements analysis has been different. Although there are procedural differences in the handling of the several types of wood products studied and even instances of apparent inconsistencies with the basic premises of the study,[15] the economic model of the study can be abstracted fairly readily.

The concept of "reasonable latitude in choice of readily available materials, including timber products" as used in the Reappraisal is interpreted in terms of an historical price relationship. While this base period is not specifically defined, it appears to be the period between the two World Wars. The estimate of potential requirements which is developed is, in essence, an estimate of the amount of wood which would be consumed if prices of wood and other materials hold in the future about the same relative positions as were typical of the period between the two World Wars.

The economic model of this analysis is extremely simple. The attempt is to determine a single point, *X*, on the demand schedule, representing the quantity which would be purchased at a price, *Y*, equal (in purchasing power) to that of the base period, with all other prices held constant at their corresponding levels. The validity of the results of such an analysis depends solely on the accuracy with which this point is determined. The significance and usefulness of the results, however, depend also on the adequacy of this economic conception.

That this is an adequate and useful model for production goal analysis can be challenged on four main grounds. These objections to the model can be introduced by reference to Figure 42–1 in which *Y* equals the base period price of the commodity for which potential requirements are to be determined, all other prices are held constant at their base period levels, and the objective is to determine the quantity which would be required or demanded at this price. This quantity will be defined by the intersection of the demand curve with the price line at level *Y*.

In Figure 42–1, *DD*, *D'D'* and *D"D"* represent only three of the infinite number of possible shapes and locations of the demand curve, and *X*, *X'* and *X"* represent corresponding possible positions of intersection between the demand curve and the price line. There is in advance no way of knowing that the desired point will fall at *X*

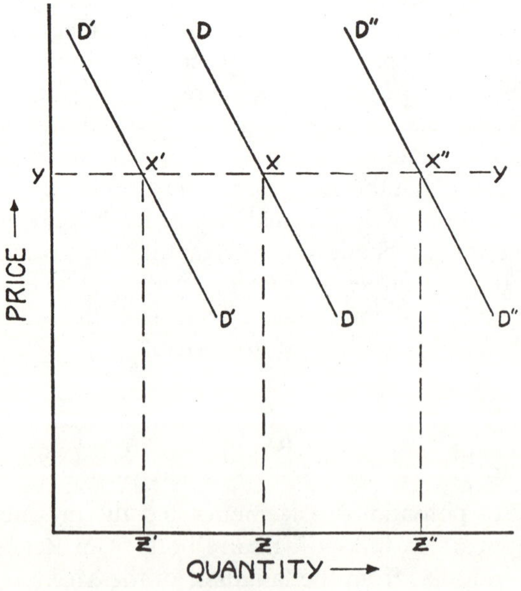

Figure 42–1.

rather than X' or X''—in fact, this is the question under investigation. Thus the assumption is necessarily involved that quantity Z, or Z', or Z'' (in other words, any quantity) can be produced at a cost equal to price Y, since if the cost were less than Y the quantity would be less than adequate to meet the demand at the cost of production while if the cost were more than Y the quantity would be more than adequate to meet the demand at the cost of production. In short, growth goal analysis based on this simple economic model of requirements involves the implicit assumption that the supply schedule can be represented by a horizontal straight line at a level corresponding to the selling price in the base period.

In the case of timber perhaps the most obvious objection to this assumption is the level at which production costs are placed. During the base period used in timber requirements analysis the supply of timber consisted almost entirely of virgin or volunteer stands, thus making it highly unlikely that the quantity of standing timber reflected any high degree of adjustment to the equilibrium requirement of equality between rate of production and rate of consumption. From this it follows that equality between cost of production and selling price of timber in this period is improbable, and so there would appear to be no reason for setting future costs of production at this level.

A more fundamental and universal objection, however, lies in the fact that this assumption implies that cost of production is completely independent of the quantity of production. This is not in accord with our general knowledge of the shape of the supply function, and there is nothing in our present knowledge of the costs of timber production to lend support to the concept of constant costs of timber production, particularly in "a national economy functioning at a high level of employment and output."

The third objection to this model is that it is static rather than dynamic. For application, it requires arbitrary selection of some particular period of time at which the balance between long-run timber supply and demand will be achieved. In the Forest Service's Reappraisal for example, the year 2020 was chosen as the target date. But the model supplies us with no means of knowing whether or not this is the optimum period over which adjustment should be made. There is no apparent reason why either 1970 or 2070 might not be just as good a target date for reaching the production goal. And the intervening patterns of production and consumption, the program needed to achieve the goal, and even the magnitude of the goal itself will vary widely, depending on the target date selected.

In more general terms, the trouble is that the model fails to recognize that both demand for and supply of timber products are continuous functions running through an infinite time continuum. A fully adequate production goal must be in terms of a continuous and not necessarily static balance. Potential timber requirements will change from decade to decade in response to basic changes in the timber supply situation, and there is no reason for selecting the requirements of one particular year or period of years as a goal in preference to those of some other period. What is needed is a solution which will reflect these dynamic features of the problem.

A final objection to this economic model lies in the assumption that all other prices remain at the same relative levels as in the base period. In a progressive economy subject to rapid technological change this is highly improbable. This problem, of course, is common to any case of specific equilibrium analysis as opposed to general equilibrium analysis.

V

The statement of the first two of these deficiencies in the economic model in itself suggests the nature of a somewhat more complete model.

The first revision of the economic model for potential requirements analysis is, as shown in Figure 42–2, simply the familiar long-run equilibrium supply-demand situation. The objective of the analysis is now seen to be the determination of the point of intersection of the long-run demand and supply schedules. Instead of involving only the determination of a single point on the demand schedule at a predetermined price, the economic model requires the determination of both supply and demand schedules in sufficient detail to enable the simultaneous solution of both quantity and price of the potential requirements estimate through estimation of the point of meeting of the two schedules.

However, even in terms of specific equilibrium analysis, this model is still incomplete. Since the movement toward the equilibrium position requires time for its completion, the model for the analysis must include provision for secular effects. In the case of forest production in which changes in productive effort require decades to become fully effective this problem is especially significant.

The problem involved here lies in the fact that the final goal to be reached will be determined in part by the path which is followed in reaching the goal. This may be illustrated by further reference to the Forst Service Reappraisal project. In analyzing the measures necessary to reach the growth goal which was set, the Forest Service[16] concluded that even if a comprehensive program of forest production were started immediately, it would be necessary to reduce consumption below present levels for a period of years in order to reach the goal. This is due to the identity between the growing stock or productive factor and the stumpage or final product of forestry enterprises. Thus, in order to increase the volume of timber which is being produced, it is necessary to increase the volume of timber held as growing stock. Obviously, one means of increasing the volume in growing stock is to decrease the volume cut during

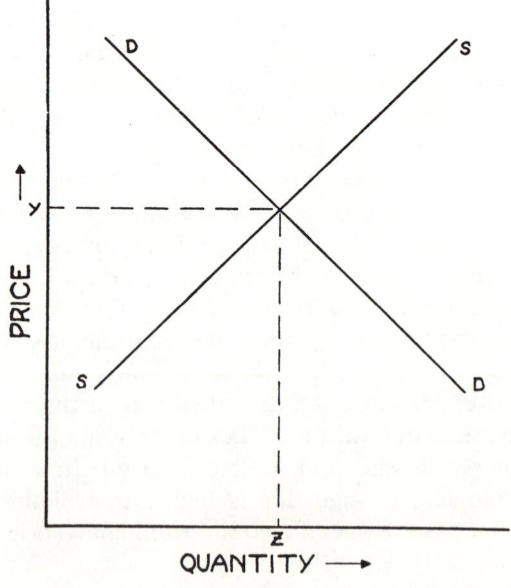

Figure 42–2.

the adjustment period. Necessarily, cut must be less than growth over any period in which growing stock is to be increased so that in time both growth and cut can be increased. Accordingly, if the growth goal is set at any level higher than the present rate of growth, then during the period of adjustment the average cut must be less than the growth goal will eventually allow. Whether or not cut must also be held below the current level of cut (which is higher than the current growth) will depend on the combined effect of the height of the eventual growth goal, the volume of the existing virgin timber reserve (which does not contribute to growth), and the effectiveness of alternative measures in increasing the growing stock.

The problem which this involves for potential requirements analysis is due to the fact that demand at any point in time is not uniquely determined but is closely related to past demand and the degree to which such past demand has been satisfied. We must think not only of potential requirements at some distant point in time at which the production adjustment will be complete but also of the entire time sequence of requirements—the amounts which would be purchased if supplies were available at the cost of production—throughout the adjustment period. To the extent to which meeting the growth goal makes it necessary to hold the cut below the level of this time sequence of requirements, timber will be in scarce supply and a scarcity price will obtain during the adjustment period. In a technologically active economy such scarcity prices can be expected to induce the development of alternate materials, modification of utilization techniques, and even changes in consumers' preferences which would not otherwise have occurred. Thus the effect of scarcity at any stage of the adjustment period will be to shift the demand curve of all future periods to the left to where it would have been in the absence of such scarcity.

For example, in Figure 42–3 let D' be the estimate of the demand curve with no allowance for the effects of intervening periods (since the extent of such scarcities cannot be estimated until the size of the growth goal is determined). Q' would then be the corresponding quantity for the growth goal. It may then be estimated that the scarcity periods in reaching this growth goal would shift the demand curve to D'', with Q'' as the corresponding growth goal. Now, however, the intervening shortages would be less than estimated for Q', so it appears that the demand curve would not be shifted as far to the left as D''. Through a series of such successive approximations the final position of the demand curve at D could be determined and the approximate growth goal Q set.

VI

In view of what has just been said of the dependence of demand in a given period on the consumption of other periods, we may logically ask, "Is the supply function in the model of Figure 42–2 subject to analogous intertemporal effects?" The answer of course is "Yes." The schedule of quantities of timber which, at some future target date, can be made available at various alternative levels of costs is related to the quantities which will be produced at all other dates between now and the target date. This is so because of the dual economic nature of standing timber—it is both capital plant and finished product. As a result, an increased (decreased) production of finished product during any period of time may lead to a reduction (accumulation) in the capital plant available for production in subsequent periods, and hence may affect costs and supply in such periods. The implications of this situation may be visualized in terms of an illustration.

Suppose that we wish to establish a timber growth goal for the year 2020, to be used as a guide to timber production programs. In conformity with the model of Figure 42–3, we attempt to forecast demand for 2020, taking account of secular demand influences in arriving at the desired function, *DD*. But to do this we must estimate consumption for each of a series of convenient time periods between now and 2020. These estimates will depend on the position of the demand and supply curves during each of the intervening time periods. Because of intertemporal supply and demand relationships, their positions will depend on the level of production in preceding periods.[17] In other words, an economic model for growth goal determination which takes account of relevant dynamic and secular effects must provide a solution of the "moving equilibrium" type in which the optimum level of production for any future date is estimated simultaneously with estimates of the optimum level of production at all other future dates. If this condition is not met, there is no guarantee that the estimated goal will increase net returns to society, much less maximize them.

A solution of the type just described eliminates the difficulty inherent in picking some particular future period as the target date for growth goal determination. It provides for a balancing of supply and demand at all future dates, and would yield goals appropriate for each and every future period. Obviously, the model would have to provide for some form of weighting, if only in recognition of the fact that contingencies which are very remote in time are less significant for public policy than those of the same order which are more immediate.

Derivation of a solution of this sort will be fraught with other difficulties both theoretical and practical. But these do not at present appear to be insurmountable. The problem is an improtant and challenging one for those who are concerned with guiding public forest policy as well as for those who are attempting to make economic analysis a more effective tool for policy evaluation.

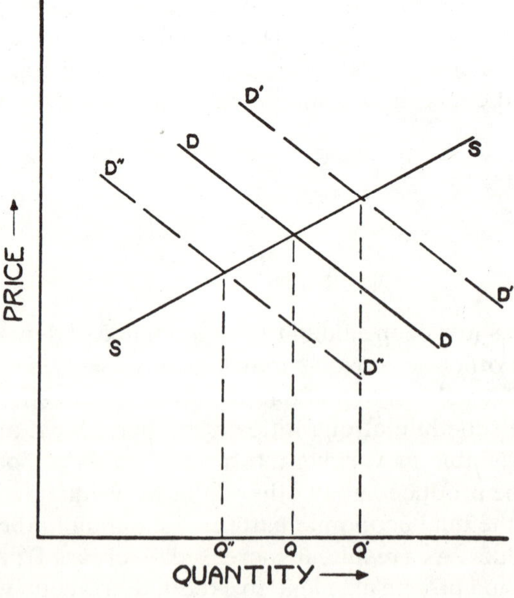

Figure 42–3.

VII

This discussion of potential requirements analysis has been developed within the specific equilibrium context. The objection that the Forest Service model assumed that all other prices remain at the same relative levels as in the base period remains to be considered. This objection could be met fully only by setting up the model in the context of general equilibrium analysis, a procedure which lies beyond the scope of this paper and of most potential requirements studies. Some recognition can be given to this problem, however, in developing the demand and supply schedules. Available evidence indicating significant trends in the relative level of prices of any of the close competitors could be developed, and suitable allowance made for the probable effects in determining the demand curve for the product under study. Allowance could similarly be made for any known trends in important cost items in developing the supply schedule.

VIII

The objection may be raised that the type of model for potential requirements analysis described here would involve so many technical difficulties on the operational level that the analysis would fall of its own weight. Certainly it must be admitted that even the rudimentary model used by the Forest Service resulted in many complexities operationally. However, it is possible that the difficulties involved in this more complete model may be exaggerated.

In any case, it is soundness rather than the simplicity of the analysis which is the more pertinent issue. It would be surprising if significant conclusions could be drawn for production policy from an analysis which ignores production costs. Only when the theoretical framework of analysis is fully understood is it possible to develop a well-balanced operational program of analysis and to evaluate adequately the results which are obtained.

NOTES AND REFERENCES

1. Act of June 4, 1897 (30 Stat. 35); section 475, title 16, U.S.C.
2. Throughout this paper "forest production" means growth of wood. It does not refer to conversion or utilization of standing timber.
3. U.S. Forest Service. 1920. Timber depletion, lumber prices, lumber exports, and concentration of timber ownership. Report on Senate Resolution 311. Washington, D.C.: U.S. Government Printing Office. Page 3.
4. E.g., in U.S. Forest Service, A national plan for American forestry, Senate Doc. 12, 73rd Cong., 1st Session. Washington: U.S. Government Printing Office, 1933; and in U.S. Forest Service, Forests and national prosperity, U.S.D.A. Miscellaneous Publication No. 688. Washington, D.C.: U.S. Government Printing Office, 1948.
5. However, the most recent nationwide drain ratios, showing drain of all classes of wood only slightly in excess of growth, have been interpreted by some groups as meaning that the forest production goal has been substantially achieved.
6. Section 9. Act of May 22, 1928 (45 Stat. 699), session 581 a-i, title 16, U.S.C.
7. U.S. Forest Service. 1940. A national plan for American forestry, work cited. R. E. Marsh and W. H. Gibbons, "Forest resource conservation," Yearbook of Agriculture, 1940. Washington, D.C.: U.S. Government Printing Office. U.S. Forest Service, Forests and national prosperity, work cited.
8. Terminology varies. Senate Document No. 12 used "requirements"; the 1940 Yearbook "possible

future consumption," and Miscellaneous Publication 688 "potential requirements." The basic concept, in each case, appears to be substantially the same.

9. U.S. Forest Service. 1943. Forestry in wartime, Report of the Chief of the Forest Service, 1942, and subsequent reports of the Chief. Washington, D.C.: U.S. Government Printing Office.
10. James C. Rettie. 1948. The purposes and basic concepts of timber consumption and requirements studies. Jour. of Forestry (April). Pages 237–242.
11. Same reference.
12. James C. Rettie and Frank J. Hallauer. 1946. Potential requirements for timber products in the United States, Report 2 from a Reappraisal of the Forest Situation, U.S. Forest Service. Washington, D.C.: Forest Service. Page 1.
13. Rettie. Work cited. Page 241.
14. John A. Zivnuska. 1949. Some aspects of the economic theory of forestry. Land Economics (May). Pages 165–172.
15. E.g., the Reappraisal analysis of construction plywood seems to abandon the basic definition of potential requirements in favor of analysis in terms of supply alone. See Rettie and Hallauer, work cited. Page 43.
16. U.S. Forest Service, Forests and national prosperity, work cited. Pages 40–45.
17. Also, where expectations influence the situation, supply and demand in a period will be influenced by expectations of market conditions in future periods.

43

LONG-TERM SUPPLY FORECASTS

Henry J. Vaux

First published in *Folia Forestalia* 101:28–34. 1971.

BIOGRAPHY

Mr. Vaux's biographical sketch appears at Item 42, page 341.

EDITORS' SUMMARY

To forecast the supply of timber or wood products (quantity related to price), it is necessary to (1) define the market area, (2) specify the time horizon, and (3) identify the influential factors in production. Such factors are production itself (being time-interrelated), timber inventory and growth rates, interest rate, forest area, and management programs. Simulation is a useful forecasting model.

QUESTIONS TO CONSIDER

(1) Empirical studies of timber supply fall into at least two categories. In one, which we may term *normative,* landowners are assumed to be rational: Government managers pursue the social optimum; private managers are profit or wealth maximizers (loss minimizers). *Positive* studies, on the other hand, assume only that quantity supplied is significantly related to independent variables; they examine managers so as to learn what the variables and relationships are. Into which category does Mr. Vaux's work fall?

(2) Suppose that one were interested in the relation of timber quantity supplied, not to timber price, but to the guiding rate of return. Would Mr. Vaux's models and methods be equally suitable? What modifications would be called for?

(3) What purposes may be served by a timber-supply forecast over a 50-year span? In the absence of a reliable forecast or means for obtaining one, are there alternative ways for serving these purposes? Are there ways that would be preferable even if a reliable forecast were at hand?

(4) Supply and demand are ordinarily regarded as independent. Mr. Vaux explains (pages 358–359) that timber supply may be a function of demand, operating through the quantity supplied. Study his analysis and restate his point in your own words. May timber demand be a function of supply, either directly or indirectly? What implications does such interdependence hold for the making and interpretation of empirical supply-demand analyses?

* * *

Long-term forecasts or projections of supply might be made in order to understand any one of a whole array of different sorts of economic problems. For example, the supply of standing timber might be forecasted for an area for dates forty, sixty, or more years in the future in order to determine whether or not additional land area should be used for timber growing. Or the supply for an area might be predicted for dates ten or twenty years in the future as one determinant of how much investment in timber processing facilities should be made. Or in an undeveloped forest area, the supply might be predicted for a five or ten year period as a guide to the rate of development of primary timber access routes. Thus, the phrase ''long-term forecasting'' embraces a good many types of problems. For each of these several types, the time horizon to be used, the selection of essential variables for study, and even the choice of methodology may be different, depending on what is appropriate for the particular problem at hand.

Thus, in a five to ten year forecast of supply for an undeveloped forest area needed to guide a program of road development, timber growth within the forecast period will be both relatively secondary in total magnitude and relatively little influenced by timber growing inputs. Within such a time period, the elasticity of supply in response to such inputs will be small. At the same time, the elasticity of supply will be much greater with respect to inputs of road construction and related transportation facilities, if they will make operable previously inaccessible portions of underdeveloped forest area.

In contrast to this problem stands the one of forecasting timber supply over a fifty-year period as a guide to timber growing policy. Over such a time span the supply response which can be obtained by physical development inputs is sharply limited by the finite size of the supply area, and indeed (except in underdeveloped areas) the elasticity of supply with respect to such inputs is negligible. But the supply response to timber growing inputs may be very substantial; it increases rapidly as the length of the forecast period approaches the length of rotation.

The preceding illustrations have been concerned with forecasting supplies of standing timber. Parallel problems in the supply of manufactured forest products could be cited to show the generality of the principle that the elasticity of supply response to different classes of inputs is highly correlated with the length of the time horizon being used so that the length of horizon influences strongly the nature of the factors affecting supply. Thus, stipulation of the time horizon is an essential initial step in long-term forecasting. This, in turn, provides the basis for analysis of input factors in order to select those whose influence on future supply may be significant.

Most frequently, long-term supply projections are undertaken in order to provide a guide to help choose between policy or program alternatives which affect wood production. Ideally, the whole range of alternatives may be projected, in which case the result is the supply function or schedule. In practice, the task of projecting may be so large that it may only be feasible to analyze a limited number of alternatives which suggest either the most likely probabilities or the range of possibilities.

DEFINING MARKET AREA

The first step in forecasting supply is to define the market area for which supply is to be predicted. The purpose for which the forecast is to be made will, of course, provide certain criteria for defining the market area. However, the nature of available physical and economic data, the geographic distribution of the forest resource itself, and the location of major centers of consumption may also be of primary importance. In general, the task of forecasting is simplified if the supply area is chosen in such a way that it is bounded by sharp discontinuities in the physical characteristics of the resource and by changes in administrative jurisdiction which permit measuring flows into and out of the market area. The detailed nature of the existing (or potential) transportation network is obviously of great importance in delineating market areas useful for the analysis of supplies of roundwood and of standing timber.

DEFINING TIME HORIZON

The time horizon to be embraced by the projection must, as has been said, be selected in relation to the particular problem in hand. Thus, a projection to be used for evaluating the economic feasibility of expanding wood processing plants would usually extend to a horizon determined by the normal economic life of such plants; but a projection used to determine the optimum scale of planting programs would usually embrace one whole rotation. Time horizons of the latter order of magnitude may thus be dictated by the logic of the problem under study. Even though there may be major uncertainties about many aspects of such distant periods, they do not justify selecting a shorter, less uncertain, time horizon in defiance of the logic of the problem.

FACTORS AFFECTING LONG-RUN SUPPLY

For any commodity, the factors influencing long-run supply are those which determine the costs of production at alternative levels of output. For roundwood these include the costs of timber growing, of timber harvesting, of extraction, and of transportation to the point where sales of roundwood are expected to take place. It should be emphasized that the relevant costs are all prospective. For example, if the time horizon appropriate for the forecast is thirty years, one must be concerned with harvesting, extraction, and transportation costs likely to be incurred thirty years from now, or at some appropriate intermediate date of harvest, not with those of the present. Similarly, the costs of timber growing must be estimated on the basis of conditions which are expected to prevail at whatever date the timber growing measures in question are applied.

Predominant elements in the cost of growing timber are those related to the capital invested in growing stock. Thus careful measurements of present growing stock and accurate projections of the potential yields from it are essential as a basis for estimating costs of capital. Of coordinate importance is the determination of the interest rate which is appropriate as a measure of the cost of using capital.

Inventory and Growth

National and regional forest inventories will usually provide information on growing stock adequate for long-term supply forecasting. Current inventories must of course be supplemented with information as to recent rates of stand reestablishment so as to insure a complete picture of the total available forest capital. Similarly, current surveys of national and regional levels of timber growth provide a starting point for estimating the physical productivity of the growing stock. Current surveys alone, however, furnish only a part of the information needed for long-term growth projections. They must be augmented by careful analysis of (1) how growth is likely to increase or decrease in the future, if the present dynamics of the forest are allowed to run their course, and of (2) how growth is likely to be modified either by variations in the future level of cutting or by the application of programs of management different from those which have produced current conditions.

The accuracy of growth information is a critical factor determining the reliability of long-run supply estimates. Because of the long time periods involved, even relatively small errors in estimates of current or prospective growth may become heavily compounded. Ordinarily, growth projections may be simplified by estimating net growth after deduction of expected mortality and losses from forest enemies.

Interest Rate

Estimation of the interest rate appropriate for calculating capital costs also is critical for reliable results because of the compounding of such costs. The rate which is appropriate will depend upon the problem in hand. For example, if the question concerns supplies of timber that private owners will place on the market in the long run, the most important guides to interest rate will be the long-term rate at which such owners can borrow capital and the marginal rate of return to capital in alternative lines of private investment. But if the question is the normative one of analyzing timber production goals for the supply area as a whole, the long-term rate on government borrowing or some overall estimate of the marginal efficiency of capital in the regional economy may be more appropriate guides.

Of particular concern is the treatment of risk and uncertainty. In short-run and market-period problems, it is often the practice to allow risk and uncertainty to be reflected in the interest rate. If used in long-term projections this practice is likely to exaggerate risk allowances to an inappropriate degree, due to the differential effects of compounding at different rates of interest. Unless there is clear evidence that risks do in fact compound annually, it is preferable to identify risk and uncertainty costs separately from the interest rate and to handle them as annual charges.

Interdependency of Outputs

When any growing forest is analyzed over a period of decades, it is apparent that growing stock available in the later years is heavily dependent on the level of cutting (or output) in the earlier years, and that allowable cut in the later years depends heavily on current growth and level of growing stock in the earlier years. Recognition of such time interdependencies is essential if costs of production are to be estimated properly. How to treat such time interdependencies constitutes a major methodological problem.

In some cases, interdependency of outputs may be treated by use of the assumption of sustained yield. Wherever this assumption is appropriate (for institutional or other reasons), it serves to specify the way in which timber output in one period is related to that in others, and the interdependency problem disappears. In many cases, however, the sustained-yield assumption may be inappropriate (for example, where the question to be studied is that of whether total growing stock should be increased, decreased, or maintained at the present level). In such cases simulation models (such as are described in "Choice of a Supply Model," below) appear to provide the most effective methodological approach because of their large capacity for analyzing intertemporal relationships.

Available Forest Area

In the long run, the area of land available for (or used for) growing timber may be subject to significant change. On the one hand, forest areas may be taken out of production in order to provide land for urban development, transportation facilities, or agriculture. Or additional land may become effectively available as a potential supplier of timber because of abandonment of agricultural land, extension of transportation into previously inaccessible areas, or for other reasons. Accordingly, an evaluation of prospective trends and potentialities in land use should be made in order to provide estimates of how forest area is likely to change over time. Such estimates must then be included as one of the independent variables in the timber supply forecast.

Forest Management Programs

Over the long run, timber growth and levels of yield are responsive to the intensity of forest management. Thus, long-term supply projections must take account of the effects of alternative levels of forestry programs concerned with protection or silvicultural improvement of the forest. Costs of such program alternatives must be estimated and their potential effect on growing stock and yield must be determined. Where the purpose of the projection is to determine the most probable level of future supplies, it may be sufficient to project a single pattern of management programs, corresponding to the level of management which is expected to prevail. Where the purpose is to evaluate management alternatives, however, it is clearly necessary to envisage several different management regions and to project the costs and changes in output which will be associated with each.

CHOICE OF A SUPPLY MODEL

For long-run supply analysis, nonstructural statistical models based on the extension into the future of regression relationships determined from historical data have limited applicability. Such models cannot take adequate account of the time interdependencies noted above. In addition, the length of long-run planning horizons is ordinarily such that many production relationships which have characterized the past are likely to be different in the future. (For example, in the short run, the impacts of increasing mechanization on the costs of output might quite properly be disregarded. But in the long run, explicit recognition of such effects is a vital part of the analysis.) Finally, in most countries, reliable statistical data covering inventories, growth, costs, and the like are available for a relatively limited period of past history. These periods are usually too short to reveal the longer run trends which are essential for the sort of supply analysis being considered here.

A more useful sort of model is based on the actual physical and economic structure of future supply. It simulates the various steps that lead to the generation of timber supply by tracing out year after year what would happen to such items as inventory, growth, and costs if specified management and cutting regimes were to be followed. Such models may be relatively simple, based on rather broad specifications as to forest growth, demand and costs of factors. Or they may be quite complex, attempting to take account of many significant details as to growing stock, products demanded, sorts of management inputs, and the like. In the latter instance, the simulation model is only practicable if a computer is available to handle the manifold calculations involved with a reasonable degree of speed.

A brief description of the principal elements in a timber supply simulation model may help to explain this method of approach. A very simple situation will be assumed in order to stress methodological aspects.

A Representative Long-Term Supply Problem

Consider a forest region of about two million hectares of accessible timberland, situated so that there are evident and substantial obstacles (either physical or political) to the import and export of forest products. The area includes substantial markets for wood pulp and sawn products. Let us assume also that all forest land is owned and operated by the central government and that only two species of timber are suitable for production in the area. (Much more complex assumptions as to ownership and species could be used without complicating the methodology.) Suppose, finally, that the central government is considering whether or not to expand output of pulp and paper products by investing in pulping facilities that will increase existing capacity by twenty-five percent. A projection of long-term supply useful in answering this question is required.

Length of Planning Period

The economics of the proposed increase in pulping facilities will depend on conditions throughout the life of the facilities in question. Therefore, the timber supply analysis must extend over a period at least as long as the expected depreciation period of the new plants. In the present case, this is judged to be 30 years.

Terminal Supply-Demand Conditions

An important (and too often unanalyzed) element in the cost of producing timber is the residual one of impact of production on more distant future conditions. Thus, in the present case, costs of increasing wood output during the next 30 years will depend in part on the state of the timber supply-demand balance at the end of the period. Other things being equal, the long-run costs of additional output during the next 30 years will be higher if they result in a severe supply-demand imbalance than if they do not.

In order to reflect properly this element in cost of production, the simulation method requires us to specify the terminal conditions of supply and demand which are expected (or desired) at the end of the planning period. Unless these terminal conditions are specified, the anticipated average unit cost of wood produced during the planning period is indeterminate and a statement of long-term supply in the economic sense cannot be made.

The specification of these terminal conditions must be made within the broader framework of forest policy considerations applicable to the problem in hand. In the present example, the most appropriate terminal condition might be an inventory sufficient to maintain output on a sustained yield basis at the level where average total unit cost of wood production equals expected average long-run price of wood. (One might readily imagine different terminal conditions which might properly be specified under particular circumstances.) Whatever the specific terminal conditions may be, however, they involve quantitative statements about the long-run conditions of (1) timber growth, (2) costs of production, and (3) the demand schedule for timber output. Each of these functional relationships must, of course, be specified in the light of conditions expected to prevail at the end of the planning period. Each must be the object of research which is carefully designed to estimate the proper function.

It may at first seem anomalous to state, as we have just done, that the long-term costs of wood production depend (in part) on the demand for wood. But this apparent contradiction is simply a reflection of the fact that a major element in the real economic cost of growing wood is the opportunity cost of forgoing consumption of wood in other time periods. The magnitude of this opportunity cost is influenced by the demand for wood in these periods. When we include the demand for wood as one of the factors defining the terminal conditions used to identify cost, we are simply recognizing the opportunity cost of using growing stock. The simulation model provides a practical way of sorting out the time interdependencies which arise out of the existence of these opportunity costs.

Initial Inventory and Growth

The simulation model must also include a detailed description of present inventory and of growth potentialities. The number of hectares of land and the volume of wood inventory, by species, site quality, and age and condition class of the forest may provide sufficient description of the inventory for supply projection purposes. In establishing the units and the standards to be used in measuring the inventory, care must be taken to insure that these are consistent with those likely to prevail in timber utilization practice throughout the 30-year planning period.

The timber supply model must also be provided with specifications as to future growth. These may be formulated either in the form of stand projections or by yield table methods. The choice may well depend on the comparative extent and reliability of the information available for use in each of these alternatives. Under either method, the growth analysis must be sufficiently precise so that the effects on future inventory and yield of a range of alternative forest management programs can be evaluated.

In the present illustration, the timber supply model provides the description of present inventory in the form of a tabulation of areas of forest land in the following forms. The tabulation is compiled separately for hardwood and for softwood:

CONDITION CLASS	AGE CLASS					
	I	II	III	IV	V	VI
A: Site I, unstocked						
B: Site I, 1–25% stocked						
C: Site I, 26–50% stocked						
D: Site I, 51–75% stocked						
E: Site I, 76–100% stocked						
F: Site II, unstocked						
G: Site II, 1–25% stocked						
Etc.						

Inventory volumes are determined from knowledge of the correlation between the tabulated condition classes and cubic volume per acre.

The effects of annual growth under a particular management regime are specified in the model by including in it estimates of the proportion of the acres in each cell which will move to some other cell at the end of one year's growth. Similarly, the effects of one year's harvest cutting of a given amount and kind are simulated by providing the model with estimates of the shifts in area from one cell to another induced by such a cut. If the relevant information is programmed for a computer, the impact on inventory of different levels of cutting and of different management programs over the next thirty years can be estimated quantitatively. This inventory/growth simulation will provide a tabulation in the above format for each year in the planning period.

Comparing Management Alternatives

Using the computing device outlined above, we may now compare the results of the various management alternatives which might be adopted over the 30-year planning period in order to provide a 25 percent increase in pulpwood output. The available management alternatives must, of course, be specifically defined for the particular problems in hand. In the present case, suppose that pulpwood output may be increased by either of the following alternatives:

(1) Planting unstocked (or understocked) softwood areas. If done immediately and on a sufficiently large scale, the terminal growing stock will be increased and cutting over the next 30 years can be expanded without violating the terminal sustained yield condition previously established. The inventory/growth simulator of the preceding section provides the information needed to determine the extent of the plantings which will be required to permit an increase of 25 percent in the pulpwood harvest.

(2) Intensifying intermediate cuttings on the more heavily stocked sites. The inventory/growth simulator provides the information needed to determine the extent of intensified intermediate cuttings needed to secure the 25 percent increase in pulpwood harvest and states the effects of this on the residual growing stock. The additional management inputs, whether of planting or other measures, needed to regain the specified terminal growing stock may then be determined.

(3) Sawnwood production may be reduced by an amount sufficient (a) to permit the 25 percent increase in pulpwood cut and (b) to build up growing stock necessary to sustain the level of yields required by the terminal conditions.

Under alternatives (1) and (2) above, the additional pulpwood output is the cost of the new plantings or other management measures needed to offset the effects of the heavier cut. Under alternative (3) the cost of the additional pulpwood output is the net income forgone (opportunity cost) as a result of curtailment of sawnwood production. Whichever of these alternatives is least costly may be regarded as the appropriate measure of the costs of the required increase in long-term supply.

Using the same simulating device and techniques described above, one may estimate the cost of other changes in output in addition to the specific case discussed above. This family of estimates, showing the cost at which different alternative levels of 30-year pulpwood output can be secured from the existing forest inventory in the area, constitutes a statement of the long-run timber supply.

44

THE LONG TREND IN WOOD USE

William A. Duerr

First published in *Proceedings of the Forest Engineering Conference, American Society of Agricultural Engineers,* East Lansing, Michigan, September 25–27, 1968. Pages 99–101. 1969.

BIOGRAPHY

Mr. Duerr's biographical sketch appears at Item 4, page 30.

EDITORS' SUMMARY

Instrumental in the long trend of wood use has been, and will be, rising labor value. The rise favors a shift in wood use away from natural forms toward reconstituted forms. In much of the twentieth century, total use has been about constant. In coming decades, the total will rise, the per-capita fall less rapidly than heretofore.

QUESTIONS TO CONSIDER

(1) In this article, the hierarchy reappears which has been referred to elsewhere. Turn again to Question 1, page 177, and Question 5, page 208. Why does this hierarchy exist? What are its consequences?

(2) What does Mr. Duerr mean when he states (page 365) that economic development not only signifies that labor has been economized, but also ensures that it will be? How can it ensure such a thing? Do not many labor-intensive activities persist in developed societies? How about the editing of books? the teaching of students? Name some other examples. Propose some counter-arguments.

(3) Look again at Question 1 for Item 38 (page 300). Do Mr. Duerr's forecasts help to describe what the lumber industry's "difficulties portend"? Is the description consistent with others' interpretation of the industry?

(4) Does this analysis of the long trend take sufficiently into account the growing convic-

tion in the United States that we must strive to restore and defend environmental quality? How would you amend the forecasts so as to give fuller recognition to environmentalism?

(5) Do professional persons have an obligation to champion the traditional elements in their professional life? Are not lumber, and surely nonreconstituted wood products in general, such an element? Should foresters not only believe that "wood is good," but also promote this belief among all consumers, so that trends such as described in this article will be changed for the better?

* * *

My purpose here is to follow the history of wood consumption in the United States from the year 1900 to the present, to make a forecast for the rest of the century, and to explain the 100-year trend as I see it.[1]

The content of our lives—the quantity and quality of our living—is, I suppose, a function of three kinds of resources: materials, ideas, and space. In our civilization, timber has been one of the major materials. The place of timber tomorrow will depend greatly on the ideas brought to bear upon its conversion and utilization. At the same time, the place of timber will largely determine the role of forest land as space. Thus the long trend in wood use is an interesting subject, both in itself and in its implications.

WHAT HAS HAPPENED TO WOOD USE

If one traces U.S. consumption of wood since the turn of the century, he discovers these historical trends (Figure 44–1, left of the vertical broken line):

1. Considering all kinds of wood products as a group, the trend in total consumption (Figure 44–1*a*) has formed, roughly, a shallow saucer: Measured in cubic feet of standing-timber equivalent and expressed as a 13-year moving average, there was a rounded height of consumption in the first decade of the century, then a slow decline to the nadir of the Depression, beyond which the line forms the other side of the saucer, rising in recent years to a level about equal to that of the early summit.
2. Clearly, with total consumption approximately constant, consumption per capita (Figure 44–1*b*) has been going down. The decline, though variable in rate, has been essentially continuous.
3. The individual wood products arrange themselves neatly into four groups with significantly different trends:
   a. Round or split products. Fire wood is the principal item in this group, representing more than three-fourths of its total in 1900, more than half today. Other items are posts, poles, piling, and hewn ties and timbers. Both total and per-capita consumption have consistently fallen off, slowly before World War II, speedily since.
   b. Sawn products—i.e., lumber, including boards and dimension, the many wooden commodities made therefrom, and sawn ties and timbers. The group's history resembles that of timber products as a whole: Total consumption sagged during the decades from 1925 to 1945, but has come back since then and held its own. Per-capita use has dwindled fairly regularly.
   c. Sliced products: veneer and plywood. Total consumption is up all the way and

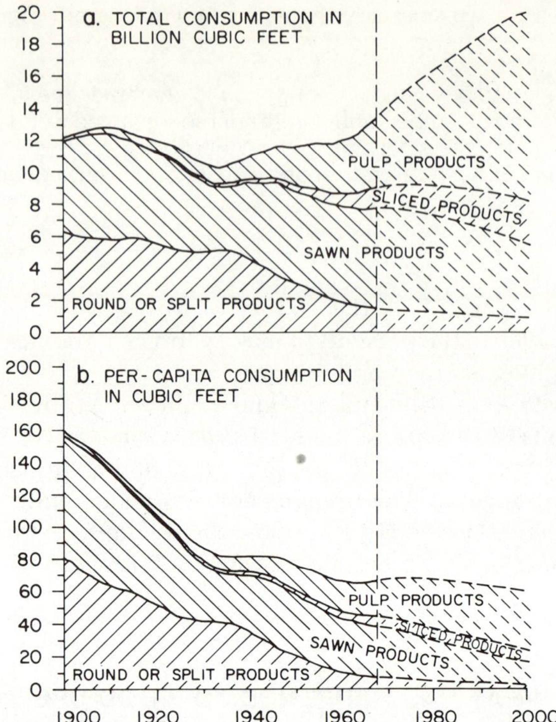

Figure 44–1. Consumption of wood products in the United States, 1900–2000: 13-year moving-average quantities expressed in terms of roundwood equivalent, including both domestic and imported wood. (Historical series are from federal government sources. Wood forecasts are the writer's; population forecasts are the Census Bureau's Series B.)

bravely up since World War II. Per-capita use also is up, but the rate of rise has been slowing off.

d. Pulp products, including paper, paperboard, all sorts of composition board, and pulp for producing rayon and acetate fibers and the like. These are the "reconstituted" wood products, in contrast to those in the first three groups, where the natural structure of the wood is preserved. They are, for the most part, chemically rather than mechanically processed. Both their total and their per-capita consumption has been rising strongly, and the rate of rise has itself been going up.

In consequence of the foregoing trends, the mix of wood products in the nation's market basket has been greatly altered:

	PERCENT IN 1900	PERCENT TODAY
Round or split products	50	11
Sawn products	48	46
Pulp or sliced products	2	43
	100	100

Lumber has held its own, the grosser products have been falling by the way, and those which pass through a more finely divided state have come on strong to occupy ground which others have vacated.

In summary, then, over the last seven decades in the United States, wood use as a whole has held generally constant in absolute terms, but has declined relatively to population and, of course, relatively to other materials. Among wood products themselves, those hardest hit have been the ones that pass through the simplest manufacturing process, reaching the consumer most nearly in their natural state: in the round or in split form. On the other hand, a comparatively good performance has been turned in by the wood products which are broken down to a greater extent in manufacture and whose natural form is more largely changed. The extreme case, the most successful group, is the most highly reconstituted: the paper and related chemical products of wood. This group is pulverized, and even reduced to separate molecules, in the course of production. Consumers have voted heavily for the resulting commodities at the same time that they have been turning their backs on other forms of wood.

ECONOMIC DEVELOPMENT AND WOOD USE

Viewed as it is here, the long trend in wood use is simply explained: It is a manifestation of social growth and notably of economic development.

Economic development means a rising trend in the output of goods and services, outstripping the increase in population. In the United States since the turn of the century, per-capita national product has climbed at a rate of about one and one-half percent per year. Man-hour output in manufacturing industries has risen annually two percent or more. The Nation's output was at the same time its income: The average person's income from wages or salary or other sources has been growing apace. That is to say, the value of human labor has been pushed up year by year. Human effort—time—has been made more precious. This is economic development.

When society raises the value of labor as it has been doing in the United States, it is not only signaling that labor has been economized, but also ensuring that labor will be economized. For in raising the value of labor, society is promoting those industries that make sparing and efficient use of human effort and suppressing those industries that make lavish use of it. Consider this process from the viewpoint of the individual industry. There is only one lasting way for the industry to meet its payrolls, and this is from the proceeds of the sale of its products. If the industry is to hold its product prices in line with the general level of prices and thus survive competition, the industry must find some way of increasing its output per man-hour of labor to keep pace with the growing cost of a man-hour of labor. The alternative is to raise prices more than the general level, lose markets to competing industries, stagnate, and ultimately perish. Thus economic development, which is an upward trend in output per man—that is, in technological efficiency—is at the same time a powerful force in defense of those lines of production which are technologically progressive and in opposition to those which are not. Indeed, it is through altering the mix of production lines in the course of such defense and opposition that economic development is achieved.

Now as for timber products: Those lines of production in which wood is kept more nearly in its natural state—say, in the form of logs or bolts or chunks—necessarily process their material piece by piece. They are involved in handling or inspecting those pieces, since one piece inescapably differs from another. They are most readily adapted to the old ways of production: to heavy use of cheap, closely supervised labor, to small-scale operation, to family, partnership, or little-corporation enterprise, to traditional management, to passive marketing, and to reliance upon public agencies for promotion, research, and development. They are technologically backward and deplore rising wage rates. They are an anachronism amidst economic development, and their days are numbered.

The long trend in wood use as a whole is the aggregate of trends in each line of wood production. It is the result of changes in product mix in obedience to the forces of economic development. It is a summary of wood industries' success, relative to that of other industries, in listening and responding to the mandate of social change in the United States.

RESPONSE OF PRODUCERS AND CONSUMERS TO DEVELOPMENT

What does one see as he examines the evidence in more detail? He sees, at each stage in production and in consumption, the forces, both actual and prospective, which shove the economic center of gravity upward through the series—

Pulp products (from particles, fibers, molecules)
↑
Sliced products (from thin pieces)
↑
Sawn products (from thick pieces)
↑
Round or split products (from chunks, logs)

What does one see at the first stage of production, the timber-growing stage? He sees many an opportunity to produce uniform stands of pulpwood from tree plantations in 10 to 20 years with little care. He contrasts this simple process with the one for producing, say, sawlogs, where the production period is relatively long and the forest, therefore, relatively varied and subject to intermediate and harvest treatments that devour skilled labor.

When harvest time comes, the young pulpwood forest may be cut over and prepared for shipment with the use of comparatively simple machinery. Logging in the older forest can be mechanized, too, but the process is necessarily selective if the values are to be preserved for which, presumably, the owner undertook the sacrifice of extra waiting and extra management outlays. A pulpwood tree can be harvested successfully with shears. However, if sawtimber is harvested in this fashion, the lowest log, which is the most valuable, may be damaged.

When one looks, next, at the transportation process, he again sees labor-saving advantages for the products of pulverized wood. Wood particles can be blown, slurried, piped, tanked, and handled in other such impersonal ways.

Coming now to the factory, one is struck by those contrasts which I spoke of earlier in summary. For example, in the lumber industry, over a period as long as one may care to name, the pace of technological development and of increase in output per man-hour has been glacial. Looking back as far as the record extends, lumber prices have been rising about two percent per year faster than the all-commodity average.

But the pulp-paper industry gives a quite different account of itself. Here, labor productivity has kept pace with the average of all manufacturing, and so have product prices.

Beyond the factory, in the course of further shipment and of marketing, the pulp and, to a lesser extent, sliced products still retain comparative advantages. They come in uniform sheets or rolls, efficiently handled, packed, stored, described, and moved.

Many of the wood products undergo further processing before they reach the final consumer. Their use in the construction industry is a major case in point. In the course of its economic development, the Nation has seen the passing of a log-cabin era, a timber era, and a lumber era in construction. We are now in a metal-stone-clay-glass era, in which plywood and pulp products play highly useful roles. A revealing small facet of this historama is to be found in the cost of installing and working with the various products. To cover 32 square feet of surface requires lifting, fitting, and fastening several pieces of lumber or a single piece of plywood or composition board. To maintain the siding on one's dwelling calls for painting and repainting wood or for washing down enameled aluminum with a hose.

And finally, what of the consumer and his preferences as they affect the use of wood? In our economically developed society, what are some of the consumer's traits? More and more, and above all, he is well-off financially and thus able to rent, borrow, or buy the legion of commodities which his way of life thrusts at him.

The consumer is adventurous and enjoys experimenting with new commodities and new substances. In his venturesomeness, he presses down the accelerator on the process of product creation, growth, ascendancy, decline, and obsolescence. So he quickens the rush of total development. So, too, he casts his votes for the innovators and against the traditionalists in respect to materials, design, and packaging. His adventurous spirit is tempered by a degree of defenselessness against the guiles of emotional advertising. Here his mentor is the industrial giant—perhaps a wood-products concern, though typically not.

The consumer is mobile, and he wants his goods light and compact. Disposables are inherently appealing to him: paper handkerchiefs, for example, and towels and clothing. He is a pampered fellow, in search of an easy life. He approves of furniture finished in plastic—wood perhaps only to the extent of the photographed grain—thus able to resist its natural predators, notably cigarettes, cocktails, and children. He wants to be warm in winter and to push a button for his heat. An open fire on the hearth delights him, but he may well find that logs pressed out of wood particles are convenient, clean, and aesthetically satisfactory.

The consumer, more and more, is well-leisured. One thing he does with his leisure is enjoy the out-of-doors. He likes the idea of setting forests aside as recreation ground. He is not always happy about the use of forest land to produce timber. He himself may own a little patch of woods, and if he does, he may have decided to keep loggers out of it.

THE FUTURE OF WOOD USE

What does all this mean for the future? As I turn to look ahead into the remainder of the twentieth century, it seems to me that in the course of explaining the past and present, I have committed myself to the outlines of a prediction. Assuming that general economic development will continue, then—

1. Total consumption of pulp products will greatly increase.
2. As a result, and despite the continued dwindling of the round-, split-, and sawn-products industries, total use of wood will increase.
3. The resurgence of total wood use will be sufficient to brake, if not to reverse, the downward trend in per-capita consumption.

These tendencies for the decades just ahead are represented conservatively in Figure 44–1, to the right of the vertical broken line. When the trends were sketched, the following were among the general considerations in view:

Cellulose and its organic associates are versatile materials, widely useful under any reasonably foreseeable technology.

Whereas the prospective supply of some of our customary materials, such as the heavier metals, is limited and uncertain, the sources of cellulose and related substances promise to be comparatively abundant.

Assuming that artificial synthesis will not be achieved commercially within the twentieth century, trees will continue to be one of the most efficient sources of plant fiber and associated compounds. A pine forest, even in a temperate climate, can produce nearly as much new organic matter per acre per year as some of the world champions, such as sugar cane plantations in Hawaii or stands of marsh grass in coastal Georgia. Its productivity is perhaps four times the world average for sugar beets, seven times that for corn, nine times that for wheat.

Furthermore, the growing and extraction of the forest product are potentially economical operations. Ideally, the forest is planted, given minimal subsequent care, and clear-cut with simple machines at the end of a rotation which is moderately short by today's standards. The harvest brings in, from a single passage over the ground, the safely stored fruits of several years' growth.

It is assumed that the relative position of foreign trade in the U.S. timber economy will hold fairly close to its recent value. The drain upon domestic forests, then, will be equal to almost nine-tenths of domestic wood consumption.

The long-run capacity of our forest lands to produce wood is limited. A limit is set by competing uses for forests and other lands: space, recreation, farming, rights-of-way, urban and suburban development, and so on. A limit is set, too, by the intensity of forest management, largely a function of the land's inherent productivity and type of ownership and of forestry technology. Given the changes now foreseeable in land use and management, long-run forest capacity in the United States is probably not much above 20 million cubic feet of wood output per year. To attain such capacity may well require more than the remaining third of the present century.

Thus technological forces will control the total use of wood just as they will control the product mix.

* * *

REFERENCE

1. This article was inspired most directly by a talk on wood consumption given some years ago to forestry students at the State University of New York, Syracuse, by Professor G. R. Armstrong.

45

DISCUSSION OF LUMBER PRICE TRENDS

Joseph Zaremba
Lee S. Settel
John A. Zivnuska
Frank A. TerBush

First published in *Journal of Forestry* 56(3):179–181; March 1958. 56(7):517–519; July 1958. 56(8):596; August 1958.

BIOGRAPHY

Mr. Zaremba is a forester-turned economist, and it was in the midst of this turning that he produced the analysis reprinted here and several other important works, including his book, *Economics of the American Lumber Industry.* In forestry-economics circles, where the producer traditionally was the object of interest, he is known as the pioneer in emphasizing the consumer. As a teacher, he began on the faculty of the Yale School of Forestry and transferred thence to the State University of New York College of Forestry and then to Fordham University. He is now Professor of Economics, State University College, Geneseo, New York.

Mr. Settel is Forester and former President of the Appalachian Southern Corporation, Ellijay, Georgia, engaged in lumber-products manufacture and marketing. He studied at the State University of New York College of Forestry and thereafter, in the early 1930s, became one of the first urban foresters, serving the City of New York. Then a varied series of assignments in public forestry brought him to Washington, where for 5 years during and after the Second World War he was associated with the Lumber Branch of the Office of Price Administration and the Lumber Division of the War Production Board.

Mr. TerBush, like Mr. Settel, is a graduate of the New York College of Forestry. After serving in the Bureau of Land Management during the early 1950s, he worked for several years as analyst for one of the lumber trade associations. He then commenced the Forest Service career which has taken him into a wide variety of positions, both in this country and on foreign assignment.

He works at present in Portland, Oregon, as Reforestation Specialist with the Division of State and Private Forestry in the Forest Service.

Mr. Zivnuska's biographical sketch is given at Item 2, page 16.

EDITORS' SUMMARY

The long-term rise in the real price of lumber in the U.S. was the result of rising per-capita income (productivity) in the economy as a whole while productivity in the lumber industry was relatively unchanging. Comments upon this thesis concern the labor theory of value, the effect of demand upon price trends, and particularly stumpage price as a lumber-price determinant.

QUESTIONS TO CONSIDER

(1) Before Mr. Zaremba's work was published, the rising trend in U.S. lumber prices was commonly explained by the migration of sawmills farther and farther away from consumption centers in the East. That is, the rising price was attributed to rising transportation cost. What is wrong with this explanation?

(2) Rising prices were at one time attributed to collusion within the lumber industry. What is wrong with this explanation?

(3) Is there any contradiction between Mr. Zaremba's thesis and the observations by Mr. Kaiser and Mr. Guttenberg (Item 24)? What has happened to real prices of lumber in the United States since the writing of Item 24? Why?

(4) Refer to Question 1 for Item 38 (page 300). Do Mr. Zaremba's ideas help identify "what lies at the root" of the lumber industry's difficulties?

(5) Do you think that Mr. Zivnuska is in fundamental disagreement with Mr. Zaremba? Can you show that the per-capita-income, or lagging-productivity, explanation of rising real price is simply a tight restatement of the generalization that supply and demand explain price?

(6) Mr. Zaremba's thesis is of basic importance to an understanding of the forest economy. Restate the thesis in your own terms.

THE TREND OF LUMBER PRICES

JOSEPH ZAREMBA

* * *

The rising trend of lumber prices in the United States has been a subject of serious discussion for many years. The fear has been frequently expressed that continued increases in lumber prices at the past rate would place the lumber industry in a progressively weaker position with respect to the manufacturers of competing materials.

Various writers have attempted to explain the cause of the rising trend of lumber prices by conditions limited exclusively to the lumber industry. Compton in 1916[1] and Bryant in 1919,[2] who were among the first to note an upward trend of lumber relative to all-commodity prices, suggested that the phenomenon was due to the gradual shifting of the center of lumber production away from the lumber-consuming centers in the East. Both Foster[3] and Hall[4] in 1946 recognized that the low efficiency of lumber mills was partially responsible for high lumber prices after World War II.

Notwithstanding the contributions of previous writers, the nature of the forces underlying the trend of lumber prices is yet to be fully explained. The purpose of this article is to show that these forces are rooted in certain relationships which exist between the lumber industry and the rest of the economy rather than in conditions limited to the lumber industry.

TREND OF LUMBER AND OF ALL-COMMODITY PRICES

The wholesale-price index of lumber in the United States from 1860 to 1955 is depicted in Figure 45–1 along with the wholesale-price index of all commodities. The latter has been relatively stable throughout this period. Apparently, in the long run, the American economy produces the goods and services demanded by consumers at fairly constant money prices.

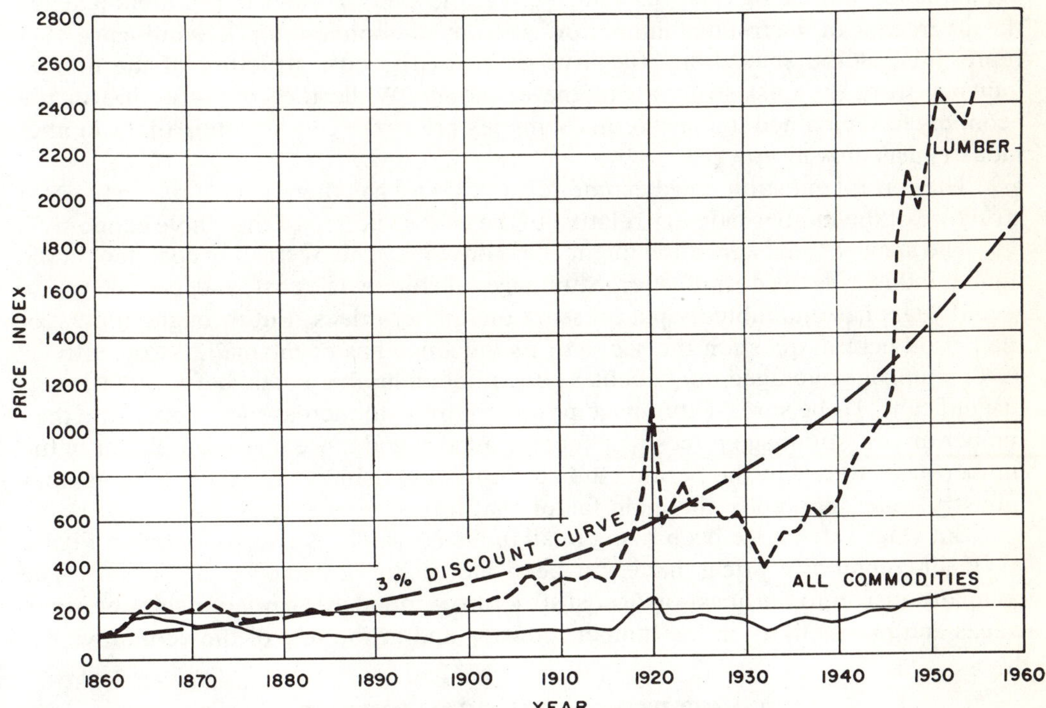

Figure 45–1. Wholesale price indices for lumber and all commodities, 1860–1955, (1860 = 100.) (These indices up to 1947 were obtained from the following source: James, L. M., 1947. The trend of lumber prices. *Journal of Forestry* 45:646. All-commodity and lumber prices since 1947 were obtained from U.S. Bureau of Labor Statistics, *Monthly Labor Review*.)

Lumber prices have followed roughly a 3-percent trend line. They have steadily departed from all-commodity prices. This departure has been particularly rapid since 1900. During the depression of the 1930s, when the lumber industry as a whole operated at a loss, the price of lumber did not decline to the previous low points of 1908 and 1915. Thus, despite a powerful downward pressure on lumber prices exerted by declining lumber demand, lumber prices remained at a relatively high level throughout the 1930s. Apparently the lumber industry produces lumber only at steadily rising prices. Why is this so?

RELATIONSHIP BETWEEN PRICE AND AVERAGE COST OF PRODUCTION

Variations in lumber prices, like variations in the prices of all commodities produced under competitive conditions, are caused by different factors according to the length of time under investigation. Price variations within a short interval are caused mainly by demand factors such as consumer income and the number of houses under construction. The longer the period under investigation, the greater is the share of attention that must be given to the influence of the cost of production on price.[5] This analysis is concerned with price variations over a period of almost a century. Consequently, the emphasis here is on the influence of the cost of production on lumber prices.

The trend of lumber prices will parallel the trend of the average costs of producing lumber since in the long run prices must cover the average costs of production. Thus the divergence of the trend of lumber prices from all-commodity prices in Figure 45–1 represents, at the same time, the divergence of the cost structure of the lumber industry from the cost structure of the economy. While the cost structure of the economy has remained stable in terms of money prices the cost structure of the lumber industry has moved upward.

The central question now becomes clearer: Why has there been a rise in the cost structure of the lumber industry relative to the cost structure of the whole economy?

The answer to this question might seem obvious. The overriding cost item in the lumber industry is the cost of labor. Stumpage, of course, is a cost, and its rapid rise in recent years has undoubtedly put pressure on lumber prices. But throughout most of the period here in question the scope of its influence has been small. As recently as 1940, stumpage averaged only about 5 percent of all lumber costs; before 1900 it was insignificant. To be sure, if stumpage prices continue to increase at a faster rate than lumber prices, stumpage prices will become a more and more important factor in the lumber price trend. In the period studied, however, rising wage rates in the lumber industry were apparently the main factor that forced lumber prices up.

But wage rates have been rising in all industries and, as already noted in Figure 45–1, all-commodity prices have not been increasing as have lumber prices. The complete answer to the question of cost structure is found in the relationships between wages and productivity in the lumber industry and in the rest of the economy.

RISING PER-CAPITA REAL INCOME— THE CAUSE OF RISING LUMBER PRICES

The correspondence between per-capita real income and per-capita productivity is generally recognized. Suppose, for example, that during a given time period the quantity of goods and services produced by the average person increases 10 percent—that is, productivity increases 10 percent. If the proportion of output allocated to investment purposes remains constant, the average person must also consume 10 percent more goods and services. Thus an increase in per-capita real income is synonymous with an increase in per-capita productivity.

During the period under investigation per-capita real income in the American economy has increased at a compound annual rate of about 2 percent.[6] Table 45–1 indicates not only that real income is increasing, but that the rate of increase itself is increasing. The present generation of Americans has inherited a remarkable economy.

Table 45–1 Annual Rate of Increase in Real Income in the United States since 1850[7]

PERIOD	RATE (PERCENT)
1850–1870	1.5
1870–1920	1.6
1920–1947	2.5
1947–1955	2.9

Changes in money wage rates have, in the long run, paralleled changes in productivity (real income) in the American economy: increases in productivity have been passed on to workers in the form of higher wage rates. When productivity in the economy increases, no sector or industry in the economy can avoid paying the resulting higher wage rates; for if an industry does not increase the wages of its labor, workers will eventually seek employment elsewhere.

When productivity in the economy increases, therefore, a given firm or industry will attempt to raise its output per worker in order to make up for its higher labor expenses. If this can be done, the firm's cost structure need not rise. If, however, the firm cannot increase its output per worker, its cost structure will shift upward.

The most exhaustive studies of productivity in the American economy have been made by S. Fabricant of the National Bureau of Economic Research.[8,9] His findings, although inconclusive, indicate that there was no appreciable change in labor productivity in the lumber industry during the period covered by his analysis, 1899 to 1939. The few studies of the industry since 1939 indicate that its productivity may have decreased since World War II.[10]

Because productivity has not increased in the lumber industry, rising money wage rates in the economy have forced the cost structure of the industry upwards. It is the fact that lumber prices in the long run are formed around this rising cost structure that explains why they have diverged from all-commodity prices.

The relationship between per-capita real income and lumber prices can now be summarized: Rising per-capita real income (productivity) in an era of rather constant all-commodity prices leads to an increase in money wage rates throughout the economy. Productivity in the lumber industry meanwhile remains about constant, and so the industry's cost structure is forced upward. Lumber prices must in the long run conform to costs. Rising per-capita real income in the American economy may thus be thought of as the cause of the upward trend of lumber prices.

RELATIONSHIPS EXPRESSED ALGEBRAICALLY

The relationships among prices, productivity, and income can be expressed algebraically. Assuming constant productivity in the lumber industry and constant all-commodity prices, the rate of increase in the lumber price trend will roughly equal the rate of increase in per-capita real income (productivity):

$$1 + p = 1 + Q,$$

where p = annual rate of change in trend of lumber prices

Q = annual rate of change in trend of per-capita real income (productivity).

If the assumptions of constant labor productivity in the lumber industry and constant all-commodity prices are dropped, we have the general proposition that the ratio of lumber prices to all-commodity prices varies directly with the ratio of the annual rate of change in productivity in the economy to the rate of change in productivity in the lumber industry:

$$\frac{1 + p}{1 + P} = \frac{1 + Q}{1 + q}$$

where P = annual rate of change in all-commodity price trend

q = annual rate of change in trend of productivity in the lumber industry.

The prices in question above are money prices. Since the real price of lumber is derived by dividing the price index of lumber by the price index of all commodities, the ratio on the left in the equation immediately above is, by definition, a measure of the rate of change in the real price trend of lumber. Thus the trend in the real price of lumber depends, fundamentally, on the relation between the trend rate of change in productivity in the economy and that in the lumber industry:

$$1 + R = \frac{1 + Q}{1 + q},$$

where R = annual trend rate of change in the real price of lumber.

COMMENTS ON STUMPAGE COSTS

LEE S. SETTEL

"The Trend of Lumber Prices" by Joseph Zaremba in the March 1958 issue of the *Journal* is a challenging paper. Rising wages without an accompanying rise in lumber productivity is a grim reality we must face up to, but no worse a reality than rising stumpage prices. Although Zaremba states, "To be sure, if stumpage prices continue to increase at a faster rate than lumber prices, stumpage prices will become a more and more important factor in the lumber price trend,'' he tends to minimize the effect of stumpage cost. He states, ''As recently as 1940, stumpage averaged only about 5 percent of all lumber costs; before 1900 it was insignificant.''

If 5 percent is to be accepted as a true average nationally for 1940 stumpage cost we must know how it was determined. National averages for stumpage costs in any one year can be very misleading. Average dollar figures for stumpage costs in any one year are influenced by the amount of short term as well as long term stumpage converted. Recessions in the lumber industry are always accompanied by declining production from the smaller mills that are dependent on short term stumpage and to a lesser degree by the large mills who own substantial acreage in fee simple or lease timber rights for 5 to 15 years. National forest shortleaf pine stumpage brought $7.00 per M bd. ft. in Arkansas during 1937 when lumber realization averaged but $21.00 per M, one-third then, of lumber costs. Two decades later, similar stumpage brought $35.00 to $50.00 per M, close to 50 percent of lumber cost. Comparable stumpage brought $10.00 in North Georgia in 1944, about 20 percent of lumber cost and $36.00 in 1957, close to 50 percent of lumber cost. Yet, 5 percent could be a valid average for the year 1940 as lumber realization had not yet risen enough to encourage the operation of sufficient short term stumpage to reflect a higher figure. If we assume $30.00 as lumber cost in 1940 and 90 percent of production from long term stumpage charged off at $1.00 per M, we have room for 10 percent short term production at a cost of $6.00 per M. This assumption can be challenged as an oversimplification, but it is not inconsistent with the oversimplification we assume when we accept any figure as a national average cost of stumpage. What was reported as stumpage cost by long term stumpage operators in 1940? March 1911 evaluations? Acquisition cost? Capitalized cost? Probably what the Bureau of Internal Revenue permitted at the time.

Nevertheless, the challenge to forestry is clear. The silviculturist must learn to grow more and higher quality per acre and our utilization men are going to have to develop lower cost harvesting methods if per-capita consumption of lumber is to be maintained or to be increased.

IN REPLY

JOSEPH ZAREMBA

According to Steer, the average stumpage price in the United States during 1940 was $2.61 per M bd. ft.,[1] and the average lumber price during the same year, f.o.b. mill, was $23.32.[2]

The latter figure, however, does not include the cost of re-manufactured lumber, such as millwork and flooring, or the cost of lumber distribution. In 1940, southern yellow pine roofers sold for about $30 per M in New York, f.o.b. cars; 2 × 4 yellow pine dimension, $44.50; and the different grades of oak flooring varied in price from $45 to $93.[3]

The average wholesale price of lumber in 1940 was therefore considerably higher than \$23.32. If the average price in 1940 was \$35 per M, the cost of stumpage as a percent of wholesale lumber prices was $7\frac{1}{2}$ percent.

But, as Mr. Settel points out, a large portion of the 1940 lumber production represented "long term stumpage," that is, stumpage purchased prior to 1940 at prices different from those prevailing in 1940. The use of 1940 stumpage prices to compute the above percentage must indeed be regarded as an oversimplification of the problem.

LUMBER PRICES AND LABOR PRODUCTIVITY

JOHN A. ZIVNUSKA

In his recent discussion of lumber prices Joseph Zaremba performs a useful service in calling attention to labor costs and labor productivity in the lumber industry. However, he does not do his argument a service when he concludes his non-algebraic presentation with the statement that "rising per-capita real income in the American economy may thus be thought of as the cause of the upward trend of lumber prices." Actually, a better summary of the main line of argument appears to be the quite different statement that "the trend in the real price of lumber depends, fundamentally, on the relation between the trend rate of change in productivity in the economy and that in the lumber industry."

However, far more unfortunate than this confusion in the presentation is Zaremba's reliance on the labor theory of value in explaining lumber prices. His explanation would be valid—though certainly not fundamental—if labor were the sole input both in lumber production and in the general economy. Since this is not the case, the conclusions tend to be misleading.

Despite its wide usage in economic discussions, the "productivity of labor" is a statistical rather than an economic concept. It is determined simply by dividing total output by the total number of labor inputs. However, production of the output requires inputs of capital (including raw material) as well as inputs of labor. Indeed, probably the most characteristic route for increasing the productivity of labor in our economy is through combining more capital with labor. This, in effect, involves replacing a labor cost with a capital cost. When the marginal cost of the capital is less than the marginal cost of the labor, methods of production are changed, the industry becomes increasingly capital intensive, and the "productivity of labor" rises. If, on the other hand, the cost of the additional capital is greater than the saving in labor cost, the change is not made, and, other things equal, the "productivity of labor" remains unchanged.

There is a lack of readily available data as to trends in "labor productivity" in the lumber industry. Thus it is not possible to determine quickly whether or not the experience of the last decade is consistent with Zaremba's thesis. Certainly his citation of a study published in 1946 and based on wartime conditions in the Southern lumber industry cannot be accepted as the basis for any conclusions concerning post-war trends. In any case, discussion of "labor productivity" in lumber involves some peculiar difficulties. The industry is engaged in the mechanical processing of a raw material, but over the years the nature of the raw material has changed. Until very recently the forest resource has been a declining resource. Manufacturing lumber under these conditions has meant going farther back and higher up in the West and shifting to scattered stands lightly stocked with small trees in the East and South. Major adjustments in technology have occurred in response to this, although for a variety of reasons the adjustments have been less capital intensive than is generally true of the American economy. There can be little doubt, however, that the modern large-scale lumber operation would require far less man-hours per thousand board feet than the similar operation of fifty years ago *if both operations were based on the same forest conditions*.

Under the generally accepted theory of price, both price and quantity of consumption are considered to be determined simultaneously by the intersection of supply and demand. It is doubtful that much meaning can be attached to discussions of price or cost which do not also consider quantity and which do not consider demand as well as supply. As has been discussed more fully elsewhere,[1] the particular pattern of volume and costs which has characterized lumber over recent decades is best explained by the combination of increases in demand with decline in a supply schedule which is markedly inelastic at the upper ranges of production experience. In these terms, Zaremba's concern is with the reasons for the decline and the inelasticity of the supply schedule. The "labor productivity" approach is simply not adequate for this task. It does not consider the fact that labor is only one of the inputs involved in the production of lumber, and fails entirely to recognize the significant changes in production context involved in any forest-based industry over the last fifty years.

It is true enough that labor is a major cost element in the production of lumber and that technological developments to reduce labor requirements per thousand board feet of output may well be crucial to profit margins in the industry. It does not follow from this, however, that such reduction in labor costs would have any particular effect on price. Supply involves both cost and volume. Careful analysis of past experience will show that within the current range of production levels the most probable effect of a reduction in costs at a particular level of output in the industry would be a very minor increase in total volume accompanied by reestablishment of practically the same price level as previously.

The reasons for this will be found in the forests which are the source of the raw material. In the past foresters have been all too prone to believe that a declining timber resource results in a "timber shortage" with direct and major impact on the ultimate consumers of the services of lumber. Instead, in a flexible economy with technological alternatives the major impact is on the lumber industry, not on society. The structure of the industry, its cost patterns, and its supply responses are direct expressions of the changes in its resource base under the conditions of wide geographic dispersion of that resource and accompanying lack of concentration of control combined with technological means for operation with low capital investment. It is in these circumstances, not in rising per capita income, that the fundamental causes of the upward trend in lumber costs must be sought. To the extent that the relative trends in "labor productivity" between lumber and the general economy do correspond to the relationship postulated by Zaremba, they must be considered as simply an alternative and less meaningful expression of the same underlying forces which explain lumber's costs.

IN REPLY

JOSEPH ZAREMBA

If I understand him, Mr. Zivnuska harbors two main objections to my analysis, one having to do with the labor theory of value, the other concerning the effect of cost changes upon lumber prices.

As for the labor theory of value, it was a live issue when it was being taught by Adam Smith in the Eighteenth Century and by Ricardo and others a little later, but today few free-world economists accept it or even find much in it to argue about. Nowhere in my analysis do I state or imply that lumber value is determined solely by labor.

On the other hand, I do use the idea of gross labor productivity, or output per worker, which Mr. Zivnuska labels as a statistical rather than an economic concept. I am interested in the concept not so much for its pedigree as for its usefulness. Comparative productivity trends are generally recognized as valid indicators of comparative long-run price trends. Let Mr. Zivnuska read, for example, Mr. Fourastie's paper on economic progress, presented in 1955 to the International Economic Association Round Table, where it is stated that: "Within one nation, the prices of two products are in inverse ratio to the productivities."[1]

As for lumber cost and price changes, Mr. Zivnuska evidently believes that holding costs down would not "have any particular effect on price." This view is not widely shared within the lumber industry itself. Indeed, one of the trade journals editorialized not long ago that "it is inconceivable" that any manufacturer of lumber "is not well aware of the fact" that an increase in his cost of production will "decrease his profits or else make it necessary for him to get a higher price for his product."[2] Whether the effect in the long run will be upon profit or price is scarcely debatable.

The remainder of Mr. Zivnuska's discussion lends helpful detail to my thesis. Improvements in logging and milling technology, he explains, have barely counteracted adverse trends in timber stocking, quality, and location, leaving productivity in the industry about constant.

Since productivity in the economy tends to increase at a progressively faster rate, the upward trend of real lumber prices will likewise steepen unless the industry raises its productivity. The benefits to be derived from increasing productivity are clearly indicated in the following quotation: "The object of the drive for increased productivity is not merely to improve the technical and economic position of firms, widen the difference between selling price and costs—in a word to increase profits. Its main goal and result is to bring down selling prices and speed up social progress and more particularly raise living standards and the purchasing power of wages and salaries."[3]

FURTHER COMMENTS ON LUMBER PRICE TRENDS

FRANK A. TERBUSH

"The Trend of Lumber Prices" . . . is the type of article that stimulates both thought and questions. Mr. Zaremba has made a good case for cost as a primary factor influencing lumber price.

Aren't there some other factors, however, that also have a bearing on this matter? In the first place, how about demand? The post-war housing boom that slowed only two years ago created a demand for lumber that certainly must have had some effect on lumber price.

Then, there is the subsidy. The F.H.A. and V.A. housing programs were and are direct subsidies to a substantial proportion (55 percent in 1955) of the housing and thus the lumber industry. This subsidy, while not mentioned in the article, must have some influence on lumber price.

I would like also to raise some questions about Mr. Zaremba's basic assumptions. One factor he brushed aside was the significance of stumpage on the cost side. In the 1957 report of the Pacific Northwest Forest and Range Experiment Station, it was pointed out that Douglas-fir stumpage prices for Federal timber in western Washington and Oregon increased from $5.20 in 1944 to $28.90 in 1955. Douglas-fir lumber prices f.o.b. mill increased from $22.20 in 1940 to $80.90 in 1955. Thus, the rate of increase was greater for stumpage than for lumber. It is hard to accept that this wasn't significant in forcing prices up. Labor costs were undoubtedly more important, but stumpage is a factor that cannot be ignored.

I would agree wholeheartedly with Fabricant's conclusions about lumber productivity for the period he studied, but I find it difficult to accept the conclusion that labor productivity has not increased significantly since 1939. The almost universal adoption of the band saw is really a post-war development. The riderless carriage, the use of electricity rather than steam, the adoption of individually powered units, rather than power from a steam-driven jack shaft, improved lumber handling systems and machines have all contributed to significant increases in lumber production per man, per mill, per shift, or productivity by any yardstick you might use. We are getting more lumber and better lumber out of the log in the 1958 mill than was possible to have gotten from the same log in a 1939 mill.

These remarks are one "dirt forester's" view of the article. I would appreciate hearing from Mr. Zaremba or any others who read and enjoyed the article as I did.

IN REPLY

JOSEPH ZAREMBA

It is important to distinguish between changes in lumber prices in the short run, as in the periods 1915 to 1920 and 1940 to 1950, and changes in lumber prices in the long run. My article was concerned with the lumber-price trend, or changes in lumber prices in the long run. The 3-percent discount curve in Figure 45–1 (my article) is a rough approximation of this trend.

Changes in lumber prices in the short period are determined primarily by changes in demand. Change in the lumber-price trend, however, is determined mainly by changes in costs of production. The post-war housing boom mentioned by Mr. TerBush unquestionably has been one of the principal demand factors which pulled lumber prices upward so rapidly immediately after World War II.

As for the cost of stumpage, it is still too early to decide whether the post-war rise in stumpage prices is a short- or long-run phenomenon. If stumpage prices should remain at their present high level, or continue to increase at a faster rate than lumber prices, I agree with Mr. TerBush that stumpage prices will become an important factor in the lumber price trend.

Reliable data on labor productivity in the lumber industry for the post-World War period is not as yet available. Thus, it is impossible to determine whether improvements in logging and milling technology have more than offset continued adverse trends in timber quality, stocking, and location. However, the fact that lumber prices remained at a very high level after the post-war housing boom subsided suggests that labor productivity actually declined after World War II.

REFERENCES

Zaremba's original article

1. Compton, W. 1916. The organization of the lumber industry. American Lumberman. Page 119.
2. Bryant, R. C. 1919. Prices of lumber. War Industries Price Bul. 43. Washington, D.C.: U.S. Government Printing Office. Pages 52–54.
3. Foster, E. 1946. Lumber snafu. Jour. of Forestry 44:396.
4. Hall, R. C. 1946. Snafu in lumber prices. Jour. of Forestry 44:685–686.
5. Marshall, A. 1920. Principles of economics. 8th ed. London: Macmillan and Co., Ltd. Page 349.
6. Fabricant, S. 1954. Economic progress and economic change. 34th Annual Rep. New York: National Bureau of Economic Res. Page 5.
7. Burck, G., and S. Parker. 1955. Productivity: the great age of 3%. Fortune 52(5):102–105, 241–242, 244, 246, 249–250. November 1955.
8. Fabricant, S. 1942. Employment in manufacturing, 1899–1939. Publication 41. New York: National Bureau of Economic Res.
9. ________. 1945. Labor savings in American industry, 1899–1939. Occasional Paper 23. New York: National Bureau of Economic Res.
10. Lowther, E. J., and R. W. Murray. 1946. Labor requirements in southern lumber production. Monthly Labor Review 63:941.

Zaremba's reply to Settel

1. Steer, H. 1952. Stumpage and log prices for the calendar year 1941. U.S. Dept. of Agriculture, Statistic Bul. No. 78. Washington, D.C.: U.S. Government Printing Office. Page 16.
2. ________. 1948. Lumber production in the United States, 1799–1946. U.S. Dept. of Agriculture, Miscellaneous Publication No. 669. Washington, D.C.: U.S. Government Printing Office. Page 24.
3. The New York Lumber Trade Journal. 1940. Volume CVIII.

Zivnuska's comments

1. Zivnuska, John A. 1955. Supply, demand, and the lumber market. Jour. of Forestry 53(8):547–553.

Zaremba's reply to Zivnuska

1. Fourastie, J. 1955. The statistical measurement of various material aspects of economic progress. Papers and Proceedings of a Round Table held by the International Economic Association. Edited by L. H. Dupriez. Page 81.
2. Southern Lumberman. July 1, 1955. Page 20.
3. Fourastie, J. 1957. Productivity prices and wages. Project 235. Paris: The European Productivity Agency of the Organization for European Economic Cooperation. Page 15.

46

CHANGING PRICE PATTERNS AFFECT SOUTHERN PINE LUMBER INDUSTRY

Clark Row
Sam Guttenberg

First published in *Journal of Forestry* 60(2):120–123. February 1962.

BIOGRAPHY

Mr. Row is Chief of Forest Economics in the Forest Service's Division of Forest Economics and Marketing Research, Washington, D.C. His position gives scope to his active interest in forest resource information and planning systems and in methods for evaluating management programs. His work includes also coordinating research in the economics of timber growing and multiple-use management at the regional experiment stations of the Forest Service. A professional forester (Duke University) and an economist (Tulane), his early research assignments were at the Southern Forest Experiment Station, New Orleans. It was there that he made his studies of southern pine lumber prices.

Mr. Guttenberg's biographical sketch appears at Item 24, page 184.

EDITORS' SUMMARY

Prices of southern pine finish lumber have been falling markedly, and of common boards slightly, in relation to dimension prices. These changes are causing the industry to saw more dimension and to work for greater productivity and fuller utilization. They are narrowing the range of log values and reducing the premium on large trees with clear boles. If these trends continue, they will affect forest management profoundly.

QUESTIONS TO CONSIDER

(1) How far can the trend continue toward narrowing the price gap between southern pine clear boards and dimension? What forces bring the two prices together? What forces pull them apart?

(2) Suppose that the trends described by the authors were to continue and that analogous trends meanwhile were to advance to the point that in any locality, all wood had the same value per cubic foot or pound. What could we deduce about the structure of the lumber industry at that time? about the structure of wood-using industry in general?

(3) Describe the principal features of a timber-management program designed for the conditions envisaged in Question 2. What might be the principal differences from current programs respecting land use and timber location? attention given to nontimber values? rotation length? intermediate cuttings? pruning? site and tree improvement? forest protection? other measures?

(4) Under the envisaged conditions, why would not a large tree still be worth more per unit than a small tree?

(5) Analyze the statement (page 386) that large landowners gave up their mills because capital gains (taxes!) were more profitable. How could this be? Explain with numerical examples. What is the crux of this interesting phenomenon?

* * *

Figure 46–1 shows wholesale prices for representative southern pine items. The curves are smoothed by 5-month binomial averaging[1] and show percentage movements for each grade equally. As finish and siding prices indicate, demand weakened substantially between 1956 and 1960; it did not recover even in the favorable 1959 market. No. 2 dimension has been steadily rising relative to No. 2 boards, long considered the bellwether item.

The relation between prices of B and Better and No. 2 boards (Figure 46–2) is now well below the 2.2 ratio that prevailed before World War II. During the war, the OPA set finish prices low to encourage production of common lumber. The ratio regained earlier levels briefly after controls were removed, but dropped during the Korean War when strong demand raised prices of common lumber. Peaks occurred during the 1949 and 1954 recessions, when common lumber dropped more sharply than clears, and again in the poor markets of 1960.

The stability of upper grades in recessions is due to the limited amounts produced and the strong finances of major producers. Lumbermen generally prefer to let inventories of clear grades build up rather than cut prices. Common grades are far more responsive to business conditions because inventories of large-volume items must be turned over to free working capital.

In constant dollars, all southern pine prices have gone down except those of dimension, which have just held their own, and those of small timbers, which have made modest gains.

* * *

Southern pine has felt the full effects of declining use of finish and has been hurt by severe competition from other species. Furthermore, the export market for high-

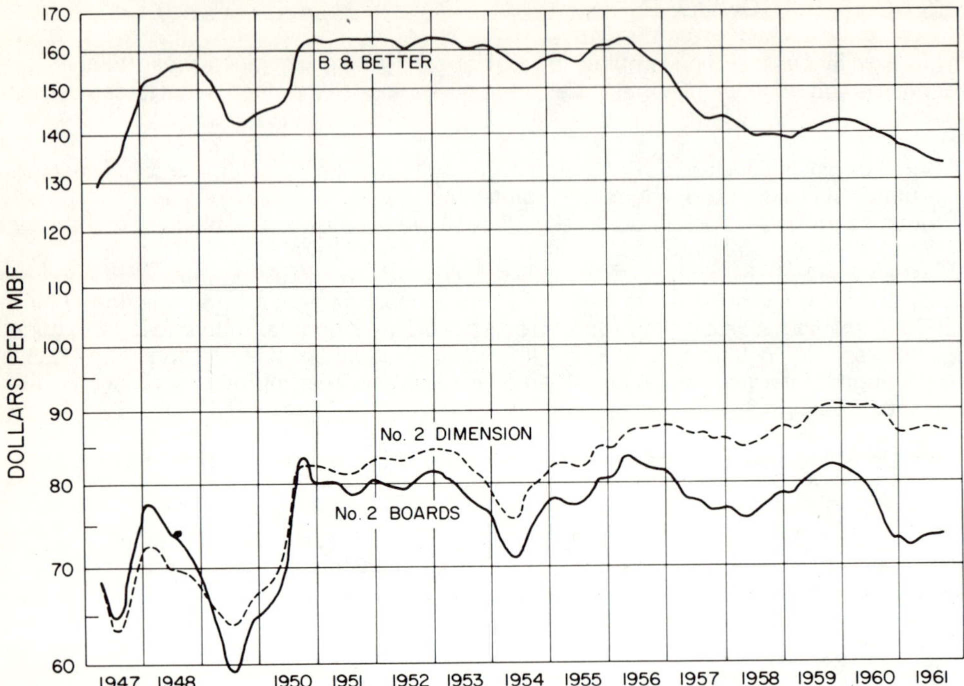

Figure 46–1. Wholesale prices for representative southern pine items. (Source: U.S. Bureau of Labor Statistics.)

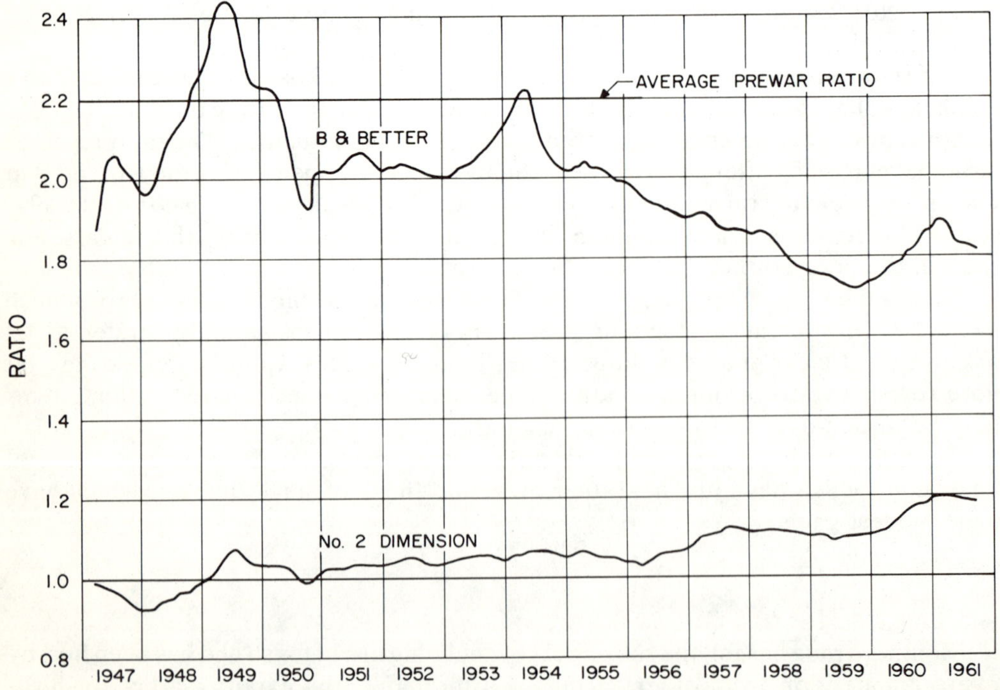

Figure 46–2. Relationships of B and Better and No. 2 dimension to No. 2 board prices.

quality southern pine, brisk in the 1920s and 1930s, has been meager since 1946. Recently it has picked up, but the 80 million feet shipped annually is only one-fourth of pre-war volume.

INDUSTRY ADJUSTS TO MARKETS

The shifting price structure has aggravated the problem of rising costs and narrow profit margins. While receipts on few lumber items have increased in the last 10 years, manufacturing costs have risen steadily. Prevailing wages are nearly twice those of 1947–1949,[2,3] and purchased stumpage has cost $30 to $40 per M board feet since 1950. Outlays for virtually every other item have also increased. To compensate for rising costs, lumbermen have been modernizing and mechanizing their operations.

According to Southern Pine Association statistics, man-hours required per M board feet at large mills have decreased an average 3.8 percent a year since 1947. The gain for the industry as a whole has been still greater, chiefly because small mills—lavish users of manpower—are much less numerous than formerly.

The best gains in productivity—5.5 percent a year—have been in logging and log transportation, both by company crews and by contractors. Mill efficiency has also increased rapidly, but manpower savings in yards and sheds, dry kilns, and shipping have been only 2.4 percent per year since 1947.

Some of the sawmill improvements have been made at a sacrifice of grade recovery. Even though finish is still worth considerably more than common lumber, today's smaller premiums no longer justify the painstaking sawing for grade that was practiced years ago. Many mills now use head saws only to make heavy slabs, flitches, and cants for further sawing by other machines. Time spent in log turning and taper-sawing has been reduced. According to major machinery distributors, at least 170 gang saws have been installed at permanent southern mills during the last 10 years. Mills with such equipment now cut 15 to 20 percent of the regional total. Sawing cants eliminates much edging without greatly affecting grade recovery; but one fifth of the gangs saw round logs, and here grade yields are reduced. Mills with tandem gangs, the first to make cants for the second, are not yet common in the South.

Mills have also changed their sawing policies to cut as much 2-inch lumber as possible. Though no statistics are available, some mill managers say that the proportion of dimension is several times that of 10 years ago. The main limitation on yields of dimension is that many second-growth logs do not meet the density requirements of the Southern Pine Inspection Bureau.[4]

Improvements in mill efficiency can be expected to continue. So far, most gains have come from modifications and expanded use of existing machine types. The development either of new processes, such as slicing logs into boards, or of products not now made from southern pine, such as rotary-cut plywood, would radically change the industry.

Increased product recovery has also helped profit margins. Narrow-kerfed sashgangs and resaws get more lumber from a log than do circular saws. Even more important, pulp chips from slabs, edgings, and trim have become a major by-product. The output of residue chips grew from nothing in 1952 to about 2.8 million cord-equivalents in 1960. The 900 southern sawmills that had installed barkers and chippers by the end of 1960 were cutting more than 50 percent of southern pine lumber.[5] In southern Arkansas, southeast Texas, southeast Georgia, and northern Florida nearly all permanent pine mills chip. Depending on the size of the logs and the minimum grade and length of lumber being cut, chips increase returns by $7 to $12 per M board feet lumber tally.[6] Chief chip production costs are depreciation on equipment, maintenance, and power. There is little or no increase in direct labor.

Large straight logs yield fewer chips per M board feet than small, low-quality ones. For instance, chip production from 17-inch logs, per M lumber tally, is about half that of 8-inch ones.[6] For small logs high chip yields partially offset low lumber values. Chipping tends to narrow the difference in value between high- and low-grade logs.

The latest development is kerf-chips from modified circular head saws.[7] Pioneered by mills in southern Georgia and northern Florida, such saws produce sawdust suitable for pulping.[8] Kerfchips may amount to one ton per M board feet, though with some reduction in lumber recovery.

Another response to the changing price structure has been a search for ways to sell clear lumber in nonstandard forms. Traditionally mills have cut only standard items for general use and have referred special orders to millwork plants. Because customers often need help in meeting grade and other requirements at minimum cost, specialties sales are more demanding than those for construction lumber. One of their advantages is that they reduce dependence on the housing market.

For many firms, the solution to the profit squeeze has been to close down or sell out. A few with large landholdings have become stumpage growers only, because capital gains on timber have been more profitable than sawmill operation. Other mills with timber holdings have been bought out by pulp and paper companies, but further acquisitions are complicated by anti-trust considerations. Mergers between lumber companies have been less widely publicized, but have occurred. Consolidations improve competitive positions by lowering overhead costs and by increasing the efficiency of management. Approximately 15 percent of southern pine is now sawn by firms with more than one permanent mill.

LUMBER PRICES AND LOG VALUES

Changing prices have influenced relative log values directly by lowering the average values of lumber in prime logs, and indirectly by encouraging manufacturing methods that have reduced yields of upper grades. The direct effect may be estimated from average yields of graded logs.

In five studies[9,10,11,12] of logs sorted by the Interim Southern Pine Grades,[13] the percentage of B and Better was:

LOG GRADE	PERCENT
1	30–60
2	20–55
3	0–20
4	0–5

The wide range in yields arises from differences in timber size, type, site, growth rate, sawmill equipment and practice, and lumber grading.

Average value per M board feet of lumber in a log, expressed as a ratio with the price of No. 2 common boards taken as 1.00, is a convenient index of relative log values. Judged by this method, the extra value for lumber from sawlogs of grades 1 and

2 has declined about one-third since 1948–1949; for ordinary logs the difference is slight:

LOG GRADE	RANGE OF QUALITY RATIOS	
	1948–1949	1959–1960
1	1.50–1.90	1.35–1.60
2	1.20–1.90	1.15–1.60
3	1.05–1.45	1.05–1.30
4	0.95–1.20	0.95–1.15

Despite the decline, clear logs are still worth appreciably more than knotty ones.

The second effect, lowering of grade recovery, varies by individual mills. It cannot be estimated from existing indirect sources, such as log-grade research, because in most recent mill studies only one-inch lumber has been cut.

Research to provide data on all production possibilities within log strata would enhance the value of log grades and aid sawmill management. With detailed volume and grade information mill managers could determine the most profitable product mix for current conditions.

EFFECTS ON FOREST MANAGEMENT

The relative decline in net realization from high-quality logs has alarmed growers of sawtimber. If all trees had equal value per M board feet, the optimum rotation would be near the culmination of mean annual increment, properly discounted. The justification for longer rotations—in southern pine over 45 years—rests in part on expectations of premium prices for clear logs. Because the extra value for quality logs has already declined by one-third, optimum sawtimber rotations have been shortened. One difficulty in estimating the effect is that data are sparse for volume yields under management, and nonexistent for grade yields.

Investments such as pruning to obtain clear logs appear less advantageous than they were 10 years ago. . . . Southern pines in adequately stocked stands normally prune themselves early. By age 20 or 30 much of the growth is on clear boles. Efforts to obtain clear wood earlier will pay only if finish lumber sells well above common grades.

The current trend toward increased output of stress-graded dimension emphasizes the need for maintaining an even growth rate throughout a rotation. To obtain such growth will require close spacing in young stands and gradual release thereafter.

WILL PRESENT TRENDS CONTINUE?

Future prices for quality southern pine lumber will largely depend on three factors.

First is the rate at which other materials will be substituted for high-quality southern pine. While the most vulnerable markets have already been lost, those that remain may be less susceptible, particularly if clear lumber becomes steadily cheaper than other materials. Product development and aggressive merchandising may alter some trends.

Second is the intensity of competition from western woods. Prime Douglas-fir stumpage is rapidly disappearing.[14] Many mills have only 5 to 10 years supply left within their hauling distances. Log production declined years ago in western Washington and is now declining in western Oregon, having reached its peak there in 1955. Furthermore, the proportion of prime logs being sawed into lumber is lessening as plywood production increases.

Less than one-third of the remaining western timber is in private ownership, and much of this is growing stock held for sustained-yield management. A growing share of the harvest comes from public lands. Forest Service sales have rapidly increased since 1957, but in some areas they are now at sustained-yield levels. Further increases must come from reappraisal of allowable cuts, and construction of access roads.

Shortages in prime timber will affect not only amounts marketed but also prices of converted products. Present trends should push western lumber prices upward, though increasing imports from British Columbia will have an offsetting effect.

Third is the regrowth of prime southern pine timber. In many areas where sawmills have extensive land holdings, high-quality stumpage is increasing. For example, in southern Arkansas pine sawtimber volume rose 41 percent from 1951 to 1959. Other areas where sawtimber is increasing are southeast Texas, southeast Louisiana, and southwest Alabama. Many industrial firms, the national forests, and a few individuals who began forest management early now have excellent second-growth stands.

The decision which these owners face is how much inventory to maintain. If the downward trend in finish prices continues, firms will reduce stocking, as some are already doing. But if the national supply of prime timber dwindles faster than the demand for clear lumber, as appears likely, price relationships for southern pine may be expected to strengthen. Growers of prime timber will have the advantage.

REFERENCES

1. Croxton, F. E., and D. J. Cowden. 1948. Practical business statistics. 2nd ed. 550 pages. Illus. New York.
2. Southern Pine Association. 1955. Southern pine wages.
3. ———. 1959. Southern pine wages. New Orleans, Louisiana: Southern Pine Association. 19 pages.
4. Southern Pine Inspection Bureau. 1961. 1961 standard grading rules for southern pine lumber. New Orleans, Louisiana: Southern Pine Inspection Bureau. 170 pages.
5. Anonymous. 1961. Crops reached new high in 1960. Pulpwood Production and Sawlog Logging 9(3):30, 32,34,36,38,40.
6. Carpenter, Roswell D. 1950. Amount of chippable waste at southern pine sawmills. U.S. Forest Service Southern Forest Exp. Sta., Occasional Paper 115, 7 pages. Illus.
7. Anonymous. 1960. A report of kerf chipping. Pulpwood Production and Sawlog Logging 8(4):6,8. Illus.
8. Martin, J. S. 1959. Pulping sawdust chips made by a coarse-feed saw. Forest Products Jour. 9:359–360.
9. Guttenberg, Sam, and Joe D. Perry. 1957. Weight-conscious lumbermen. Southern Lumberman 195(2430):54,56. Illus.
10. ———, Donald Fassnacht, and William C. Siegel. 1960. Weightscaling southern pine sawlogs. U.S. Forest Service Southern Forest Exp. Sta., Occasional Paper 177, 6 pages. Illus.
11. McClay, T. A. 1954. Lumber grade yields in the loblolly-shortleaf pine type by the Southern pine log grades. U.S. Forest Service Southeastern Forest Exp. Sta., Station Paper 37, 12 pages. Illus.
12. ———, and H. J. Pawek. 1955. The southern pine log grades: their application in a stumpage appraisal. Southern Lumberman 190(2379):70, 72, 74, 76. Illus.
13. Southern Forest Exp. Sta., et al. 1953. Interim log grades for southern pine. U.S. Forest Service Southern Forest Exp. Sta. 18 pages. Illus.
14. Young, Charles E. 1961. The economics of chips from mill residuals—supply and demand and mutual benefits. Southern Pulp and Paper Manufacturer 24(2):55,56,58,60.

47

SOME ECONOMIC ASPECTS OF FOREIGN TRADE IN FOREST PRODUCTS

H. R. Josephson

Presented at the Western Technical Division of the American Pulpwood Association in Tacoma, Washington, October 11–12, 1967.

BIOGRAPHY

Mr. Josephson spent his early professional years in California: as a student of forestry and of economics at the University of California, as a member of the university's forestry faculty, and as an employee of the California Forest Experiment Station. After a tour with the Air Force in World War II, he joined the Washington Office of the Forest Service, where he worked in the Division of Forest Economics and Marketing Research. He retired in 1973 from the directorship of the Division, after some 22 years in the position. He is known for his leadership of the 1962 and 1970 nation wide studies of the timber situation in the United States. He is known, too, for his many analyses of foreign forestry and of international trade, of which the article reprinted here is an example.

EDITORS' SUMMARY

Log exports to Japan from the Pacific Northwest are a complex issue. Wood-using interests and conservationists are generally opposed to the exports; free-trade advocates and those concerned with payments balances are generally in favor. Our policies and laws, too, favor the status quo. Exporting, as a log-supply problem, is better met by intensifying timber management than by instituting controls.

QUESTIONS TO CONSIDER

(1) What, exactly, *was* the Japanese log-export issue at the time of Mr. Josephson's address?

(2) Japan has long been a good customer for West Coast softwood timber. In the early years of the trade, timber was shipped in the form of huge cants ("Jap squares") for remanufacture at destination. Later, timber was shipped as long logs, often small and even crooked and of formerly despised species. What happened to produce such changes? Why was timber not shipped to Japan as finished lumber, plywood, and other end products—i.e., why, in this case, was the United States' relation to Japan characteristic of a colony-mother country relationship?

(3) In order to understand the issue which Mr. Josephson discusses, one surely must have a grasp of the United States economy, including the forest economy. How much more widely must one's grasp extend? to the situation in Japan? in still other nations? Explain and, if possible, provide the missing information.

(4) How does one explain the fact (pages 392–393) that the domestic income from exporting logs was greater than from manufacturing the logs into lumber?

(5) One solution to the log-export problem, as Mr. Josephson points out, is to grow more timber. What are other solutions? How acceptable are they?

* * *

In this discussion I'd like to refer particularly to log exports as an example of the varied problems that arise in the area of international trade.

The present controversy over export of logs to Japan is only the latest in a series of foreign trade issues of concern to the forest industries of the Pacific Northwest. Just prior to World War II, congressional hearings were held to consider legislation that would have prohibited export of Douglas-fir peeler logs. A few years ago the lumber industry made a strenuous attempt to obtain some curtailment of competing softwood lumber imports from Canada. And over the years, various attempts have been made to limit the flow of imports of other products such as hardboard and hardwood plywood.

The current controversy over log exports is a difficult economic and political issue. It is one of major concern to state legislatures, the Congress, and various agencies including the Forest Service.

Economists are generally agreed that free international trade, at least of non-strategic items, is of mutual benefit to the trading partners. But business groups often take a position, as in other countries such as Canada, that we should limit exports of raw materials, encourage exports of manufactured products, and protect domestic manufacturers from foreign competition. This conflict of basic policies is clearly evident in the log export situation.

TRENDS IN LOG EXPORTS

The recent buildup of log exports to Japan has rapidly developed into a major business. In 1960 we exported about 100 million board feet of logs to Japan, but by 1966 exports had climbed to more than a billion board feet. In the first half of this year, exports were up another 40 percent above a year ago—an annual rate of around 1.4 billion board feet. Hemlock and true fir comprise about 82 percent of these log exports to Japan; Douglas-fir 15 percent; and Port Orford cedar 3 percent.

Trade in logs between the United States and Canada is also of some significance in this picture, although a minor factor compared with the Japanese log trade. Last year, for example, log exports to Canada from the Pacific Northwest totaled 58 million board feet, while imports amounted to 42 million board feet—a net export of 16 million board feet to Canada.

This export surplus in logs is in marked contrast to the U.S. position as a major net importer for all forest products combined. Last year, for example, imports of lumber, pulp, newsprint, plywood, veneer, and other wood-based products were equivalent to 2.2 billion cubic feet of round timber. This was nearly 4 times the volume of timber exports equivalent to 0.6 billion cubic feet. Net imports of 1.5 billion cubic feet equivalent made up 13 percent of the total consumption of forest products in the United States.

Softwood log exports from western Oregon and Washington make up a sizable proportion of the total harvest of logs in this area—over 9 percent of last year's cut of around 12 billion board feet. These exports equaled only 2 percent of the total output of timber in the United States. But in local areas contiguous to ports where procurement of export logs is concentrated, the growing log export business must be considered to have major impacts.

FUTURE JAPANESE MARKETS

The prospective trend in Japanese demand for logs is necessarily conjectural, but the trend will undoubtedly continue upward. Articles in Japanese trade papers and comments of Japanese spokesmen suggest that log exports from the Pacific Northwest might easily double within the next decade. In addition to an expansion of log imports for production of lumber in Japan, there is the additional possibility that the Japanese might develop a softwood plywood industry based upon Douglas-fir or other species from the Pacific Northwest.

The growing export of chips to Japan will also have an effect on the wood supply situation in this region. In the first half of this year, chip exports reached 166,000 cords. Each month chip exports climb as more specialized vessels designed to carry chips go into service. It would seem to be only a matter of time before essentially all plant residues in this region, and sizable volumes of logging residues and thinnings as well, will be chipped for either the domestic or the export market.

Increased shipments of chips from British Columbia to Japan may also lead in time to a marked reduction in the movement of chips from B.C. to U.S. mills in the Northwest. Last year such imports amounted to about 822,000 tons.

Japan consumed about 32 billion board feet of timber last year. Two-thirds of this was obtained from the forests of Japan, and one-third was imported. North America supplied about 25 percent of the timber imports and the Soviet Union about 16 percent. Other Pacific countries supplied 59 percent of the imports, mainly tropical hardwoods. It is of interest that Japanese timber trade with the U.S.S.R. has been expanding under recent trade agreements, including exchange of Siberian logs and lumber for Japanese-built pulp mill equipment.

PRICE IMPACTS

The rise of log exports to Japan undoubtedly is a factor behind the growing problems of log procurement and rising prices for stumpage in the Pacific Northwest, although a factor that is difficult to quantify. Between 1960 and 1966 national forest stumpage prices for hemlock, for example, roughly doubled from $10.50 to $20.60 per M board feet.

The principal group adversely affected by log exports seems to be made up of the smaller nontimber-owning sawmills in areas of concentrated log procurement, but export competition is not the only problem faced by these mills. The number of sawmills operating in the Pacific Northwest has dropped sharply since 1950, a trend evident long before the recent rise in log exports. Some of the difficulties encountered by the nontimber-owning mills thus stem to a large extent from competition and continuing adjustments within the evolving timber industries of the area.

The lumber industry as a whole in the Douglas-fir region, though comprised of fewer and fewer mills, has increased output significantly since 1960 during the period of rising log exports. Output of peeler logs for the domestic plywood industry also increased by some 2 billion board feet during this period. As yet there appear to be no sizable impacts of log exports on the pulp and paper industry of the region nor on the plywood industry.

POSSIBILITIES FOR INCREASING LUMBER EXPORTS

Some of the opposition to log exports to Japan rests upon the assumption that a reduction in log exports would lead to an increase in domestic lumber production for the Japanese market. This is a doubtful assumption.

Japanese spokesmen have stated that they prefer to import logs instead of lumber in order to obtain lumber products manufactured to Japanese specifications. They assert that U.S. producers cannot or will not cut lumber to Japanese standards, although this problem does not seem to be of as much concern to lumber producers in British Columbia, nor is it of significance in the export of cants and squares. The major advantage to the Japanese in importing logs or squares lies in the generation of employment and income from the manufacture of lumber and other wood products obtained from the imported material. Because of such factors, in 1966 the Japanese imported only about 26 million board feet of lumber from the Pacific Northwest, 174 million board feet from the United States as a whole, and about 270 million board feet from Canada.

Thus it seems probable that if U.S. log exports were curtailed, Japan would attempt to increase log imports from other sources such as Siberia and New Zealand, and perhaps increase imports of squares and other lumber items from British Columbia, Siberia, and Alaska.

RELATIVE INCOME FROM EXPORTS VERSUS DOMESTIC MANUFACTURE

The question of whether log exports provide as much income to the regional economy as domestic lumber manufacture has been a subject of much discussion. One study of this question by the Pacific Northwest Forest and Range Experiment Station showed that a sample of export logs produced approximately $100 in income per thousand board feet for landowners, loggers, longshoremen, and exporters. The same logs manufactured into lumber at domestic mills would have produced an estimated value of approximately $95 per thousand board feet.

The direct employment provided, on the other hand, was estimated at about 8 man-hours per thousand board feet for log exporting, compared with about 13 man-hours in domestic lumber manufacture. Such differences in employment would not be as great, however, if indirect employment in providing capital facilities were taken into account.

There are additional secondary or indirect benefits beyond the sawmill from domestic manufacture of logs, such as in transportation and marketing of lumber. But such values are also generated in handling goods imported from Japan or other countries in exchange for export logs.

During recent years, moreover, when timber supplies have been adequate for both export and domestic markets, the income generated in the Northwest by log exports must be considered as an addition to, rather than in lieu of, income from domestic lumber manufacture.

ADEQUACY OF TIMBER SUPPLIES

If log exports were curtailed, it is possible that the Pacific Northwest would supply somewhat more lumber for the U.S. market. At the present time, any such increase could well be at the expense of less competitive areas such as parts of the Rocky Mountains.

Within the next decade or so, on the other hand, it is likely that rising demands for timber products in the United States will require all of the available cut in both the Pacific Northwest and in other western areas. Thus we can expect that log exports will be in more and more direct competition with domestic manufacture of timber products, and controversies over log exports very likely will be accentuated.

On most public lands in the Northwest cutting is at or above the level that can be sustained permanently with present forest management efforts. Increased intermediate cutting will be required even to maintain the levels recently achieved. Cutting on private lands is expected to continue the downward trend of recent years. Hence the desirability of intensifying forest management in the region seems clearly evident.

There are some current possibilities of easing the pressures for timber in the Pacific Northwest by shifting part of the log export demand to the Rocky Mountain area. This could amount to as much as 300 million board feet annually.

Alaska is also mentioned as an alternative source of logs, but in view of the expanding Alaskan production of cants and squares for the Japanese market and policies to promote local industry, prospects are not promising for increased log supplies from this area.

FOREIGN TRADE POLICY CONSIDERATIONS

Proposals to curtail log exports encounter an objection that the United States is committed to encouragement of international trade and general reductions of trade barriers. Under the Trade Agreements Program started in 1934, the United States has aggressively moved to lower tariffs and to foster trade throughout the world. The

General Agreement on Tariff and Trade (GATT) signed by the United States and some 60 other countries is dedicated to reduction of tariffs and other trade barriers. To curtail log exports would conflict with this basic policy.

An adverse balance of international payments further complicates the matter of log exports. Last year, for example, total receipts from exports of goods and services from the United States amounted to $43 billion, while expenditures for imports of goods and services were valued at $37.9 billion. This favorable balance of goods and services was more than offset, however, by grants to foreign countries, U.S. capital investments in foreign countries and other payments. As a result the United States suffered a balance of payments deficit of about $1.4 billion. Such continuing deficits in our balance of payments have lead to a rather precipitous drop in U.S. gold stocks from a post-war peak of around $25 billion to a current figure of about $13 billion.

The value of log exports to Japan, amounting to $94 million in 1966, is relatively small compared with the total U.S. balance of payments deficit. Nevertheless, there is understandable reluctance to sacrifice any business that helps balance our international trade position.

Still another factor is the special position of Japan as a U.S. trading partner in a highly strategic part of the world. Japan is the second best customer of the United States, with purchases in 1966 amounting to some $2.3 billion and sales to the United States about $3 billion. Both the economic and the political importance of our growing international trade with Japan must be considered when appraising any proposal that might begin a series of retaliatory actions.

EXPORT CONTROL ACT OF 1949

Effective action to curtail Japanese log exports would presumably have to be based upon new legislative authority, even though the Export Control Act, as amended, is still in effect. This Act empowers the Administration through the Department of Commerce to impose export controls on specified commodities on grounds of "national security," "foreign policy," or "to protect the domestic economy from the excessive drain of scarce materials and to reduce the inflationary impact of abnormal foreign demands."

The latter provision of this Act was used in 1964 to establish quotas on exports of black walnut logs. After a year this control was terminated, however, despite continuing strong pressures from industry and congressional sources for continued limitation of walnut log exports. In the case of walnut, as in the current matter of softwood log exports from the Pacific Northwest, the Department of Commerce has concluded that the Export Control Act of 1949 is not an appropriate authorization for conserving a natural resource that may be in short supply.

Part of the difficulty in establishing controls on log exports stems from the fact that the GATT requires that an exporting nation placing controls on exports of an exhaustible natural resource must also restrict domestic consumption. The Export Control Act does not include authority to impose such domestic controls. Also, it's a pretty safe assumption that controls on domestic production or consumption would not be welcomed by U.S. timber industries. And restrictions on volumes of timber cut would certainly be opposed by many landowners and other economic groups who benefit from log exports.

OTHER LEGISLATION

Among other attempts to limit log exports, you will also recall a request made to the Secretary of Agriculture in 1964 to use the authority of the Sustained Yield Act of 1944 to establish a federal "sustained yield unit" made up of 13 national forests in the Pacific Northwest. Such action was proposed to assure that national forest timber cut in this area would be available only for domestic manufacture. This proposal was rejected by the Secretary, however, on the grounds that a sustained yield unit was not necessary for the purposes of the Act, nor an appropriate way of meeting problems of international trade.

The set-aside program under the Small Business Act also has been suggested as a means of curbing log exports, even though this program was designed only to aid small business obtain a fair share of available public timber. In the Puget Sound area, where much of the export log procurement is concentrated, nearly half of the timber recently sold by the Forest Service has been included in set-asides for small businesses. Under SBA regulations at least 70 percent of the timber in these set-asides must be processed domestically.

The present set-aside procedure would not in itself preclude the export of a major part of the logs being produced on federal lands. And even if set-asides of public timber were further increased, such action would not constitute an effective export control as long as there are unrestricted cutting and disposal of the major flow of logs produced from private and state lands.

Other proposals have been made to limit exports of logs cut on national forests under authority of the Act of June 4, 1897, which provides that national forest timber must be used in the state of origin. This policy is being followed in Alaska to require local processing of national forest timber.

Restriction of interstate and export shipments of logs produced in national forests in Oregon and Washington, for example, could have far-reaching adverse impacts on the timber industries in these states. As in the case of other proposals relating only to public lands, moreover, log export controls could not be effective unless applied to both public and private timber.

Some groups have suggested that the Japanese voluntarily impose quotas on log imports from the Pacific Northwest, or accept a mixed quota of logs, lumber, and plywood. Such proposals to date have not been acceptable to the Japanese. Nor would action by Japan to limit log imports be entirely consistent with the General Agreement on Trade and Tariff or with related U.S. trade policy. Nevertheless, as a result of suggestions made by Secretary Freeman at a meeting of the Joint U.S.-Japan Committee on Trade and Economic Affairs held in Washington in mid-September, that Committee has agreed to a joint examination of the problem of reconciling conservation and trade interests in the use of timber resources of the Pacific Northwest and Alaska. This forthcoming joint review of the log export situation could lead to at least a partial solution of the log export problem.

CONCLUSIONS

In these remarks I have outlined some of the complex economic and political issues involved in the difficult controversies that arise in the field of foreign trade.

These relate, in the case of softwood log exports, for example, to the clash of opposing domestic interests in lumber manufacture versus log exporting, the changing economic pattern of the timber industries, international trade commitments, the problem of balance of payments, the strategic importance of U.S.-Japanese trade, and the lack of adequate legislative authority to restrict log exports.

The log export problem shows no signs of going away. It is a matter that is being given continuing serious attention by the various federal departments that are concerned, as well as by members of Congress and other groups.

One immediate step that can be taken is to improve the total timber supply situation on the West Coast to better meet both domestic and export demands. This means accelerating those forestry programs that will grow more timber for the future. It means providing more access roads for intermediate cutting and other forest management practices. It means stretching the available timber supply through more complete utilization in both the woods and the mills of the region. Such measures could well be the most significant long-run response to the log export problem.

48

RISK AND THE LUMBER FUTURES MARKET

Vernon L. Robinson

First published in *Forest Products Journal* 20(12):11–16, December 1970.

BIOGRAPHY

Mr. Robinson has, since 1957, been Research Forester in the Southeastern Forest Experiment Station, stationed at Asheville, North Carolina and at Athens, Georgia. He is the author of professional papers on a range of subjects in forestry and in resource economics, including marketing, as illustrated here. A native Iowan, he earned bachelor's and master's degrees in forestry at Iowa State University, then a Ph.D. at the University of Georgia. He is an active member of the American Agricultural Economic Association, the Forest Products Research Society, and the Society of American Foresters.

EDITORS' SUMMARY

The futures market for white fir dimension lumber at the Chicago Mercantile Exchange permits many lumber manufacturers, dealers, and users to hedge against losses and to enjoy other advantages. Because of the nature of probable gains and risks, large-scale western lumbermen can hedge to good advantage; hardwood lumbermen cannot; southern pine firms are intermediate.

QUESTIONS TO CONSIDER

(1) What is "hedging"? What is its purpose? How is it made possible by a market for futures?

(2) How is it possible for a firm in one industry (e.g., southern pine) to hedge in the futures market for the product of a different industry (e.g., western fir)? What conditions would have to hold in order for a lumberman to be able to hedge successfully in the market for wheat futures? Why can there not be just one large, composite market for futures, in which anyone who wants to hedge may do so?

(3) If a holder of stocks negotiates a short hedge by selling futures and if the market later falls off, the hedger is protected, but someone else has paid for his protection. Who? Why should anyone be willing to provide such a service?

(4) Would not a firm which hedged short and then witnessed a large price increase feel that it had made a mistake by depriving itself of a fortuitous gain?

(5) All in all, and certainly if the sample in this book is representative, students of wood-using industry have given a great deal of attention to lumber—possibly more than to pulp-paper, though the latter promises to be of greater significance. How do you explain this emphasis upon lumber?

* * *

The advent of futures trading in dimension lumber at the Chicago Mercantile Exchange on October 1, 1969, has stimulated questions from those in forestry circles who are unfamiliar with such a market. What is it? How does it work? What can it do for the lumber industry? This paper describes the essentials of futures trading with respect to lumber, presents a few examples of how the industry might benefit from this activity, and explores the trading potential with respect to various segments of the industry.

THE LUMBER EXCHANGE

A lumber futures exchange is an organized market place where buyers and sellers may conduct trading in contracts for the future delivery of lumber. These contracts are agreements between the two parties to deliver or accept delivery of a specific species, grade, and size of lumber at a specified time and place at a definite price. Only members of the exchange are allowed to do the actual buying and selling, but they, for the most part, act as brokers for lumber producers, wholesalers, retailers, builders, and speculators. These customers establish the future prices by their competitive bidding, which reflects their estimates of future supply and demand conditions. The prices change from time to time as these expectations are altered in light of current and prospective market influences. The Chicago Mercantile Exchange provides the market place, supervises the trading with established rules, and acts as a clearing house in guaranteeing the fulfillment of all contracts. It also disseminates the market information instantly throughout the country and thereby encourages wider participation in the price formulation process.

THE LUMBER CONTRACT

To facilitate trading, futures contracts are bought and sold in terms of standardized trading units for delivery in certain specified months. For lumber, the trading unit is 40,000 board feet (car-load lots) of kiln-dried (the 1971 contract is being written for trading units of 90,000 board feet of kiln- and air-dried) construction-grade white-fir nominal 2 by 4s of random length, manufactured in California, Idaho, Montana, Nevada, Oregon, or Washington.[1] No other species, grade, or size of lumber is deliverable under this contract. The fact that the lumber contract is so restrictive with respect to species, grade, and size is no great disadvantage to most potential Exchange participants, however, because the delivery option would be exercised only under rare circumstances. Experience in other commodities indicates that only 2 percent or less of the contracts traded result in delivery.[2] The normal situation is for a given contract position to be cancelled by taking an opposite position of equal size in the same future prior to delivery.[3]

Each delivery month in which trading takes place is called a future. At the time of this writing, the trading months are July, September, and November 1970 and January, March, and May 1971.

THE HEDGE

The basic reason for the existence of a futures market is to provide the facilities which can help business firms protect themselves against financial losses arising from adverse commodity price movements.[4,5] The Exchange participant is able to transfer most of his price risk to the speculator by counter-balancing his inventory or forward price commitments with investments in futures contracts—a process known as hedging. There are two types of hedges, the short hedge and the long hedge. The short hedge is used by lumber producers and distributors to protect the value of their inventory. This hedge is initiated by selling futures. The long hedge is used to protect the forward price commitments of producers, distributors, and builders. It is initiated by buying futures.

The essential feature of hedging, whether it is a short or long hedge, is that the trader simultaneously holds opposite positions in the cash and futures markets. If the lumber producer holds an inventory, then he should sell lumber future contracts. When the inventory is sold, the commitment in the futures market is simultaneously liquidated. This is done prior to actual delivery by buying an equivalent number of contracts of the same future. To the extent that the prices in the two markets move in unison, the gains in one market will offset the losses in the other market.

To illustrate the arithmetic of a short hedge, let us assume that a white-fir lumber producer wants to buy a tract of timber on October 1. The tract is expected to yield 400,000 board feet of 2 by 4s at a cost of $77 per M.B.M. (thousand board measure) for stumpage, manufacturing, and overhead. Because he wishes to avoid the price risk associated with the sale of his lumber 5 months hence, the producer calls his local broker and finds that March lumber futures are selling at $80 per M.B.M. when adjusted to his cash market location. The futures price is generally considered to be for delivery in Portland, Oregon, but it should be adjusted to represent the individual trader's cash market by adding the appropriate transportation costs. The producer decides that this is a desirable price for his lumber, so he buys the stumpage and simultaneously instructs his broker to sell 10 future contracts for March delivery. A commodity account has previously been opened for the producer at his local stockbrokerage firm, and he has deposited the $2,000 initial margin for the 10 contracts.

During the next 5 months, while the hedged position is held, the price of the March future can be expected to fluctuate. To insure the integrity of each transaction against a price change, the Exchange requires a maintenance margin of $100 per contract. This means that if the future price rose so that the initial margin declines by more than $100 per contract ($2.50 per M.B.M.), the producer would be expected to bring the margin back up to $100 per contract.

On March 3, the lumber is ready for sale, but 2 by 4 prices have fallen to $74 per M.B.M. in both the cash and futures market. The producer decides to sell the lumber and instructs the broker to buy ten March contracts. This closes out the producer's commitment in the futures market. His selling price is the $74 per M.B.M. obtained in his normal market channels plus a $6 per M.B.M. gain obtained in the futures market, equaling the $80 per M.B.M. he had hedged for back in October, Table 48–1. The cost of the hedge was $300 in broker commission fees ($30 per contract) plus the interest given up on the margin deposit, a total of less than $1 per M.B.M. The net selling price was therefore about $79 per M.B.M. If the producer had not established this hedge, he would have obtained $2,000 less on the transaction.

Table 48–1. The Short Hedge on a Declining Market

TIME	MARKET	
	Cash	Futures
Oct 1 Mar 3	Buy @ $77 Sell @ $74	Sell @ $80 Buy @ $74
	Loss $ 3	Gain $ 6
Selling Price $74 + $6 = $80		

Table 48–2. The Short Hedge on a Rising Market

TIME	MARKET	
	Cash	Futures
Oct 1 Mar 3	Buy @ $77 Sell @ $82	Sell @ $80 Buy @ $82
	Gain $ 5	Loss $ 2
Selling Price $82 − $2 = $80		

The hedging procedure not only protects the producer from major losses due to downward price movements, but it also prevents him from obtaining major gains from price advances. If instead of falling, the cash and futures prices had risen to $82 per M.B.M. on March 3, the producer would have gained $5 per M.B.M. in the cash market, but he would have lost $2 of this in buying back the futures contracts, leaving him again with a selling price of $80 per M.B.M. (Table 48–2).

The results of a long hedge are exactly the opposite of those of the short hedge described above. For example, a contractor who hedged on October 1 by buying contracts at $80 per M.B.M. would have lost $6 per M.B.M. on his futures transaction. He could have bought the lumber at $74 per M.B.M. on March 3 had he not hedged (Table 48–3). While this hedge may be considered a poor speculative decision in light of the outcome, the contractor was able to make other decisions—a firm forward commitment, for example, with the assurance that his lumber costs were going to be about $80 per M.B.M.

Table 48–3. The Long Hedge on a Declining Market

TIME	MARKET	
	Cash	Futures
Oct 1 Mar 3	Sell @ $80 Buy @ $74	Buy @ $80 Sell @ $74
	Gain $ 6	Loss $ 6
Buying Price $74 + $6 = $80		

Table 48–4. A Hypothetical Short Hedge of Douglas-fir 2 by 4s

TIME	MARKET	
	Cash	Futures
June '69 July '69	Buy @ $110 Sell @ $ 96	Sell @ $95 Buy @ $80
	Loss $ 14	Gain $15
Selling Price $80 + $15 = $95		

HEDGING BENEFITS

As previously mentioned, the principal benefit of a futures market lies in its ability to help participants avoid the uncertainty of major price changes over time. Lumber producers, distributors, and builders incur this uncertainty whenever they maintain an inventory of stumpage, logs, or lumber or plan for purchases of these items in advance of their need. It arises from the probability of a decrease in the value of an inventory or an increase in the cost of the forward commitments due to unfavorable price movements. For example, between June and July of last year, the wholesale price of construction-grade Douglas-fir 2 by 4s dropped $14 per M.B.M. This resulted in a substantial decrease in the value of inventories of this lumber item. Had these Douglas-fir 2 by 4s been hedged on the white-fir contract in June and sold in July, the producer would have gained $1 per M.B.M. rather than losing $14 per M.B.M. (Table 48–4).

In addition to risk avoidance, the futures market can also provide other benefits to the lumber industry, depending upon the type of hedge used. The trader, using the long hedge, for example, can buy white-fir 2 by 4s for delivery at some future date with a down payment of only the margin requirements. This reduces the amount of capital tied up in inventory and eliminates or reduces the fixed investment in storage facilities and maintenance costs.[3] The contractor using this hedge can fix his lumber costs well in advance, thus removing the uncertainty of lumber price changes from his forward contracting.

The trader who establishes a short hedge in a distant future when lumber is selling at a premium to the cash market can earn a carrying charge from the hedge. This type of hedging manifests itself in the form of basis trading where the basis is defined as the difference between the cash and future prices. These cash-future price spreads are recognized as a function of stocks in the case of seasonally produced commodities that need to be stored in significant quantities, such as wheat, corn, and oats.[6,7] This price-of-storage theory may also apply to lumber and help lumber products and distributors earn a carrying charge on their drying-yard and inventory holdings.

Finally, the short hedge also increases the hedger's capacity to borrow money. Banks tend to lend less money than the current value of the collateral pledged as security for a loan because they wish to protect the value of these assets against changes in their market price. When the collateral is hedged, they generally recognize that the price risk has been diminished and are therefore willing to lend a larger percentage of its current value. For example, if a bank will lend a firm 50 percent of the current value on an unhedged inventory, then it may be willing to lend it up to 70 percent, or more, on a hedged inventory.[8,9] This enables the hedger to handle a larger volume of the commodity on a given initial investment and thereby reduces his unit costs of operation. With constant or increasing returns to scale, such a hedge may be worthwhile even though it results in a loss on the futures transaction.

TRADING POTENTIAL

When opposite positions are taken in the cash and futures markets, the hedger will be protected against major changes in price, provided the cash price of the commodity traded parallels the futures price. Therefore, if price protection is the main benefit to be obtained from futures trading, then the incentive for participating in the futures market will hinge largely upon the magnitude of the price risk associated with lumber production, distribution, and use, and upon the gain, i.e., the reduction in risk that can be obtained from hedging. Those firms confronted with a low price risk or a small hedging gain would not be expected to participate, unless other hedging motives controlled their actions. Thus, the trading potential can be established in broad, general terms by showing the relative risk and potential gain arising from various segments of the lumber industry.

THE PRICE RISK

Historically, lumber prices have fluctuated widely over time and as a result have been an important element in the uncertainty facing the lumber industry. Some form of the dispersion of these price fluctuations can therefore be taken as a measure of risk. For example, risk can be cast in terms of the difficulty lumbermen have in formulating their price expectations over various intervals of time. Figure 48–1 shows the 95-percent confidence interval containing the expected change in price for white-fir dimension lumber for time periods of 1 to 6 months. As the time period between successive price comparisons lengthens, the gap between positive and negative price changes widens, indicating the greater degree of uncertainty associated with longer planning horizons. Although there is a slight positive bias, negative price changes can be expected with nearly the same frequency as positive ones. The acuteness of this risk is documented by the fact that the lumber and wood products industry has a lower profit rate and a much larger profit fluctuation than either the average of all manufacturing industries or the competing stone, clay, and glass industries.[10]

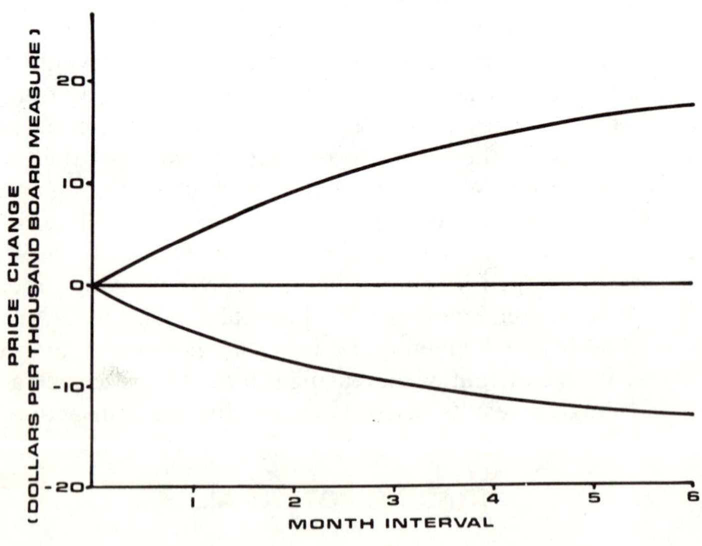

Figure 48–1. A 0.95 confidence interval containing the expected change in price for white-fir dimension for time intervals of 1 to 6 months.

Table 48–5. Coefficient of Variation and Correlation on Lumber Prices, Bureau of Labor Statistics Commodities, for the 10-Year Period, 1959–1968.

BUREAU OF LABOR STATISTICS COMMODITY	COEFFICIENT OF Variation	COEFFICIENT OF Correlation*
	Percent	
DOUGLAS-FIR	15	0.971
Flooring, C, and Better	18	0.641
Drop Siding, C, and Better	13	0.650
Dimension, Construction, Dried	14	0.984
Dimension, Const. 25 Pct., Std., Green	17	0.974
Boards, Construction, Dried	15	0.924
Boards, Const. 25 Pct., Std., Green	16	0.894
Timbers, Construction, Green	18	0.896
Dimension, Utility, Green	24	0.927
Boards, Utility, Green	17	0.880
Timber, Utility, Green	20	0.943
SOUTHERN PINE	12	0.918
Flooring, B, and Better	10	0.813
Finish B, and Better	10	0.804
Drop Siding, C, Grade	13	0.785
Dimension No. 1	11	0.928
Dimension No. 2	13	0.939
Boards, No. 2	13	0.921
Boards, No. 3	16	0.917
Timbers, No. 1	9	0.863
Timbers, No. 2	11	0.887
OTHER SOFTWOOD	16	0.975
Ponderosa Pine, Boards, No. 3	21	0.924
Ponderosa Pine, Boards No. 4	23	0.881
Ponderosa Pine, Shop, No. 2	17	0.774
OTHER SOFTWOOD (Cont'd.)		
Larch, Douglas-fir, Dim., Std./Better	17	0.988
White Fir, Dim., Std., and Better	18	1.000
Eastern White Pine, Boards, No. 3, Com.	7	0.539
Redwood Boards, F. G., Green	29	0.830
Redwood, Bevel Siding, Clear All Heart	10	0.767
Redwood, Boards, Clear, F. G., Dry	11	0.750
Cypress, C Select, Finish	3	0.198
Cypress, No. 2 Common	5	0.604
Western Hemlock, Dimension	20	0.927
Cedar, Bevel Siding, Clear	12	0.819
HARDWOOD LUMBER		
Oak, Red, Flooring, Select	11	0.626
Oak, Red, No. 1 Common	18	0.701
Oak White	7	0.344
Gum, No. 1 Common	16	0.223
Gum, No. 2 Common	15	0.581
Maple, Flooring	13	0.708
Maple, No. 1 Common	7	0.417
Poplar, No. 1 Common	10	0.643
Poplar, No. 2-B Common	7	0.357
Cottonwood, No. 2 Common	7	0.469
Basswood	7	0.572
Birch, No. 1 Common	6	0.500
Beech, No. 2 Common	11	0.682

*With white fir, dimension, standard, and better used as a surrogate for the near-month future price.

When Bureau of Labor Statistics wholesale price data[11] and a 3-month time interval are used, the coefficient of variation shows that significant differences occur in the magnitude of the risk among individual lumber items (Table 48–5). For example, the relative stability of southern pine prices suggests that southern pine lumbermen are not as likely to use the futures market for hedging against price risk as western lumbermen.

Price risk can also be defined as the standard deviation in income resulting from holding a lumber inventory from one time period to another. The risk of holding x units is then given by the product xs where s is the standard deviation of the probability distribution for price change between the two time periods. (Johnson[12] shows the derivation of this measure when risk is defined as the variance in expected income. Here the standard deviation of ex-post price changes was employed rather than assuming a distribution of subjective price changes.) The risk of holding different quantities of white-fir dimension over various intervals of time is shown in Figure 48–2. For example, a firm holding a 10 million-board-foot inventory of white-fir dimension for 1 month has a price risk of about $25,000. If the inventory is held a longer time, say for 3 months, the price risk rises to $57,000. Also, the larger the inventory, the greater the price risk. Since the risk rises as the quantity of lumber held increases and as the planning horizon lengthens, large firms would have a greater incentive for participating in the lumber futures market than small firms. When various species were compared, it was found that the risk of holding southern pine dimension was considerably less than that of holding the western species (Figure 48–3).

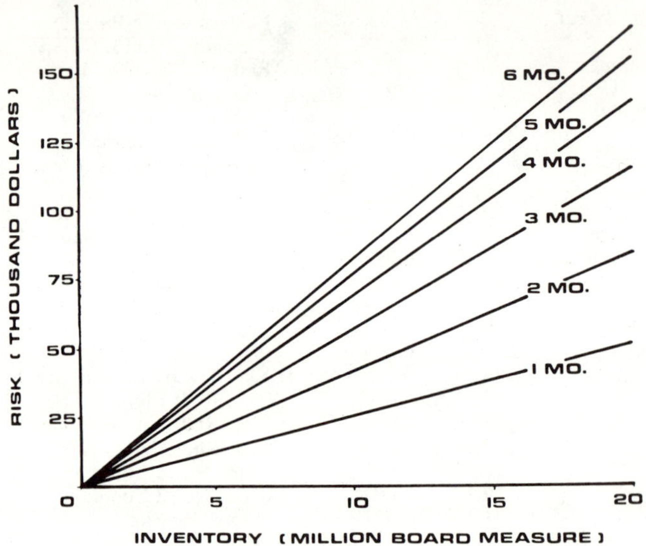

Figure 48–2. Risk associated with holding white-fir dimension by size of inventory is shown for time intervals of 1 to 6 months.

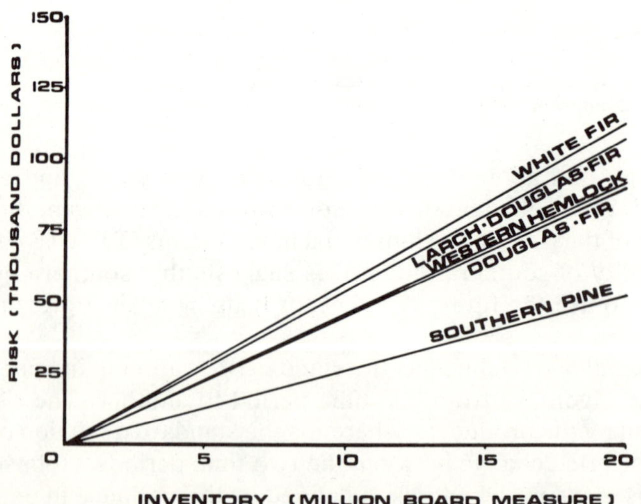

Figure 48–3. Risk of holding dimension lumber over a 3-month interval by species and size of inventory.

THE HEDGING GAIN

Hedging effectiveness is based upon the assumption that the future and cash commodity prices move more or less up and down together. Thus, if there is a high positive correlation between the two prices, then any loss in the cash market is compensated for by a corresponding gain in the futures market, and vice versa. The gain to be derived from hedging is therefore a function of the price correlation between the commodity being traded on the Exchange and the commodity being hedged. Specifically, the price risk of a hedged position designed to minimize the standard deviation in income of holding an inventory is given by xsk where k is the coefficient of alienation between the two prices. (The coefficient of alienation is defined as $\sqrt{(1 - r^2)}$, where r is the correlation coefficient.) The first part of this expression, xs, is the risk of an unhedged position. The difference between the risk of a hedged and unhedged position is the hedging gain (Figure 48–4). Since the percentage gain, with respect to the risk, increases at an increasing rate as the correlation coefficient increases, a high positive correlation is desirable.

The correlation between the futures price and the price of the lumber item to be hedged is, therefore, a useful criterion for judging both the gain to be expected from the hedge and, consequently, the potential participation in futures trading by various segments of the lumber industry. When these computations were made with white-fir dimension as a surrogate for the near-month futures price and 47 other Bureau of Labor Statistics lumber price series, the correlations on all the hardwood and most of the softwood lumber items were found to be too low for effective hedging (Table 48–5).

The highest correlations were on dimension, timber, and board lumber items, which account for most of the softwood lumber produced. But, again, significant species differences may exist. The correlation on the composite Douglas-fir series, for example, was 0.97, while southern pine was only 0.92. Individual item comparisons show that the southern pine correlations were lower than those on the western items in 5 out of 8 cases and none of the southern pine correlations were above 0.94. Since a correlation of 0.95 or greater is recommended for effective hedging,[13] western lumbermen may have a distinct hedging advantage over southern lumbermen due to differences underlying regional supply and demand conditions.

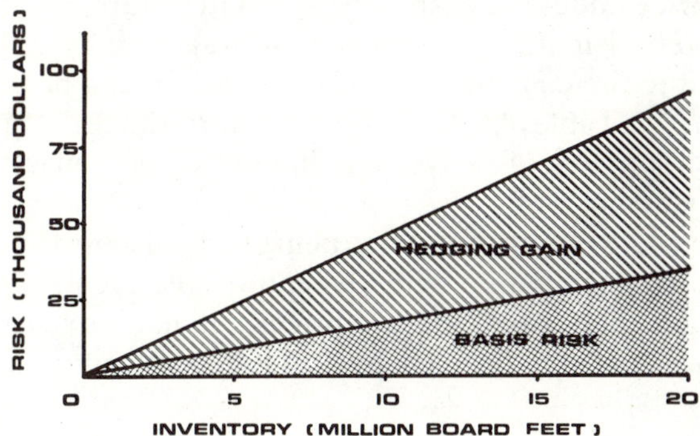

Figure 48–4. Ex-post hedging gain and basis risk associated with holding a hedged inventory of western hemlock dimension over a 3-month interval.

The idea that successful hedging depends upon a high correlation between the cash and futures price is in keeping with traditional theory. Another theory, however, suggests that a high correlation is an improper standard by which to test the effectiveness of hedging.[14] It is based upon the assumption that cash-future price spreads are less variable and more predictable than price changes. Hence, according to this theory, successful hedging depends only upon the seasonal basis pattern being reasonably repetitive over time.

These price correlations [of Table 48–5], however, were low in comparison with those found in another study where price series were analyzed with the publication "Random Lengths" as the data source.[15,16] Here it was shown that various western softwood lumber prices correlate so closely that one futures contract could effectively apply to other types of lumber with respect to species, size, grade, and moisture content. Perhaps the Bureau of Labor Statistics data are too aggregated for meaningful correlation comparisons. In any event, intelligent use of the futures market requires the individual participant to know his own cost and market-price situation in relation to the specifications called for in the futures contract. It is the correlation between these prices that will determine the gain an individual can expect from hedging.

A high positive price correlation is also desirable because when a futures position is initiated the trader assumes another type of risk called the basis risk (Figure 48–4). The smaller the (absolute) value of the correlation coefficient, the greater the basis risk. The basis is defined as the difference between the current cash and futures prices. These cash-future price spreads arise due to specification, location, and time factors.[17] For a given location, the future price of white-fir 2 by 4s normally can be expected to be greater than their current cash price by at most the costs of storage, insurance, and interest required to carry the lumber through time to that future date. The longer the period of time before the contract matures, the greater will be the futures price over the cash price. As the delivery month approaches, however, one of the two prices must move relative to the other, because the delivery provision in the contract should prevent the futures price from being greater than the cash price plus the cost of delivery at contract maturity. If the short trader has accurately anticipated the basis movement in this normal market situation, the hedge will usually provide him with full price protection and it may also earn a carrying charge for him (Table 48–1).

While a maximum premium exists for a distant future over the cash or nearer-month future, the reverse does not hold true.[18] There is no limit to the discount of a more distant future price under the cash or near-month future price. The short hedger should view this inverted market situation with great caution. Since the hedge is initiated at a discount to the cash market to begin with, it can be effective only if the cash price falls sharply (Table 48–4). Should a short squeeze develop, the forced buying of the limited supplies of deliverable lumber could cause the short hedger's position to steadily worsen.

Also, the more remote the lumber items being hedged from the item defined in the contract, i.e., different specification with respect to species, size, grade, or moisture content, the greater the chance that the two prices will not move together in a regular manner.[19] The risk arising from these changes in the basis cannot be hedged against.

CONCLUSIONS

The Chicago Mercantile Exchange has provided the lumber industry with a new dimension for making its marketing management decisions. The existence of two markets, one for now and one for the future, makes hedging possible. This marketing tool can help the industry to forward-price, minimize its inventory risk, earn a carrying charge on its inventory, and enlarge its inventory financing.

In general, all levels of the marketing channel can benefit from this activity. But an analysis suggests that the strongest trading potential exists among large firms producing, distributing, and using western dimension, timber, and board lumber items. The price risk is high, and these large-volume items seem to correlate strongly enough with a surrogate of the futures contract to insure a hedging gain. It appears, however, that southern pine lumber producers, distributors, and users do not have as strong an incentive for hedging as their western counterparts. Southern pine prices tend to be more stable and their correlations are lower than those on the western construction species. Hardwood lumber probably cannot be effectively hedged because the supply and demand factors underlying its prices are significantly different from those influencing the price of white-fir 2 by 4s.

An analysis of the type presented above is obviously subject to a number of limitations. In the first place, the impact of seasonal and cyclical price movements and trends were not removed prior to making the risk computations. This tends to exaggerate the magnitude of the risk, as lumbermen can formulate their price expectations with reference to specific points in time. Secondly, a surrogate for the futures price had to be employed and the price series used may have been too aggregated for a detailed analysis of the trading potential. However, this analysis has focused attention on some of those salient features of futures trading which have not been expressed in the more popular articles appearing in forestry literature. Futures trading can be a real aid to management decision-making, but a large part of the industry may not be able to take advantage of it.

In the final analysis, each individual firm will have to determine its own trading potential. Futures trading exists for broilers, live cattle, frozen orange juice, cocoa, copper, cotton, eggs, live hogs, mercury, frozen pork bellies, potatoes, propane gas, silver, sugar, and most of the grain crops, such as wheat, corn, oats, rye, and soy beans. The fact that futures trading is so widely practiced suggests that benefits are being derived from hedging. This opportunity now exists for many firms in the lumber and plywood industries as well.

REFERENCES

1. Chicago Mercantile Exchange. 1969. Rules and specifications for lumber futures. Chicago, Illinois. 6 pages.
2. Ragsdale, J. M. 1964. Nature of the futures contract. *In* Chicago Board of Trade. Pages 14–24. Proc., Agric. Writers Conference, Chicago, Illinois, September 17–18.
3. Gold, G. 1959. Modern commodity futures trading. New York: Commodity Research Bureau, Inc. 244 pages.
4. Arthur, H. B. 1966. Economic risk, uncertainty and the futures market. *In* Futures Trading Seminar, A Commodity Marketing Forum for College Teachers of Economics. Pages 213–242. Volume 3. Madison, Wisconsin: Minir Publishers, Inc.
5. Gold, G. Work cited.
6. Ehrich, R. L. 1969. Cash-futures price relationships for live beef cattle. American Jour. of Agric. Economics 51(1):26–40.
7. Working, H. 1949. The theory of price of storage. American Economic Review 39:1254–62.
8. Gold, G. Work cited.
9. Powers, M. 1968. Hedging, forward contracting and agricultural credit. A study of factors in making loans to farmers and firms. South Dakota Agric. Exp. Sta. Bul. 545. 27 pages.
10. Craig, G. A. 1961. Profits and risk in the lumber industry. Jour. of Forestry 59(8):580–589.
11. U.S. Dept. of Labor, Bureau of Labor Statistics. Wholesale prices and price indexes. Washington, D.C. Monthly.

12. Johnson, L. L. 1959. The theory of hedging and speculation in commodity futures. Review of Economic Studies 27:139–151.
13. Harris Trust and Savings Bank. 1969. Feasibility study. Lumber and plywood futures. Chicago, Illinois.
14. Working, H. 1953. Hedging reconsidered. Jour. of Farm Economics 35:544–561.
15. Same reference.
16. Random Lengths Publishing Co. Random lengths yearbook. Eugene, Oregon. Annual.
17. Gold, G. Work cited.
18. Same reference.
19. Dow, J. C. R. 1939. A theoretical account of futures markets. Review of Economic Studies 7:185–195.

INDEX OF AUTHORS

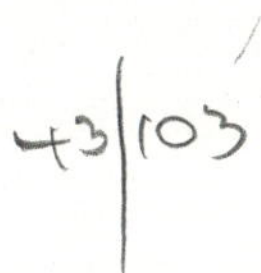